NATO ASI Series

Advanced Science Institutes Series

A series presenting the results of activities sponsored by the NATO Science Committee, which aims at the dissemination of advanced scientific and technological knowledge, with a view to strengthening links between scientific communities.

The Series is published by an international board of publishers in conjunction with the NATO Scientific Affairs Division

A **Life Sciences**	Plenum Publishing Corporation
B **Physics**	London and New York
C **Mathematical and** **Physical Sciences**	Kluwer Academic Publishers Dordrecht, Boston and London
D **Behavioural and** **Social Sciences**	
E **Applied Sciences**	
F **Computer and** **Systems Sciences**	Springer-Verlag Berlin Heidelberg New York
G **Ecological Sciences**	London Paris Tokyo
H **Cell Biology**	

The ASI Series Books Published as a Result of
Activities of the Special Programme on
SENSORY SYSTEMS FOR ROBOTIC CONTROL

This book contains the proceedings of a NATO Advanced Research Workshop held within the activities of the NATO Special Programme on Sensory Systems for Robotic Control, running from 1983 to 1988 under the auspices of the NATO Science Committee.

The books published so far as a result of the activities of the Special Programme are:

Syntactic and Structural Pattern Recognition

Edited by

Gabriel Ferraté

Instituto de Cibernética, Diagonal, 647, 2.ª planta
E-08028 Barcelona, Spain

Theo Pavlidis

Department of Electrical Engineering, SUNY
Stony Brook, NY 11794, USA

Alberto Sanfeliu

Instituto de Cibernética, Diagonal, 647, 2.ª planta
E-08028 Barcelona, Spain

Horst Bunke

Institut für Informatik und angewandte Mathematik
Universität Bern, Länggassstrasse 51
CH-3012 Bern

Springer-Verlag Berlin
Heidelberg GmbH

Proceedings of the NATO Advanced Research Workshop on Syntactic and Structural Pattern Recognition, held in Barcelona-Sitges, Spain, October 23–25, 1986.

ISBN 978-3-642-83464-6   ISBN 978-3-642-83462-2 (eBook)
DOI 10.1007/978-3-642-83462-2

© Springer-Verlag Berlin Heidelberg 1988
Originally published by COPYRIGHT in Springer-Verlag Berlin Heidelberg New York 1988
Softcover reprint of the hardcover 1st edition 1988

2145/3140-543210

About the editors of this volume

Gabriel Ferraté, Polytechnical University of Catalonia, Barcelona, Spain
Was born in Reus, Spain in 1932. He received the Dr. Eng. degree in 1958 from the Higher Technical School of Engineering of Barcelona (ETSIIB). He is Professor of Automatic Control at the Polytechnical University of Catalonia (Instituto de Cibernética). From 1969 to 1972 he was Chairman of the ETSIIB and since then President of the Polytechnical University. Presently he is also responsible for the research policy of the Government of Catalonia. He has been a member of the Executive Committee of IFAC and now he is President of the Spanish Committee of Automatic Control.

Theo Pavlidis, SUNY, Stony Brook, NY, USA
Is leading Professor of Computer Science at SUNY at Stony Brook since 1986. He was previously with AT&T Bell Laboratories at Murray Hill and Princeton University. He is the author of three books and over 100 papers. He served as the editor-in-chief of the IEEE Transactions on Pattern Analysis and Machine Intelligence (PAMI) from 1982 to 1986. His research interests include pattern recognition, computer vision, image processing, and graphics. Recent research has been centered on the recognition of text and graphics together and the development of a graphics editor using image analysis techniques.

Alberto Sanfeliu, Instituto de Cibernética, Barcelona, Spain
Received the diploma and Ph.D. from the Polytechnical University of Catalonia (UPC) (Spain), in 1978 and 1982, respectively. From 1975 to 1979 he was with the Instituto de Cibernética (UPC-CSIC) de Barcelona. For a period of two years (1979-1981) he visited Purdue University (USA) in the department of Prof. K.S. Fu. He is now Professor Titular at the Polytechnical University of Catalonia and is doing research at the Instituto de Cibernética. He is the present chairman of the IAPR Technical Committee on Syntactic Pattern Recognition and the President of the Spanish Working Group for Pattern Recognition. His area of research interests includes syntactic and structural pattern recognition and artificial intelligence, mainly related to robotics and computer vision areas.

Horst Bunke, University of Berne, Berne, Switzerland
Received the diploma and the Ph.D. in Computer Science from the University of Erlangen, West Germany, in 1974 and 1979, respectively. From 1976 to 1984 he was with the University of Erlangen. In 1980-81 he was on leave visiting Purdue University, West Lafayette, IN. In 1983 he was temporarily with the University of Hamburg, West Germany. Since 1984, he has been Professor of Computer Science at the University of Berne, Switzerland. From 1982 to 1986 he was the chairman of the IAPR Technical Committee on Syntactic Pattern Recognition. He is an editor in charge of the International Journal of Pattern Recognition and Artificial Intelligence, and a member of several scientific organizations. His research interests include syntactic and structural pattern recognition and artificial intelligence.

Acknowledgments

The authors gratefully acknowledge the financial support and encouragement provided by Dr. Mario di Lullo who most unfortunately passed away in 1986 before the Workshop was initiated, and by Dr. L.V. da Cunha who had to manage the new situation once he became the Program Director of the NATO Scientific Affairs Division. We want also to acknowledge the financial support of the Commisió Interdepartamental de Recerca e Innovación Tecnològica (CIRIT) and the help provided by M. Carmen Rodriguez and other people of the Instituto de Cibernética (UPC-CSIC) whose valuable aid allowed the Workshop to be organized.

October 15, 1987
Gabriel Ferraté
Theo Pavlidis
Alberto Sanfeliu
Horst Bunke

PREFACE

Thirty years ago pattern recognition was dominated by the learning machine concept: that one could automate the process of going from the raw data to a classifier. The derivation of numerical *features* from the input image was not considered an important step. One could present all possible features to a program which in turn could find which ones would be useful for pattern recognition. In spite of significant improvements in statistical inference techniques, progress was slow. It became clear that feature derivation was a very complex process that could not be automated and that features could be symbolic as well as numerical. Furthermore the spatial relationship amongst features might be important. It appeared that pattern recognition might resemble language analysis since features could play the role of symbols strung together to form a word. This led to the genesis of *syntactic pattern recognition*, pioneered in the middle and late 1960's by Russel Kirsch, Robert Ledley, Nararimhan, and Allan Shaw. However the thorough investigation of the area was left to King-Sun Fu and his students who, until his untimely death, produced most of the significant papers in this area. One of these papers (syntactic recognition of fingerprints) received the distinction of being selected as the best paper published that year in the IEEE Transaction on Computers.

Therefore syntactic pattern recognition has a long history of active research and has been used in industrial applications. There is at least one industrial inspection system relying on syntactic pattern recognition (developed by Joe Mundy and R.E. Joynson at General Electric).

It was realized early that the methodology of formal languages may not have been appropriate for all problems and one could investigate techniques that emphasize feature analysis and description of their relations using other mathematical tools, such as graph matching. This area of investigation came to be known as *structural pattern recognition*. I used the term in 1971 in the title of a paper published in *Frontiers of Pattern Recognition*, edited by S. Watanabe: Structural Pattern Recognition: Primitives as Juxtaposition Relations. The use of the word "structure" meant to emphasize that pictures of objects can be seen as being built from simpler elements and the recognition process involves both the recognition of the elements and investigation of their relationships. The classification technique was not as relevant. If the features were mapped in a finite set of symbols and were linearly arranged, then syntactic techniques were clearly appropriate. The desire to relax the requirement of mapping into a finite alphabet led to the investigation of attributed grammars where each symbol may have numerical attributes.

One weakness of both structural and syntactic pattern recognition is the difficulty of *automatic* inference. Given N sets of samples from each of N classes we would like to find automatically the parameters of a discriminant set. For example, if the samples are strings over a finite alphabet we would like to find N grammars, such that each one of them would accept exactly the string of one class. This problem has a trivial and uninteresting solution but no one has found solutions that are both interesting and can be found efficiently. This has been a serious impediment in the practical application classifier. Another new direction is offered by the combination of *artificial intelligence* and syntactic or structural pattern recognition. It seems that successful combinations are the challenge for the fourth decade of pattern recognition.

The workshop at Sitges covered not only topics in the mainstream of syntactic and structural pattern recognition but also in the main adjacent areas. The conference papers and the later discussion in the working groups showed that the meaning "structural" in *syntactic and structural pattern recognition (SSPR)* plays an important role in this area, mainly concerning matching of strings, trees and graphs. These techniques, which are widely used in other related areas, are one of the important topics in SSPR. It seems that successful combinations are the challenge for the fourth decade of pattern recognition.

The papers in this volume cover matching and parsing techniques mainly in error correcting graphs, grammatical inference under constraints that simplify structural combinations, and logic-syntactic integration. From the application point of view, the papers show some significant work done in speech recognition (using hybrid techniques), recognition of drawings, feature identification, cryptosystems, and applications to histo-pathology. Moreover there are several papers on image understanding where SSPR is one of the main applied techniques.

In addition to the regular sessions there were *Working groups* on 2D and 3D Image Understanding ; Speech and Waveform Recognition ; Hybrid Methodologies ; and Models and Inference. The summaries of the group discussions are incorporated in the volume. One of the interesting aspects treated in the working groups was the relation between representation and techniques, which will probably be the heart of the new generation of pattern recognition techniques.

A final discussion was handled in the Panel, a summary of which is included in the book. We hope that this coverage will help other researchers to meet the challenges of the fourth decade.

Theo Pavlidis
Stony Brook, NY, USA
September 1987

TABLE OF CONTENTS

VIII.   HYBRID APPROACHES II

Session Chairman: H. Bunke, University of Bern, Bern (Switzerland)

IX. WORKING SESSIONS

2D and 3D Image Understanding

Chairman: R. Mohr, Centre de Recherche en Inf. de Nancy (France)

Waveform and Speech Recognition

Chairman: R. De Mori, McGill Univ. Montréal, (Canada)

Hybrid Techniques

Chairman: H. Bunke, University of Bern, Bern (Switzerland)

Models and Inference

Chairman: T. Pavlidis, SUNY, Stony Brook, (USA)

   Chairman: G. Ferraté, Instituto de Cibernética (Spain)

   Panelists: R. De Mori (Canada)
                  T.C. Henderson (USA)
                  R. Mohr (France)
                  T. Pavlidis (USA)
                  A. Sanfeliu (Spain)
                  L. Shapiro (USA)

I. MATCHING AND PARSING I

A STRING CORRECTION METHOD BASED ON THE CONTEXT-DEPENDENT SIMILARITY

E. Tanaka

Utsunomiya University
Dept. of Information Science
Faculty of Engineering
Utsunomiya 321-31
Japan

Abstract: Primitive sequences of patterns, phoneme sequences and spellings are essentially context-dependent. This paper describes the string-to-string correction problem based on the context-dependent edit operations. The edit operations investigated allow substituting symbols of a string into another symbols, deleting symbols from a string, and inserting symbols into a string whose costs determined depending on their contexts. The invariance of the similarity to parallel transformations, a separation theorem and topological equivalence are described.

1. Introduction

A metric or similarity measure between two strings $A = a_1 a_2 \ldots a_m$ and $B = b_1 b_2 \ldots b_n$, which are composed from an alphabet Σ, is quite useful. Levenshtein [1](1965) defined first this kind of metric for $\Sigma = \{0,1\}$, when he discussed a synchronization error correcting code. Suppose that k_i symbols of A are erroneously substituted, n_i symbols of A are deleted and m_i symbols are inserted to A to transform A to B. He defined the distance between A and B as the minimum value of "$k_i + m_i + n_i$" among possible numbers of triple (k_i, m_i, n_i). However the calculating method of the distance was not described in his paper. The metric is called the Levenshtein metric. After that, Sakoe and Chiba [2](1971) defined a new metric between strings for time normalization problem of speech, and calculated it using the dynamic programming. Sankoff [3](1972) proposed a similarity measure between different length strings and its calculating method in relation to the problem to infer phylogenetic diagrams from DNA sequences composed of $\Sigma = \{A,C,G,T\}$. The calculating method of the Levenshtein metric could be easily obtained from his calculating method. In speech and hand printed character recognition, we need to correct phoneme sequences and the sequences of recognized characters. In connection with this problem, Okuda, Tanaka and Kasai [4](1972)

NATO ASI Series, Vol. F45
Syntactic and Structural Pattern Recognition
Edited by G. Ferraté et al.
© Springer-Verlag Berlin Heidelberg 1988

defined a metric and proposed its calculating method. This metric is to minimize "$p \times k_i + q \times m_i + r \times n_i$" where p, q and r are nonnegative weights for a substitution, an insertion and a deletion, respectively. Later, this metric is called the one dimensional weighted Levenshtein metric (1WLD). 1WLD is quite important to the theory of synchronization error correcting code [5](1976). The idea of the Levenshtein metric was considered by Wagner and Fischer [6](1974), and Sellers [7](1974), independently. Lowrance and Wagner [8](1975) extended the set of edit operations to include the operation of interchanging the positions of two adjacent symbols. Many results on this subject are collected in Sankoff and Kruskal [9](1983).

Vowels in continuous speech are highly context-dependent. For instance, in Japanese /u/ in /aua/ is in most cases identical with /w/, and /u/ in /aue/ is sometimes recognized as format transition. That is, /aue/ means /ae/. The context-dependent phonological rules in English are found in [11](1975). For instance, a schwa between two voiceless stops is deleted and this can be written as follows.

$$[\partial] \;\rightarrow\; \phi \left/ \left(\begin{array}{c} C \\ + \text{ stop} \\ - \text{ voice} \end{array} \right) - \left(\begin{array}{c} C \\ + \text{ stop} \\ - \text{ voice} \end{array} \right) \right.$$

Examples are multiply ([mʌltəplay] → [mʌltplay]) and potassium ([pətæsiəm] → [ptæsiəm]). This paper describes a context-dependent similarity measure (Chapter 3) between strings which is a generalization of 1WLD. The context-dependent similarity measure enables us to discuss the invariance of similarity to parallel transformation (Chapter 4), a separation theorem (Chapter 5) and topological equivalence between strings (Chapter 6). Two sequences "*abacc****" and "***abacc**" are equivalent, if we are not concerned about "*". If we can decide their equivalence in a similarity system, the system can be called to be invariant to parallel transformation. In matching of recorded voices, we have to adjust their initial points. However, this system does not need such adjustment. Consider character recognition using chain codes. Let α and β be standard characters, and A_α and B_β be their chain codes. Since there are many variations for each character, we have many different chain codes, say $\tilde{A}_\alpha = \{A_\alpha, A_{\alpha 1}, A_{\alpha 2}, \ldots\}$ and $\tilde{B}_\beta = \{B_\beta, B_{\beta 1}, B_{\beta 2}, \ldots\}$. For any $A_{\alpha i}$ and $B_{\beta j}$ ($i = 1, 2, \ldots$; $j = 1, 2, \ldots$) the similarity relation $S(A_\alpha, A_{\alpha i}) < S(A_\alpha, B_{\beta j})$ must be satisfied, where $S(A_\alpha, A_{\alpha i})$ denotes the similarity from A_α to $A_{\alpha i}$. The separation theorem states the condition to obtain the above relation in general. In speech, vowels are sometimes pronounced longer than their standard length. For instance, the pronounciation /vèriii looon/ is an emphasized expression of /veri lon/ (=very long). From the point of string comparison, these two pronounciations are topologically equivalent. Another example of topological equivalence is found in character recognition using chain codes. This paper is mainly based on the previous report [18](1984) which is an extension of [19](1984) and the one dimensional version of [16](1984) and [20](1985). Recently, Kohonen[2](1986) discusses the same problem and shows experimental results in speech recognition.

2. Brief review of the one dimensional weighted Levenshtein metric

In this section we review the one dimensional weighted Levenshtein metric [4]. Let $A = a_1 a_2 \ldots a_m$ and $B = b_1 b_2 \ldots b_n$ be two finite strings of symbols from a given alphabet Σ. Define a mapping M_s from A to B. M_s is a set of pairs of integers (i,j) which satisfies the following conditions, where i and j are for a_i and b_j, respectively.

  1) $1 \leq i \leq m$, $1 \leq j \leq n$.
  2) For (i_1,j_1), $(i_2,j_2) \in M_s$,
    a) $i_1 = i_2$ iff $j_1 = j_2$.
    b) $i_1 < i_2$ iff $j_1 < j_2$.

Let u_s be the number of elements in M_s such that $a_i \neq b_j$ and $(i,j) \in M_s$. Let $v_s = n - |M_s|$ and $w_s = m - |M_s|$. u_s, v_s and w_s are considered as the number of substitutions, that of insertions of extra symbols and that of deletions to transform A to B, respectively, where $|M_s|$ denotes the number of elements in M_s. The one dimensional weighted Levenshtein distance from A to B, denoted by $D(A,B)$, is defined as

$$D(A,B) = \min_s \{ p \times u_s + q \times v_s + r \times w_s \}, \tag{2-1}$$

where p, q and r are nonnegative weights assigned to substitutions, insertions and deletions, respectively. 1WLD has following properties:

  1) $D(A,B) \geq 0$, with equality iff $A = B$.
  2) $D(A,B) + D(B,C) \geq D(A,C)$.
  3) $D(A,B) = D(B,A)$, if $q = r$. $\tag{2-2}$

$D(A,B)$ can be computed by the following recurrence relation.

$$d(i,j) = \min \{d(i-1,j)+r, \; d(i-1,j-1)+p(i,j), \; d(i,j-1)+q\} \qquad (2\text{-}3)$$

where $d(i,0) = i \times r$, $d(0,j) = j \times q$ and

$$p(i,j) = \begin{cases} p, & \text{if } a_i \neq b_j. \\ 0, & \text{if } a_i = b_j. \end{cases} \qquad (2\text{-}4)$$

Then

$$D(A,B) = d(m,n). \qquad (2\text{-}5)$$

The computational complexity to compute $D(A,B)$ is proportional to mn.

3. A one dimensional single side context-dependent similarity

Let λ be the null string. Let Σ' and Σ^* be $\Sigma' = \Sigma \cup \{\lambda\}$ and the set of all strings of symbols from Σ including the null string λ, respectively. Consider $A = a_1 a_2 \ldots a_m$ and $B = b_1 b_2 \ldots b_n$. Let $K(a_i)$ be a substring of A including a_i, that is, $a_{i-h} a_{i-h+1} \ldots a_i \ldots a_{i+k}$ ($1 \leq i-h$, $i+k \leq m$). If $K(a_i) = K(b_j)$, it is called that a_i matches b_j with context $K(a_i)$, or $K(a_i)$ matches $K(b_j)$. If A is not a substring of B, that is, $B \neq \xi A \zeta$ ($\xi, \zeta \in \Sigma^*$), write $A \nrightarrow B$. Otherwise, write $A \rightarrow B$. Note that if $A \nrightarrow B$, there is at least one substring A_i of A such that $A_i \neq B_j$ for any substring B_j of B. In other words, we can find at least one context $K(a_i)$ of a_i such that a_i does not match any b_j in B. Let $\bar{A} = \$A\$$ and $\bar{B} = \$B\$$, where $\$$ is an end mark. Each symbol x in $\bar{A}$ has at least one context $K(x)$ such that x does not match any y in $\bar{B}$, even if $A \rightarrow B$. Let $K(\xi)$ and $c(\xi, \zeta : K(\xi))$ be a context of ξ and the cost to replace ξ to ζ under the condition $K(\xi)$, respectively, where $\xi, \zeta \in \Sigma^*$. $c(\xi, \zeta : K(\xi))$ is called "a cost concerning ξ" and is a (single side) context-dependent cost. Without $K(\xi)$, c is a context-independent cost. Hereafter, we use the notation $\alpha \xi \beta$ for a context of ξ, where $K(\xi) = \alpha \xi \beta$, and $\alpha, \beta \in \Sigma^*$. $c(\xi, \zeta : \alpha \xi \beta)$ is interpreted as follows:

If $\xi \neq \lambda$ and $\zeta \neq \lambda$, c is the cost of substitution of ζ for ξ.

If $\xi \neq \lambda$ and $\zeta = \lambda$, c is the cost of deletion of ξ.

If $\xi = \lambda$ and $\zeta \neq \lambda$, c is the cost of insertion of ζ.

[Example 1] Consider $A = abac$ and $B = abc$. "a" with context ab and "b" with context ab in A match "a" and "b" in B, respectively. However, "a" with context aba does not match "a" in B. Let $C = abcd$.

Then $\bar{B}$ = \$abc\$ and $\bar{C}$ = \$abcd\$. "a" in B does not have its context that does not match K(a) in C. However, "a" in $\bar{B}$ with context \$$\underline{a}$bc\$ does not match "a" in $\bar{C}$.

One of the possible ways to determine cost $c(\xi,\zeta:K(\xi))$ is to use the probability p to transform ξ to ζ in the context $K(\xi)$, that is, $c(\xi,\zeta:K(\xi))$ = $-\ln(p)$. Assume that the probability to substitute "f" **for** "a" in $\underline{a}$b and that to substitute "g" for "b" in b$\underline{c}$ are p_1 and p_2, respectively. That is, $c(a,f:\underline{a}b)$ = $-\ln(p_1)$ and $c(b,g:b\underline{c})$ = $-\ln(p_2)$. Then the probability to transform A = abc to B = fgc is p_1p_2. The cost to transform A to B is $-(\ln(p_1)+\ln(p_2))$. This means that the smaller the cost is, the easier the transformation is.

[Example 2] Consider A = efefeg (= $a_1a_2a_3a_4a_5a_6$) and B = hefeg (= $b_1b_2b_3b_4b_5$).

(i) Assume that x is substituted for a_1, a_3 and b_2 with probability p_1, and x is substituted for a_5 and b_4 with probability p_2, where $p_1 \neq p_2$. Then, we can set

$$c(e,x:\underline{e}f) = -\ln(p_1), \quad c(e,x:\underline{e}g) = -\ln(p_2).$$

(ii) If there is inconsistency in a cost system, we can dissolve the inconsistency to lengthen the contexts. Consider a cost system which includes the following cost functions:

$$c(e,x:\underline{e}f) = -\ln(p_1), \quad c(e,x:f\underline{e}) = -\ln(p_2).$$

Both cost functions can be applied to a_3. This inconsistency can be removed using the following cost functions:

$$c(e,x:\underline{e}fef) = -\ln(p_1), \quad c(e,f:f\underline{e}f) = -\ln(p_1),$$
$$c(e,x:f\underline{e}g) = -\ln(p_2).$$

(iii) Note that we can determine any symbol of any string uniquely in a set of strings by using symbol \$. Consider D = ab (= d_1d_2) and E = fab (= $e_1e_2e_3$). Assume that x is substituted for d_1 and e_2 with probability p_1 and p_2, respectively. The cost functions concerning "a" can be defined as follows:

$$c(a,x:\$\underline{a}) = -\ln(p_1), \quad c(a,x:f\underline{a}) = -\ln(p_2).$$

In the following, we use the following types of cost functions for A = $a_1a_2\ldots a_m$.

$$c(a_i, d : \alpha_i \underline{a}_i \beta_i) \quad \text{and} \quad c(a_i, \lambda : \alpha_i \underline{a}_i \beta_i)$$

$$(1 \leq i \leq m. \quad \text{If } i = 1, \ \alpha_1 = \lambda. \quad \text{If } i = m, \ \beta_m = \lambda.)$$

$$c(\lambda, d : \underline{\lambda} a_1 \ldots),$$

$$c(\lambda, d : \ldots a_i \underline{\lambda} a_{i+1} \ldots) \quad (1 \leq i < m),$$

$$c(\lambda, d : \ldots a_m \underline{\lambda}), \tag{3-1}$$

where $d \in \Sigma$. We assume that there is no inconsistent cost function in a given cost system.

[Example 3] Consider the transformation from $A = a_1 a_2 a_3 a_4 a_5$ to $B = b_1 b_2 b_3 b_4$ in Fig.1. Assume that the costs concerning symbols in A are

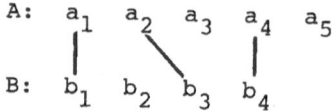

Fig.1. A transformation from "A" to "B".

defined when $|\alpha_i| = 1$ and $|\beta_i| = 1$ in (3-1), where $|\alpha|$ is the length of α. b_1, b_3 and b_4 are substituted for a_1, a_2 and a_4, respectively. a_3 and a_5 are deleted. b_2 is inserted. The total cost C_T of this transformation is

$$C_T = c(a_1, b_1 : \underline{a}_1 a_2) + c(a_2, b_3 : a_1 \underline{a}_2 a_3) + c(a_3, \lambda : a_2 \underline{a}_3 a_4)$$
$$+ c(a_4, b_4 : a_3 \underline{a}_4 a_5) + c(a_5, \lambda : a_4 \underline{a}_5) + c(\lambda, b_2 : a_1 \underline{\lambda} a_2).$$

The correction action applies all primitive corrections simultaneously and in parallel to the entire string.　　　　　　　　　　　　　　　　　　　■

Let C be the set of costs. If we need to express the cost system C explicitly, the notation $S_C(A,B)$ is used instead of $S(A,B)$. The one dimensional single side context-dependent similarity (1SCDS) from $A = a_1 a_2 \ldots a_m$ to $B = b_1 b_2 \ldots b_n$ can be computed by the following recurrence relation.

$$s(i,j) = \min \begin{cases} s(i-1,j) + c(a_i, \lambda : \alpha_i \underline{a}_i \beta_i), \\ s(i-1,j-1) + c(a_i, b_j : \alpha_{i'} \underline{a}_i \beta_{i'}), \\ s(i,j-1) + c(\lambda, b_j : \alpha_{i''} \underline{\lambda} \beta_{i''}), \end{cases} \tag{3-2}$$

where

$$s(i,0) = \sum_{k=1}^{i} c(a_k, \lambda:\alpha_k \underline{a}_k \beta_k),$$

$$s(0,j) = \sum_{k=1}^{j} c(\lambda, b_k:\underline{\lambda} a_1 a_2 \ldots). \qquad (3\text{-}3)$$

Then

$$S(A,B) = s(m,n). \qquad (3\text{-}4)$$

Based on (3-2)∿(3-4), we obtain the following algorithm.

Algorithm

```
s(0,0) := 0;
for i := 1 to |A|  do
  s(i,0) := s(i-1,0) + c(a_i,λ:α_i a_i β_i);
for j := 1 to |B|  do
  s(0,j) := s(0,j-1) + c(λ,b_j:λa_1 a_2 ...);
for i := 1 to |A|  do
  for j := 1 to |B|  do
    begin
      Δ1 := s(i-1,j) + c(a_i,λ:α_i a_i β_i);
      Δ2 := s(i-1,j-1) + c(a_i,b_j:α_i, a_i β_i,):
      Δ3 := s(i,j-1) + c(λ,b_j:α_i„ λβ_i„);
      s(i,j) := min {Δ1,Δ2,Δ3};
    end;
S(A,B) := s(m,n).
```

The computational complexity of the algorithm is proportional to mn.

[Example 4] Assume that the cost functions for symbols in C = abca are as follows:

$$c(a,x:\underline{ab}) = 0 \text{ if } x = a, \text{ and } = 1 \text{ otherwise,}$$
$$c(b,x:a\underline{bb}) = 0 \text{ if } x = b, \text{ and } = 2 \text{ otherwise,}$$
$$c(c,x:b\underline{c}a) = 0 \text{ if } x = c, \text{ and } = 1 \text{ otherwise,}$$
$$c(a,x;c\underline{a}) = 0 \text{ if } x = a, \text{ and } = 2 \text{ otherwise,}$$
$$c(a,\lambda:\underline{ab}) = 0.5, \quad c(b,\lambda:a\underline{bc}) = 0.5, \quad c(c,\lambda:b\underline{ca}) = 0.5,$$
$$c(a,\lambda:c\underline{a}) = 3, \quad c(\lambda,y:\underline{\lambda}ab) = 2, \quad c(\lambda,y:a\underline{\lambda}b) = 1,$$
$$c(\lambda,y:b\underline{\lambda}c) = 1, \quad c(\lambda,y:a\underline{\lambda}b) = 1, \quad c(\lambda,y:a\underline{\lambda}) = 2,$$

where x,y ε {a,b,c}.

Let D = acc. Fig. 2 is the diagram to compute S(C,D). The route drawn
by the thick line is the shortest route from P to Q, that indicates
S(C,D). Hence S(C,D) = 2.5. To transform C **to D**, "b" is deleted and
"c" is substituted for "a".

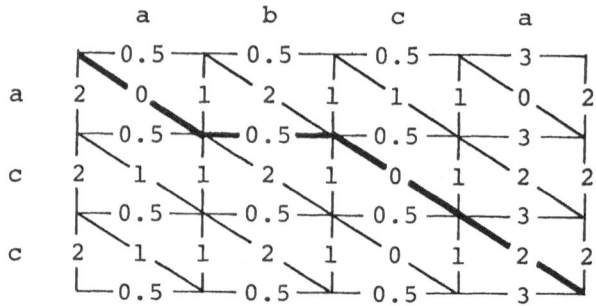

Fig.2. Computational diagram.

In the following, assume c(a,a:K(a)) = 0 for any "a" in Σ without
stating it explicitly.

[Example 5] Let A = a, B = b and C = c. Assume that c(a,b) = 1,
c(a,c) = 5, c(b,c) = 1 and all other cost functions are equal to 10.
Then, S(A,B) = 1, S(A,C) = 5 and S(B,C) = 1, i.e., S(A,B)+S(B,C) < S(A,C).
That is, the context-independent cost function does not satisfy the triangular axiom. In other words, the context-independent cost function is not
a distant function. Hence, the context-dependent cost function is also not
a distant function.

This fact does not reduce the importance of the context-dependent matching.
The followings are reasons: (1) Words and phoneme sequences are highly
context-dependent. For instance, "q" of a English word is always followed
by "u", and the examples of phoneme sequences are described in Introduction.
(2) Cost functions for patterns are not always distance functions [14],[15].
The point is that the importance of a cost function for pattern recognition
lies not in being a distance function but in the ability to classify
patterns. Usually, a similarity measure is defined as S(A,A) = 1 and
S(A,B) = 0, where B is one of the most different strings from A. If the
conventional definition is required, we can transform S(C,D) to S̄(C,D)

such that $\bar{S}(C,D) = 1/\{S(C,D)+1\}$.

4. Isomorphism - the invariance of the similarity to parallel transformation -

Let $\#$ denote a special symbol, for instance a blank. A substring $\hat{A}$ of a string A is called the character part of A, if the leftmost symbol of $\hat{A}$ is the leftmost non-$\#$ symbol of A and the rightmost symbol of $\hat{A}$ is the rightmost non-$\#$ symbol of A. If $S_C(\hat{A},\hat{B}) = k$ follows $S_C(A,B) = k$, then a given cost system C is called to be invariant to parralel transformation. If C is invariant to parallel transformation and $S_C(A,B) = 0$, it is called that A is isomorphic to B. The following 4 types of costs are called the boundary cost functions:

$$c(\#^n, \zeta_1 : \underline{\#}^n\hat{A}), \quad c(\#^n, \zeta_2 : \hat{A}\underline{\#}^n), \quad c(\lambda, \zeta_3 : \underline{\lambda}\hat{A}), \quad c(\lambda, \zeta_4 : \hat{A}\underline{\lambda}),$$

where n is a nonnegative integer, ζ_1, ζ_2, ζ_3, $\zeta_4 \in \Sigma^*$, $\#^n$ is the n-times concatenation of $\#$, and $\#^n\hat{A}$ means that $\#^n$ is connecting to the left of $\hat{A}$.

[Lemma 1] Assume that B is a parallel transformation of A, and assume the following boundary cost functions:

$$c(\#^n, \boldsymbol{\lambda} : \underline{\#}^n\hat{A}) = 0, \quad c(\#^n, \lambda : \hat{A}\underline{\#}^n) = 0,$$

$$c(\lambda, \#^n : \underline{\lambda}\hat{A}) = 0, \quad c(\lambda, \#^n : \hat{A}\underline{\lambda}) = 0, \tag{4-1}$$

where n is a nonnegative integer. Then, $S(A,B) = 0$ for any cost assignments to symbols in $\hat{A}$.

Proof. From the assumption, $\hat{A} = \hat{B}$. Assume that $A = \#^s\hat{A}\#^t$, $B = \#^u\hat{A}\#^v$, $u \leq s$ and $t \leq v$, without loss of generality, where s, t, u and v are nonnegative integers. Since $c(\#^{s-u}, \lambda : \underline{\#}^{s-u}\hat{A}) = 0$ and $c(\lambda, \underline{\#}^{v-t} : \hat{A}\#^{v-t}) = 0$, we have

$$S(A,B) = S(\#^s\hat{A}\#^t, \#^u\hat{A}\#^v)$$

$$= S(\#^{s-u+u}\hat{A}\#^t, \#^u\hat{A}\#^{v-t+t})$$

$$= S(\#^u\hat{A}\#^t, \#^u\hat{A}\#^t)$$

$$= 0.$$

Note that $c(\#^n, \lambda : \underline{\#}^n\hat{A}) = 0$ can be decomposed into $c(\#, \lambda : \#\underline{\#}^{n-1}\hat{A}) = 0$, $c(\#, \lambda : \#\#\underline{\#}^{n-2}\hat{A}) = 0$, ..., and $c(\#, \lambda : \#^{n-1}\underline{\#}\hat{A}) = 0$. Other three types of costs of (4-1) also can be expressed as combinations of types of costs such as $c(\xi, \zeta : K(\xi)) = 0$, where ξ , $\zeta \in \Sigma'$. Therefore, we can decide whether A is isomorphic to B or not using the algorithm described in chapter 3.

[Example 6] Let $C = \#ab\#c\#\#\#$, $D = \#\#\#ab\#c\#$ and $E = \#abc\#\#\#\#$. Then $\hat{C} = ab\#c$. The boundary cost functions of C can be written in simpler forms than (4-1). That is,

$$c(\#^n, \lambda : \underline{\#}^n a) = 0, \quad c(\#^n, \lambda : c\underline{\#}^n) = 0,$$

$$c(\lambda, \#^n : \#\underline{\lambda} a) = 0, \quad c(\lambda, \#^n : c\underline{\lambda}\#) = 0. \tag{4-2}$$

Then, we have $S(C,D) = 0$ and $S(C,E) \neq 0$, where any of the other costs concerning symbols or strings is supposed not to be 0. Consider 1WLD. Let $q_\#$ and $r_\#$ be the cost of insertion of # and that of deletion of #, respectively. If $q_\# = r_\# = 0$, then $D(C,D) = 0$ and $D(C,E) = 0$. Therefore, 1WLD can not be invariant to parallel transformation by any weight assignment

5. A separation theorem

Using context-dependent cost functions, we can cluster a set of strings based on the similarity measure.

[Lemma 2] Assume that $A \nleftrightarrow B$. Then, we can make $S(A,B) > U$ for any U by adjusting the costs concerning symbols in A.

Proof. From the assumption, there is a substring A_i of A that does not match any possible substring B_j of B. Put

$$c(u, x : K(u)) = U \quad (x \in \Sigma', \quad x \neq u),$$

$$c(\lambda, y : \underline{\lambda} a_i) = U \quad (y \in \Sigma),$$

$$c(\lambda, y : a_i \underline{\lambda}) = U,$$

for each symbol u in A_i , where $K(u) = A_i$. Using these cost assignments, the substitution of x ($\neq$ u) for u in K(u), the deletion of u and the insertion of y into B cost U. In other words, A_i can not be transformed to any B_j at the cost less than U. Therefore, we have $S(A,B) \geq U$. ∎

Note that even if $A \rightarrow B$ ($A \neq B$), we can make $S(\bar{A}, \bar{B}) > U$ for any U by adjusting the costs concerning symbols in A, since $\bar{A} \nrightarrow \bar{B}$. Let $\overset{o}{A}$ be a set of contexts in cost functions of (3-1). Let $M(A,B) = \overset{o}{A} - \overset{o}{B}$, i.e. it is the set of contexts of $\overset{o}{A}$ which are not in $\overset{o}{B}$. Then, $M(A,B)$ is a set of consistent contexts such that (i) any $K(a_i)$ in $M(A,B)$ does not match any substring of B, and (ii) each a_i has only one context $K(a_i)$ in $M(A,B)$.

[Lemma 3] If $M(A,B) - M(A,C) \neq \phi$ (the null set), we can make $S(A,B) > S(A,C)$ by adjusting the costs concerning symbols in A.

Proof. Let K_A be an element in $(M(A,B) - M(A,C))$. Put

$$c(u,x:K(u)) = U \quad (x \in \Sigma', \quad x \neq u),$$

$$c(\lambda,y:\underline{\lambda}u) = U \quad (y \in \Sigma),$$

$$c(\lambda,y:u\underline{\lambda}) = U,$$

$$U > S(A,C), \tag{5-1}$$

for each symbol u in K_A, where $K(u) = K_A$.
Then, the cost to transform K_A to any substring of B is at least U. Therefore,

$$S(A,B) \geq U > S(A,C).$$

Hence, we have $S(A,B) > S(A,C)$.

[Example 7] Consider $A = abac$, $B = bac$ and $C = ababbc$ and assume the followings.

$$M(A,B) = \{\underline{ab}, a\underline{b}, ab\underline{a}\}, \quad M(A,C) = \{a\underline{c}\}.$$

Clearly, $\underline{ab} \in (M(A,B)-M(A,C))$. Assume that the maximum cost of all other costs concerning symbols in A except $c(a,x:\underline{ab})$ $(x \in \Sigma')$ is V, and set $c(a,x:\underline{ab}) = 10V$. Then

$$S(A,B) \geq 10V > S(A,C).$$

Consider 1WLD and assume that $q = r$ (*1) for the weights of 1WLD. Since $D(A,B) = r$ and $D(A,C) = 2q$, then $D(A,C) \geq D(A,B)$ under the condition (*1). That is, we can not obtain $D(A,B) > D(A,C)$.

Let $\tilde{A} = \{A_1, A_2, \ldots, A_m\}$ and $\tilde{B} = \{B_1, B_2, \ldots, B_n\}$. Let

$$S(\tilde{A},\tilde{B}) = \min \{S(A_i,B_j)\}$$

and

$$M(\tilde{A},\tilde{B}) = \bigcap_{i=1}^{m} \bigcap_{j=1}^{n} M(A_i,B_j) - \bigcup_{i=1}^{m} \bigcup_{j=1}^{m} M(A_i,A_j)$$

[Theorem 4 (Separation theorem)] If $M(\tilde{A},\tilde{B}) \neq \phi$, we have

$$S(\tilde{A},\tilde{B}) > S(A_i,A_j), \quad S(\tilde{A},\tilde{B}) > S(B_s,B_t),$$

by adjusting the costs concerning symbols of strings in $\tilde{A}$, where $1 \leq i,j \leq m$ and $1 \leq s,t \leq n$.

Proof. Let $\hat{A}$ be an element in $M(\tilde{A},\tilde{B})$. Define the cost functions as follows:

$$c(u,x:K(u)) = U \quad (x \in \Sigma', x \neq u)$$

$$c(\lambda,y:\underline{\lambda}u) = U \quad (y \in \Sigma),$$

$$c(\lambda,y:u\underline{\lambda}) = U,$$

$$U > \bar{U},$$

for each symbol in $\hat{A}$, where $\bar{U} = \max \{S(A_i,A_j), S(B_s,B_t)\}$ $(1 \leq i,j \leq m$ and $1 \leq s,t \leq n)$. Then $S(A,B) \geq U$. Therefore, we have

$$S(\tilde{A},\tilde{B}) \geq U > \bar{U} \geq S(A_i,A_j),$$

$$S(\tilde{A},\tilde{B}) \geq U > \bar{U} \geq S(B_s,B_t).$$

Note that the separation theorem can not hold for 1WLD.

6. A one dimensional both sides context-dependent similarity
   - topological equivalence -

If A can be transformed to B by the mapping $x^m \to x$ and $x \to x^n$ for any x in Σ and any positive integers m and n, A is called to be topologically equivalent to B, and that is denoted by $A \overset{t}{=} B$.

[Example 8] Let C = abbbcaa, D = abcccaa and E = abcbca. C can be transformed to D by $b^3 \to b$ and $c \to c^3$. Therefore $C \overset{t}{=} D$. However, C can not be transformed to E by the mapping in the forms $x^m \to x$ and $x \to x^n$. Hence, C is not topologically equivalent to E.

Let $c(\xi,\zeta:K_A(\xi),K_B(\zeta))$ be the cost to replace ξ to ζ under the conditions $K_A(\xi)$ and $K_B(\zeta)$, where $K_A(\xi)$ and $K_B(\zeta)$ denote the context of ξ in A and that of ζ in B, respectively. This cost is called the both sides context-dependent cost. The minimum sum of costs to transform string A to string B is the one dimensional both sides context-dependent similarity (1BCDS, in abbreviation) from A to B. In general, it is not easy to compute 1BCDS for given strings. However, in a special case such as to decide topological equivalence, we can easily compute 1BCDS. In the following, let $K_A(x)$ and $K_B(\underline{\lambda}x)$ denote that x is connected with another x in A and λ is connected with x in B, respectively. Define costs $c(x,\lambda:K_A(x),K_B(\underline{\lambda}x))$ and $c(\lambda,x:K_A(\underline{\lambda}x),K_B(x))$ for x in Σ as follows:

$$c(x,\lambda:K_A(x),K_B(\underline{\lambda}x)) = \begin{cases} 0, & \text{if } K_A(x) \text{ and } K_B(\underline{\lambda}x) \text{ are satisfied.} \\ r_x, & \text{otherwise. } (r_x \neq 0) \end{cases} \qquad (6\text{-}1)$$

$$c(\lambda,x:K_A(\underline{\lambda}x),K_B(x)) = \begin{cases} 0, & \text{if } K_A(x) \text{ and } K_B(\underline{\lambda}x) \text{ are satisfied.} \\ q_x, & \text{otherwise. } (q_x \neq 0) \end{cases} \qquad (6\text{-}2)$$

The cost functions $c(x^m,x) = 0$ and $c(x,x^n) = 0$ are obtained from (6-1) and (6-2), respectively. Let other cost functions be arbitrary. The similarity from A to B computed using the cost functions (6-1) and (6-2) is called the topological similarity from A to B, and it is denoted by $T(A,B)$. The recurrence formula to compute the topological similarity from $A = a_1 a_2 \ldots a_m$ to $B = b_1 b_2 \ldots b_n$ is as follows:

$$t(i,j) = \min \begin{cases} t(i-1,j) + c(a_i,\lambda:K_A(a_i),K_B(\underline{\lambda}a_i)), \\ t(i-1,j-1) + c(a_i,b_j), \\ t(i,j-1) + c(\lambda,b_j:K_A(\underline{\lambda}b_j),K_B(b_j)), \end{cases} \qquad (6\text{-}3)$$

where

$$t(u,0) = \sum_{k=1}^{u} c(a_k,\lambda), \quad t(0,v) = \sum_{k=1}^{v} c(\lambda,b_k). \qquad (6\text{-}4)$$

Then,

$$T(A,B) = t(m,n). \qquad \blacksquare \quad (6\text{-}5)$$

We can use the following formula instead of (6-3).

$$t(i,j) = \min \begin{cases} t(i-1,j) + c(a_i,\lambda), \\ t(i-1,j-1) + c(a_i,b_j), \\ t(i,j-1) + c(\lambda,b_j), \\ t(i-u,j-v), & \text{if } x^u = a_{i-u+1}a_{i-u+2}\ldots a_i, \text{ and} \\ & \qquad x^v = b_{j-v+1}b_{j-v+2}\ldots b_j, \\ t(i-u,j), & \text{if } x^{u+1} = a_{i-u+1}a_{i-u+2}\ldots a_{i+1}, \\ t(i,j-v), & \text{if } x^{v+1} = b_{j-v+1}b_{j-v+2}\ldots b_{j+1}. \end{cases} \qquad (6\text{-}6)$$

where $x \in \Sigma$.

7. Concluding remarks

We have introduced a context-dependent similarity measure for strings and presented its computing algorithm which is analogous to the computing method of 1WLD. This similarity could be invariant to parallel transformation by adjusting cost functions. The separation theorem and topological equivalence between two strings are described.

Acknowledgement

The author is grateful to Ms. K. Tanaka for typing the manuscript.

References

[1] Levenshtein, B.I. Binary codes with correction of deletions, insertions and substitutions of symbols, Dokl.Acak.Nauk. SSSR, 163(1965), 845-848.

[2] Sakoe, H. and Chiba, S. Recognition of continuously spoken words based on time normalization by dynamic programming, J.Acoust.Soc.Japan, 29,9 (1971), 483-490.

[3] Sankoff, D. Matching sequences under deletion/insertion constraints, Proc.Natl.Acad.Sci.USA, 69,1 (1972), 4-6.

[4] Okuda, T., Tanaka, E. and Kasai, T. A garbled word correcting method by an extended metric, Ann.Joint Meeting of Elect. and Electro. Eng. Tokai District, Japan, 18a-B-6 (1972).
Okuda, T., Tanaka, E. and Kasai, T. A method for the correction of garbled words based on the Levenshtein metric, IEEE Trans.Comput., C-25,2 (1976), 172-178.

[5] Tanaka, E. and Kasai, T. Synchronization and substitution error-correcting codes for the Levenshtein metric, IEEE Trans.Inf.Theory, IT-22 (1976), 156-162.

[6] Wagner, R. and Fischer, M. The string to string correction problem, JACM,21,1 (1974), 168-173.

[7] Sellers, P. An algorithm for the distance between two finite sequences, J.Comb.Theory Se.A,16 (1974), 253-258.

[8] Lowrance, R. and Wagner, R.A. An extension of the string-to-string correction problem, JACM,22,2 (1975), 177-183.

[9] Sankoff, D. and Kruskal, J.B. Time warps, string edits, and macromolecules: The theory and practice of sequence comparison, Mass.,USA, Addison-Wesley (1983).

[10] Abe, K. Distances between strings of symbols --Review and remarks, 6th ICPR,München (1982).

[11] Oshika, B., Zue, V., Weeks, V., Neu, H. and Aurbach, J. The role of phonological rules in speech understanding research, IEEE Trans ASSP, ASSP-23,1 (1975), 104-112.

[12] Yokota, M., Akizawa, K. and Kasuya, H. Automatic identification of vowels in connected speech uttered by multiple speakers, Trans.IECE, Japan,J65-D,1 (1982), 134-135

[13] Tanaka, E. and Kikuchi, Y. A metric between pictures, Trans.IECE, Japan,63-D,12 (1980), 1018-1024.

[14] Isomichi, Y. and Ogawa, T. A pattern matching by dynamic programming, J.Inf.Proc.Soc.Japan,16,1 (1975), 15-22.

[15] Yamada, H., Saito, T. and Mori, S. An improvement correlation method
     -locally maximized correlation-, Trans.IECE,Japan,64-D,10(1983),
     970-976.
[16] Tanaka,E. A context-dependent similarity measure for pictures,
     7th ICPR, Montreal (1984).
[17] Peterson, J.L. Computer programs for detecting and correcting
     spelling errors, CACM,23,12 (1980),676-687.
[18] Tanaka, E. A one dimensional context-dependent similarity measure,
     A paper of technical group on pattern recognition and understanding,
     IECE,Japan (1984).
[19] Tanaka, E. A context dependent similarity between strings, Trans.IECE,
     Japan,67-A,6,612-613 (1984).
[20] Tanaka, E. A two dimensional context-dependent similarity measure,
     Trans.IECE,Japan,E68,10,667-673 (1985).
[21] Kohonen, T. Dynamically expanding context, with application to the
     correction of symbol strings in the recognition of continuous speech,
     8th ICPR,Paris (1986).

AN ERROR-CORRECTING PARSER FOR A CONTEXT-FREE LANGUAGE BASED ON THE CONTEXT-DEPENDENT SIMILARITY

M. Ikeda(*), E. Tanaka(**), O. Kasusho(*)

*(*) The Ins. of Sci. and Ind. Res.
Osaka Univ., Ibaraki-Shi, 565 Japan
(**) Fac. of Engineering, Utsunomiya Univ.
Utsunomiya-Shi, 321 Japan*

ABSTRACT Patterns are essentially context-dependent. This paper describes an error-correcting parser for a context-free language, which finds most similar sentences to an input sentence based on the context-dependent similarity(CDS). The proposed algorithm is obtained by modifying the Lyon's error-correcting parser. Possible application are to the problem of pattern recognition, speech recognition and language processing.

1. Introduction

Several error-correcting parsers have been developed for syntactic pattern recognition[1], natural language processing, error correction in linguistic deta transmission, error correcting compilers, and so on. The first error-correcting parser for a context-free language(CFL) was proposed by Aho and Peterson[2]. It is an extension of the Earley's parser[3] for CFL. Since their parser uses error-productions, both computing time and memory space are enormously large. This deficiency has been removed by Lyon[4]. Another approach[5],[6] is based on the Cocke-Kasami-Younger's bottom-up parser(CKY) for CFL[7]. Recently, Tanaka[8] has developed an ECP based on the Graham-Harison-Ruzzo's parser(GHR)[9] which combined the CKY's and the Early's.

These error-correcting parsers are the least error correction methods, which finds most similar sentences based on the Levenshtein distance(LD)[10] or the weighted Levenshtein distance(WLD)[11]. WLD is a context dependent measure. However, patterns such as speech and printed characters have context-dependent deformation. Therefore, WLD cannot treat them correctly.

In this paper , we propose an error-correcting parser for a context-free language based on both the one dimensional context-dependent similarity[12] and the Lyon's error-correcting parser.

2. Definitions

In this section, we will give some definitions for formal languages and a similarity measure.

2.1 Grammars and Languages.

A phrase structure grammar is a quadruple $G=(N,T,P,S)$, where N, T,P and S are a finite set of nonterminals , a finite set of terminals, a finite set of productions, and the start symbol, respectively. A production is a rewriting rule in the form $\alpha \rightarrow \beta$, where $\alpha \in V^* N V^*$, $\beta \in V^*$ and $V=N \cup T$. V is called a set of vocabularies, and V^* is a set of all finite sequences composed of elements in V including the empty sequence λ. If $\alpha \in N$, G is called a context-free grammar. If $\alpha \rightarrow \beta \in P$, then $\xi \alpha \eta$ => $\xi \beta \eta$. That is, $\xi \beta \eta$ is derived directly from $\xi \alpha \eta$. If $\alpha_1 => \alpha_2$, $\alpha_2 => \alpha_3$,•••, $\alpha_{k-1} => \alpha_k$, then α_k is derived from α_1 and write $\alpha_1 \overset{*}{=>} \alpha_k$. The language generated by G is defined as $L(G)=\{x \mid S \overset{*}{=>} x, x \in T^*\}$. If G is a context-free grammar, $L(G)$ is called a context-free language.

2.2 The one dimensional context-dependent similarity.

Let $\alpha = a_{(1)} a_{(2)} \cdots a_{(m)}$ and $\beta = b_{(1)} b_{(2)} \cdots b_{(n)}$ be two sequences. Let $K(a_{(i)})$ be a subsequence of α including $a_{(i)}$, that is, $a_{(i-h)} a_{(i-h+1)} \cdots a_{(i)} \cdots a_{(i+k)}$ $(1 \le i-h, i+k \le m)$. We call this sequence a context of $a_{(i)}$. If $K(a_{(i)})= K(b_{(j)})$, it is called that $a_{(i)}$ matches $b_{(j)}$ with context $K(a_{(i)})$. Let $c(\xi, \eta : K(\xi))$ be the cost to replace ξ with η under the condition $K(\xi),$ where $\xi, \eta \in T^*$. Without $K(\xi)$, c is a context-independent cost. Hereafter we use the notation $\alpha \underline{\xi} \beta$ for a context of ξ, where

$K(\xi)=\alpha\xi\beta$, and $\alpha,\beta \in V^*$.

$c(\xi,\eta:\alpha\xi\beta)$ is interpreted as follows.

If $\xi\neq\lambda$ and $\eta\neq\lambda$, c is the cost of substitution of η for ξ.

If $\xi\neq\lambda$ and $\eta=\lambda$, c is the cost of deletion of ξ.

If $\xi=\lambda$ and $\eta\neq\lambda$, c is the cost of insertion of η.

Let C be the set of cost functions. We use the notation $S(\alpha,\beta)$ to express the one dimensional context-dependent similarity from α to β. $S(\alpha,\beta)$ can be computed by the following recurrence relation.

$$s(i,j)= \min \begin{cases} s(i-1,j)+c(a_{(i)},\lambda:\alpha_{(i)},\lambda:\alpha_{(i)}\underline{a_{(i)}}\beta_{(i)}) \\ s(i-1,j-1)+c(a_{(i)},b_{(j)}:\alpha_{(i')}\underline{a_{(i')}}\beta_i), \\ s(i,j-1)+c(\lambda,b_{(j)}:\alpha_{(i'')}\underline{\lambda}\beta_{(i'')}), \end{cases}$$

where

$$s(i,0)=\sum_{k=1}^{i}c(a_{(k)},\lambda:\alpha_{(k)}\underline{a_{(k)}}\beta_{(k)}),$$

$$s(0,j)=\sum_{k=1}^{j}c(\lambda,b_{(k)}:\underline{\lambda}a_{(1)}a_{(2)}\cdots).$$

Then $S(\alpha,\beta)=s(m,n)$.

The computation complexity of the algorithm is proportional to mn.

2.3 Error-correcting parsing problem based on the one dimensional context-dependent similarity.

We now define the error-correcting parsing problem discussed in this paper. Assume that the length of context is not greater than 1. If we consider an error-correcting parsing problem without this assumption, an algorithm becomes so complex that it requires enormous time and memory space.

[Definition] Assume that a context-free grammar G= (N,T,P,S) , a set of cost functions, an input sentence ω and the maximum error cost W_{max} are given. The error-correcting parser finds sentences $\tilde{\omega}$s which satisfy the following two conditions,

    (1) $\tilde{\omega} \notin L(G)$,

    (2) $S(\tilde{\omega},\omega) \leq S(\omega',\omega) \leq W_{max}$ for any sentence ω' in $L(G)$.

If there exists no such sentence, then ω is rejected.

3. An error-correcting parsing(ECP) algorithm.

First, we define the error-handling rule as follows;

    (1) match : $X \rightarrow (mat)$

    (2) substitution error : $X \rightarrow (sub)/K(X)$

    (3) deletion error : $X \rightarrow (del)/K(X)$ and

    (4) insertion error : $X \rightarrow X(ins), X \rightarrow (ins)X/K(X)$,

where $X \notin T$ and $K(X)$ is a context of X. The set of error-handling rules P_E is defined depending on the set of cost functions C as described in §3.2.

Let $G=(N,T,P,S)$ be a context-free grammar and $\omega=a_{(1)}a_{(2)} \cdots a_{(n)}$ be an input sentence. The ECP algorithm proposed in this paper constructs a triangular parse table T, whose entry is denoted by $t_{i,j}$ for $0 \leq i \leq n-1$ and $0 \leq j \leq n-i$. Each $t_{i,j}$ is a set of items $[\psi,\delta,\zeta,w]$. ψ is a dotted rule of the form $X \rightarrow \alpha \cdot \beta/\theta$,where "$\cdot$" is a metasymbol not in V. Dotted rules are classified into two groups. If $X \rightarrow \alpha\beta \in P$ and $\theta=\phi$, then $X \rightarrow \alpha \cdot \beta$ is called a dotted rule. If $X \notin T$, then $X \rightarrow \alpha \cdot \beta/\theta$ is called a dotted error-handling rule, where θ is a context. δ and ζ have the information about the context and are written in the form (x,y) ,where $x,y \notin T$. We use a notation ε to indicate (ϕ,ϕ), where ϕ is the empty symbol. The last component w is error cost. An item which has a dotted rule of the form $X \rightarrow \alpha \cdot$ is called a final-item and an other item is called a stranded-item.

We will describe a relationship between an item and error-correcting parsing. Let $\omega=a_{(1)}a_{(2)} \cdots a_{(n)}$ be an input sentence. Assume that $[X \rightarrow \alpha \cdot \beta,(l,r),(L,R),w]$ is in $t_{i,j}$. This means that there exists at least one derivation $\alpha \overset{*}{=}> \xi=Rb_{(1)}b_{(2)} \cdots b_{(m)}l$ and the context-dependent similarity between ξ and $\omega_{(i+1,i+j)}$ is w on the assumption that r is on the right of ξ and L is on the left of ξ, where $\omega_{(i+1,i+j)}=a_{(i+1)} \cdots a_{(i+j)}$ (see Fig.3-1), and it is denoted by $S(L[\xi]r,\omega_{(i+1,i+j)})$.

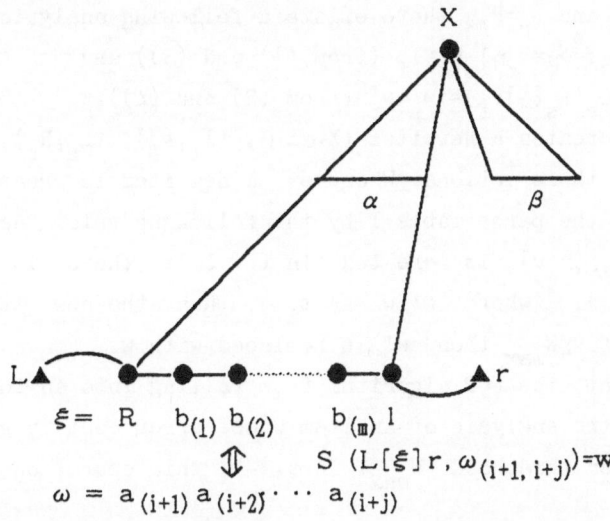

Fig.3-1 A relationship between items and trees

3.1 Parsing mechanism

Parsing mechanism of the ECP algorithm is similar to that of the Earley's parser. The Earley's parser is composed of three procedures, these are , SCAN, CREATE and COMPLETE.

In the ECP algorithm, SCAN is replaced with the error-handling procedure(ERROR) described in the next section. In this section , we give an informal explanation of CREATE and of COMPLETE.

CREATE: Assume that $[X \rightarrow \alpha \cdot Y\beta, \delta, \zeta, w]$ is in $t_{i,j}$, where $X, Y \in N$, $\alpha, \beta \in V^*$, CREATE adds a new item $[Y \rightarrow \cdot \gamma, \delta, \epsilon, 0]$ to $t_{i+j+1,0}$ for all $Y \rightarrow \gamma$ in P.

COMPLETE: COMPLETE will be activated when a final-item is created. Assume that a final-item $E_f = [Y \rightarrow \gamma \cdot, (l_f, r_f), (L_f, R_f), w]$ be created in $t_{p,q}$. There must be a stranded-item $E_s = [X \rightarrow \alpha \cdot Y\beta, (l_s, r_s), (L_s, R_s), w]$ in $t_{i,j}$, where $i+j=p$ and $\alpha \neq \lambda$.

The item E_f means that there exists a following analysis;

    (1) $Y \overset{*}{=}>_f R_f \xi l_f$ and

    (2) $S_c(L_f[R_f \xi l_f]r_f, \omega_{(p+1,p+q)}) = w_f$.

The item E_s indicates that there exists a following analysis;

    (3) $\alpha \overset{*}{=}>_s R_s \eta l_s$ and

    (4) $S_c(L_s[R_s \eta l_s]r_s, \omega_{(i+1,p)}) = w_s$.

If $l_s=L_f$ and $r_s=R_f$, there exists a following analysis ;

(5) $\alpha Y \overset{*}{=>}\mu=R_s\eta l_s R_f \xi l_f$ (from (1) and (3)) and

(6) $S_c(L_s[\mu]r_f)= w_s+w_f$ (from (2) and (4)).

COMPLETE creates a new item $[X \rightarrow \alpha Y \cdot \beta, (l_f,r_f), (L_s,R_s), w_s+w_f]$. Fig.3-2 illustrates these actions. Whenever a new item is created, the item is stored into the parse table T by the following rule. Assume that a new item $[X \rightarrow \alpha \cdot \beta, \delta, \zeta, w]$ is created in $t_{i,j}$. If there is a similar item $[X \rightarrow \alpha \cdot \beta, \delta, \zeta, w']$, where $w' \leq w$ in $t_{i,j}$, then the new item is rejected. Otherwise, if $w \leq W_{max}$ then w' is replaced with w.

To prevent the ECP algorithm from falling into an infinite loop, we will stop the analysis of an item whose error cost is greater than the pre-determined threshold W_{max}. In §.3-3, this operation is denoted by @.

3.2 Error-correction mechanisms.

Assume that $[X \rightarrow \alpha \cdot b\beta, (l,r), (L,R), w]$ is in $t_{i,j}$, where $X \not\in N$, $b \not\in T$ and $\alpha, \beta \not\in V^*$. This means that $\alpha \overset{*}{=>}\gamma$, $Sc(L[R\gamma l]r, \omega_{(i+1,i+j)})=w$ and b is derivable on the right of γ. If $r=\phi$ or $r=b$, then the right context assumption r is satisfied. The procedure ERROR acts upon such an item.

First, we describe error-correcting procedures for the cases of match hypothesis and substitution-error hypothesis.

[Match]

If $b=a_{(i+j+1)}$, the procedure creates an item $[b \rightarrow (mat) \cdot, (b,\phi), (\phi,r), 0]$ in $t_{i+j+1,1}$.

[Substitution]

If a cost function $c(b,a_{(i+j+1)}:x\_y)=e$ is in C, check whether the context of b can be derived or not. Since the analysis proceeds from left to right in the ECP, we can check the left context of b but cannot right one. Therefore, the ECP goes on analyzing on the assumption that the right context y will be derived on the right of b. After checking on the condition, create a new item $[b \rightarrow (sub) \cdot /x\_y, (b,y), (x,r), e]$ in $t_{i+j+1,1}$.

Next, let us consider deletion-error hypothesis.

[Deletion]

If there is a cost function $c(b,\lambda:x\_y)=e$ is in C, a new item $[b \rightarrow (del) \cdot /x\_y, (b,y), (x,r), e]$ is created in $t_{i+j+1,0}$.

There are two cases in the insertion-error handling. One is for the case that a symbol is inserted with both side contexts or with only left context and the other is for the case that a symbol is inserted with only right context.

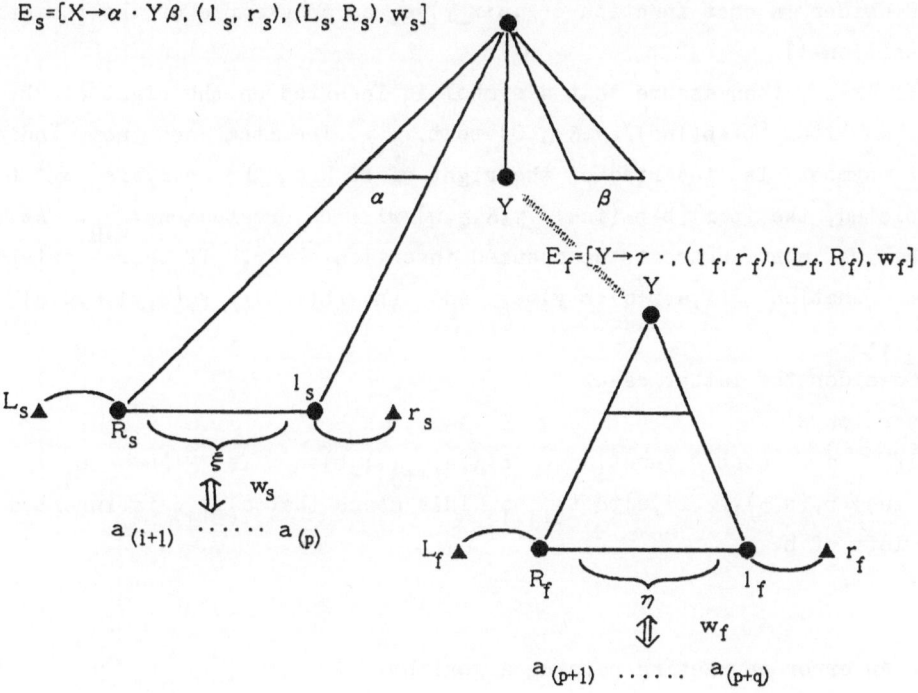

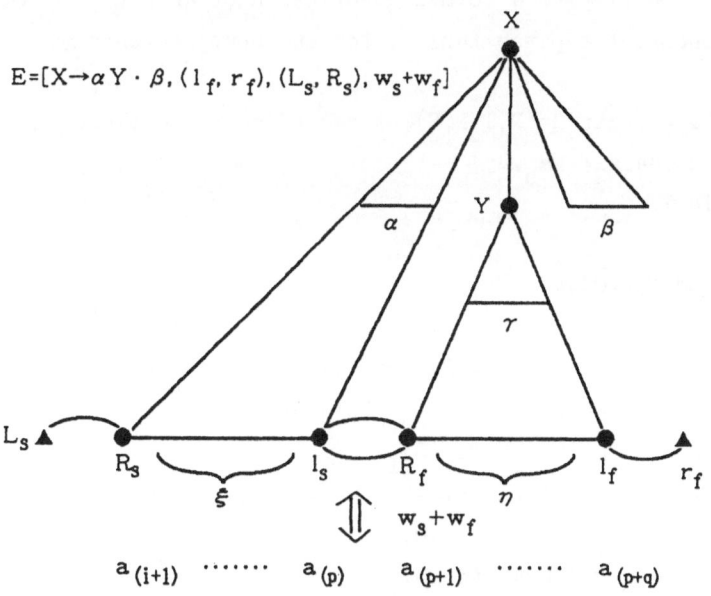

Fig.3-2   The procedure COMPLETE.

Consider a cost function $c(\lambda,u:x\_y)=e$, as an example of the former.
[Insertion-1]

' If x=b , then assume that a symbol is inserted on the right of b and add an item $[b\rightarrow \cdot b(ins)/\_y,\delta,\varepsilon,0]$ to $t_{i+j,0}$. The item does not indicate what symbol is inserted on the right of b. When the analysis of b is completed, the item $[b\rightarrow b\cdot(ins)/\_y,\delta,\zeta,w]$ will be created on $t_{p,q}$. At this time, we must retreat the assumed insertion-error. If there exists a cost function $c(\lambda,a(p+q):b\_y)=e$, add $[b\rightarrow b(ins)\cdot/\_y,(b,y),\zeta,w+e]$ to $t_{p,q+1}$.

Consider the latter case.
[Insertion-2]

If a cost function $c(\lambda,a_{(i+j)}:\_b)=e$ is in C, add $[b\rightarrow(ins)\cdot b,(b,\phi),(\phi,r),e]$ to $t_{i+j,1}$. This means that $a_{(i+j)}$ is inserted on the left of b.

3.3 An error-correcting parsing algorithm.

In this section, we will give a formal description of an ECP algorithm. Algorithm-1 is to construct a parse table T for an input sentence ω.
[Algorithm 1]
Input: A context-free grammar $G=(N,T,P,S)$, a set of cost functions and an input sentence $\omega=a_1a_2\cdots a_n$.
Output: A parse table T.
Method:

     [1] $t_{0,0}@[S'\rightarrow\cdot S\#,\xi,\xi,0]$.

     [2] i:=0

     [3] Repeat procedures (1),(2),(3) and (4) until no new item can be created in $S_{(i)}$, where $S_{(i)}=\{t_{p,q}|p+q=i\}$.

       (1) { The procedure CREATE for $S_{(i)}$ }
Assume a stranded item $[X\rightarrow\alpha\cdot Y\beta,\delta,\zeta,w]$ is in $t_{p,q}$, where p+q=i. For all $Y\rightarrow\gamma\langle P$, $t_{i,0}@[Y\rightarrow\cdot\gamma,\delta,\varepsilon,0]$.

       (2) { The procedure ERROR1 for $S_{(i)}$ }
Assume that a stranded item $[X\rightarrow\alpha\cdot b\beta,(l,r),(L,R),w]$ is in $t_{p,q}$, where p+q=i, r=b or r=ϕ and $X\langle V$.

       (2-1){ For match hypothesis }
If $b=a_{(i+1)}$, then $t_{i,1}@[b\rightarrow(mat)\cdot,\varepsilon,\varepsilon,0]$.

(2-2){ For substitution-error hypothesis }

If $b{\neq}a_{(i+1)}$, $c(b,a_{(i+1)}{:}x\_y)=e{\notin}C$ and $l=x$,

then $t_{i,1}@[b{\to}(\text{sub}){\cdot}/x\_y,(b,y),(x,r),e]$.

(2-3){ For deletion-error hypothesis }

If $c(b,\lambda{:}x\_y)=e{\notin}C$ and $l=x$,

then $t_{i,0}@[b{\to}(\text{del}){\cdot}/x\_y,(b,y),(x,r),e]$.

(2-4){ For insertion-error hypothesis(1) }

If $c(\lambda,a_{(i+1)}\_b)=e{\notin}C$,

then $t_{p+q,0}@[b{\to}(\text{ins}){\cdot}b,(1,\phi),(\phi,r),e]$

(2-5){ For insertion-error hypothesis(2) }

If $c(\lambda,u{:}b\_y){\notin}C$,

then $t_{p+q,0}@[b{\to}{\cdot}b(\text{ins})/\_y,(1,r),(\phi,r),0]$.

Assume that a stranded item $[b{\to}b{\cdot}(\text{ins})/\_y,(b,y),(L,R),w]$ is in $t_{p,q}$, where $p+q=i$.

(2-6){ For item created by (2-5) }

If $c(\lambda,a_{(i+1)}{:}b\_y)=e$ is in C,

then $t_{p,q+1}@[b{\to}b(\text{ins}){\cdot}/\_y,(b,y),(L,R),w+e]$.

(3){ The procedure COMPLETE1 for $S_{(i)}$ }

Assume that a final item $[b{\to}\gamma{\cdot}/\theta_f,(1_f,r_f),(L_f,R_f),w_f]$ is in $t_{p,q}$, where $p+q=i$ and $b{\notin}T$. Search an item of the form $[X{\to}\alpha{\cdot}b\beta/\ \theta_s,(1_s,r_s),(L_s,R_s),w_s]$ in $S_{(p)}$. For each item in $t_{u,v}$, for $u+v=p$, check consistency between these items, that is, if $r_s=R_f$ and $L_s=L_f$ or not. If the condition is satisfied, carry out the following operation;

(3-1)In case that $\beta=(\text{ins})$, $X=b,\theta_s=\_y$ and $\alpha=\lambda$.

If $y=r_f$, then $t_{u,v+q}@[b{\to}b{\cdot}(\text{ins})/\theta_s,(b,y),(L_s,R_s),w_s+w_f]$

(3-2)Otherwise,

$t_{u,v+q}@[X{\to}\alpha b{\cdot}\beta/\theta_s,(1_f,r_f),\zeta,w_s+w_f]$,

where $\zeta=\zeta_f$ if $\alpha=\lambda$, and $=\zeta_s$ otherwise.

(4){ The procedure COMPLETE2 for $S_{(i)}$ }

Assume that a final item $[Y{\to}\gamma{\cdot},(1_f,r_f),(L_f,R_f),w_f]$ is in $t_{p,q}$, where $Y{\notin}N$ and $p+q=i$. Search a stranded-item of the form $[X{\to}\alpha{\cdot}Y\beta/\theta_s,(1_s,r_s),(L_s,R_s),w_s]$ in $S_{(p)}$.

For each item in $t_{u,v}$, where $u+v=p$, check consistency between these items, that is, if $r_s=R_f$ and $l_s=L_f$ or not. If the conditions are satisfied, carry out the following operation;

$t_{u,v+q}@[X{\to}\alpha Y{\cdot}\beta,(1_f,r_f),\zeta,w_s+w_f]$,

where $\zeta=\zeta_f$ if $\alpha=\lambda$, and $=\zeta_s$ otherwise.

[4] i:=i+1

[5] If i≤n, then goto [2]

[6] For all items of the form $[S' \to S \bullet \#, (1, \phi), \varepsilon, w]$ in $t_{0,n}$, $t_{0,n} @ [S' \to S\# \bullet, \varepsilon, \varepsilon, \varepsilon, w]$

S' and # are the extended symbols not in V. The number to be given for the production rule $S' \to S$ is 0. After the construction of T has been completed, if there exist items of the form $[S' \to S\# \bullet, \varepsilon, \varepsilon, w]$, then ω is accepted with error cost w. Otherwise, ω is rejected.

[Example 3-1]

Let us consider the grammar G and a set of cost functions C.

G=(N,T,P,S), N={S,A,C}, T={a,b,c},

P={ (1) S→Sa, (2) S→bA, (3) A→Ca, (4) S→a, (5) C→c }.

C={ c(λ,b:_a)=1, c(a,b:_a)=1, c(c,λ:b_a)=0.5 }

For C, error-handling rules are defined as follow,

(20)	$a \to (\text{ins})\, a \cdot ,\ \varepsilon,\ \varepsilon, 1$		
(21)	$S \to S\, a \cdot ,\ (a, \phi),\ \varepsilon, 1$		
(22)	$S \to b\, A \cdot ,\ (a, \phi),\ \varepsilon, 0.5$		
(23)	$S \to a \cdot ,\ \varepsilon,\ \varepsilon, 1$		
(24)	$S' \to S \cdot \#,\ (a, \phi),\ \varepsilon, 0.5$		
(25)	$S' \to S \cdot \#,\ \varepsilon,\ \varepsilon, 1$		
(26)	$S' \to S\# \cdot ,\ \varepsilon,\ \varepsilon, 0.5$		
(5)	$b \to (\text{mat}) \cdot ,\ (b, \phi),\ \varepsilon, 0$	(13)	$a \to (\text{mat}) \cdot ,\ (a, \phi),\ \varepsilon, 0$
(6)	$a \to (\text{ins}) \cdot a,\ \varepsilon,\ \varepsilon, 1$	(14)	$a \to (\text{mat}) \cdot ,\ (a, \phi),\ (\phi, a), 0$
(7)	$a \to (\text{sub}) \cdot /\_a,\ (a, a),\ \varepsilon, 1$	(19)	$A \to C\, a \cdot ,\ (a, \phi),\ (b, \phi), 0.5$
(8)	$S \to b \cdot A,\ (b, \phi),\ \varepsilon, 0$		
(9)	$S \to a \cdot ,\ (a, a),\ \varepsilon, 1$		
(10)	$S' \to S \cdot \#,\ (a, a),\ \varepsilon, 1$		
(11)	$S \to S \cdot a,\ (a, a),\ \varepsilon, 1$		
(1)	$S' \to \cdot S\#,\ \varepsilon,\ \varepsilon, 0$	(12)	$A \to \cdot C\, a,\ (b, \phi),\ \varepsilon, 0$
(2)	$S \to \cdot S\, a,\ \varepsilon,\ \varepsilon, 0$	(15)	$C \to \cdot c,\ (b, \phi),\ \varepsilon, 0$
(3)	$S \to \cdot b\, A,\ \varepsilon,\ \varepsilon, 0$	(16)	$c \to (\text{del}) \cdot /b\_a,\ (c, a),\ (b, \phi), 0.5$
(4)	$S \to \cdot a,\ \varepsilon,\ \varepsilon, 0$	(17)	$C \to c \cdot ,\ (c, a),\ (b, \phi), 0.5$
		(18)	$A \to C \cdot c,\ (c, a),\ (b, \phi), 0.5$

Fig.3-3 A parse table constructed by Algorithm-1 in Example 3-1.

(6) a→(ins)a, (7) a→(sub)/_a, (8) c→(del)/b_a,

(9) x→(mat), for any x∉T .

The parse table T for an input sentence ω=ba is given in Fig.3-3. The items (21),(22) and (23) are candidates for the final solution. The parsing trees corresponding to the items (21), (22) and (23) are shown in Fig.3-4 (a),(b) and (c), respectively. By step [6], the item (26) with least error cost is selected and the item (26) is created.

Algorithm-2 is to make a right parse from the parse table T.

[Algorithm 2]

Input:A cycle-free CFG G=(N,T,P,S), a set of cost functions C, an input sentence ω and a parse table T.

Output: A right parse π for ω, or message "reject".

Method: Number the rules in $P \cup P_E$.

Search $[S' \rightarrow S\#\cdot, \varepsilon, \varepsilon, w]$ in $t_{0,n}$, do the procedure $R(S' \rightarrow S\#\cdot, \phi, \phi, w, 0, n, \lambda)$.

The procedure R is defined as follow:

procedure $R(A \rightarrow \alpha\cdot, L, R, w, i, j, \pi)$;

 var k,l: integer;

 begin

  [1] Let π be h followed by the previous value of π, where h is the number of production A→α.

  [2] If $\alpha = X_1 X_2 \cdots X_m$, then k:=m ; v:=j;

  [3] Find an item $[X_k \rightarrow \beta\cdot, (L,R), (L',R'), w_f]$ in $t_{p,q}$ for p and w_f such that $[A \rightarrow X_1 X_2 \cdots X_{k-1}\cdot X_k \cdots X_m, (L',R'), (L,R), w-w_f]$ is in $t_{i,j-p}$. Call $R(X_k \rightarrow \beta\cdot, L', R', ws, p, q, \pi)$.

   k:=k-1; v:=p;

  [4] Repeat [3] until k=0;

 end;   ⊞

[Example 3-2]

Let us apply Algorithm-2 to the parse table of Example 3-1. Initially, we execute $R(S' \rightarrow S\cdot, \phi, \phi, 0.5, 2, \pi)$. The sequence of calling the procedure R is described in Fig. 3-5. After the completion of this call, we obtain the right parse π=9859320.

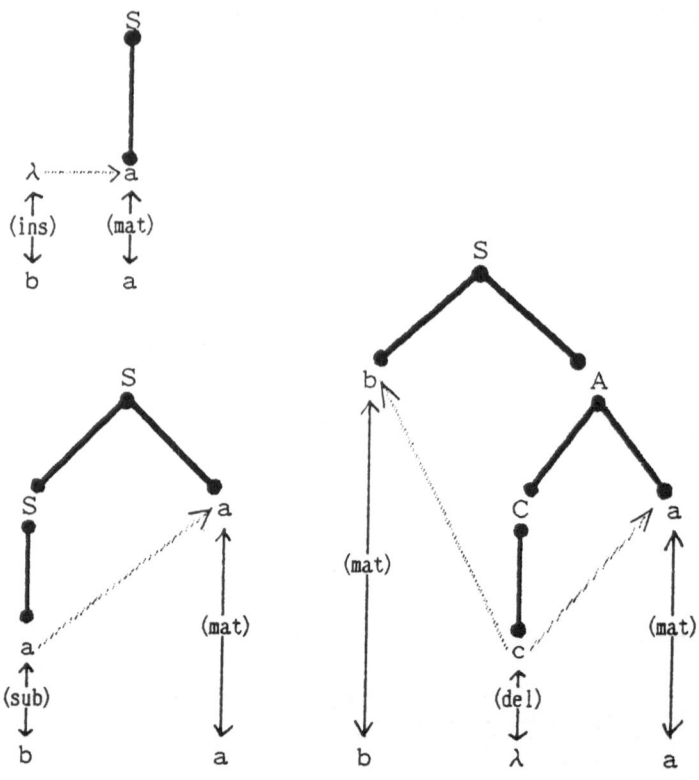

Fig.3-4   Parsing trees in Example-1.

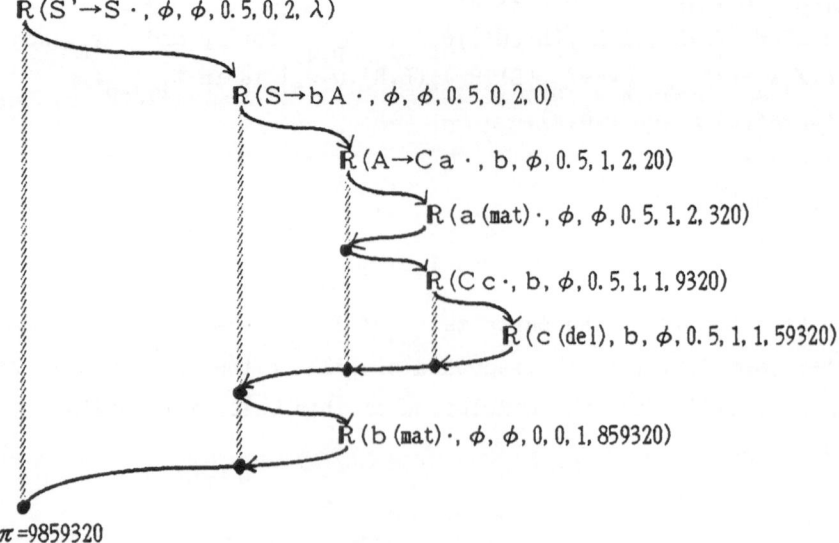

Fig.3-5   The diagram of a sequence of calling the procedure **R** in Example 3-2.

3.4 Properties of the algorithm

In the following, we will give the space and time complexities of the Algorithm-1. Let us begin with upper bounds of the space complexity. An item has four components ψ, δ, ζ and w as described at the beginning of this chapter. Assuming that a context-free grammar G and a set of cost functions C are fixed. The maximum number of strings that can appeare on (ψ, δ, ζ) is $K*|P \cup P_E|*(|T|+1)^4$, where K is the maximum length of left hand side of rule in P P_E and $|P \cup P_E|$ denotes the number of elements in $P \cup P_E$. For each triple (ψ, δ, ζ), only one item (ψ, δ, ζ, w) is stored in $t_{i,j}$. Therefore, each entry on T has a fixed number of items. Then, the upper bounds of the space complexity is $O(n^2)$.

Next, consider upper bounds of the time complexity. It takes a fixed amount of time to create an item in both CREATE and ERROR. Since there are $O(i)$ items on $S_{(i)}$, the total space complexity is $O(n^2)$. Both COMPLETE-1 and -2 carry out the search procedure for all the final-item in $S_{(i)}$. The search procedure passes over $S_{(1)}$ through $S_{(i-1)}$. Since it takes fixed time to check context relationship between two items, the time complexity of the search procedure is $O(i^2)$. If the condition is satisfied, it takes fixed time to create a new item. Therefore, the total time complexity is $O(n^3)$.

4. Concluding remarks

We have presented an error-correcting parsing algorithm for a context-free language based on the one dimensional context-dependent similarity. The algorithm is a modified version of Lyon's error-correcting parser for a context-free language. Compared with the Lyon's, the proposed algorithm has following advantages ;

(1) Since the error-correcting mechanism is based on the one dimensional context-dependent-similarity, the ECP can be applied to pattern recognition with context-dependent deformation.

(2) The space and time complexity is as same as those of the Lyon's.

Acknowledgement

The authors would like to thank Dr.Kamata who is an assistant professor of Utsunomiya University.

References
(1) Fu,K.S.: "Error-correcting parsers for syntactic pattern recognition" in DATA Structure, Computer Graphics and Pattern Recognition, Klinger et.al. Eds. Academic,N.Y.(1976).
(2) Aho,A.V. and Peterson,T.G.: "A minimum distance error-correcting parser for context-free languages", SIAM J.Comput.,1,pp.305-312 (1972).
(3) Earley,J.:"An efficient context-free parsing algorithm", CACM, 13, pp.94-102 (1970).
(4) Lyon,G.:" Syntax-directed least-errors analysis for context-free languages; A practical approach", Comm.ACM, 17, pp.3-14(1974).
(5) Fung,L.W. and Fu, K.S.: "Maximum-likelihood syntactic decoding", IEEE Trans. Inf. Theory, IT-21, pp.423-430(1975).
(6) Yamasaki,S. and Tonomura,T.:"On a bottom-up least error correction algorithm for context-free languages", J.Inf.   Process.Soc.Jpn., 18,pp.781-788(1977).
(7) Cocke,J. and Schwarts,Y.T.:"Programming language and their compilers", Courant Inst. of Mathematical Science, N.Y.(1967).
(8) Tanaka,E. : "An improved error-correcting parser for a context-free language", Trans. IECE, Jpn, E-67, 7, pp.379-385(1984).
(9) Graham,S.L., Harrison,M.A. and Ruzzo, W.L. :" An improved context-free recognizer", ACM Trans. Program. Lang. Syst.,2,pp.415-462(1980).
(10) Levenshtein,B.I. : " Binary codes with correction of deletion, insertion and substitution of symbols", Dokl.Acak.Nauk.SSSR,163, pp.845-848(1965).
(11) Okuda,T., Tanaka,E. and Kasai,T.:" A garbled word correcting method by an extended metric", Ann. Joint Meeting of Electrical and Electronics Eng. Tokai Distinct,Jpn.,18a-B-6(1972).
     Okuda,T., Tanaka,E. and Kasai,T.:" A method for the correction of garbled words based on Levenshtein metric", Trans. Comput., C-25,20, pp.172-178(1976).
(12) Tanaka,E. : "A context-dependent similarity measure for strings", Trans. IECE, Jpn, J-67-A, 6, pp.612-613(1984).

ORDERED STRUCTURAL MATCHING

L. Shapiro

Dept. of Electrical Engineering
University of Washington
Seattle, WA 98195
U.S.A.

1. Introduction

A structural description of an object is a representation of the object that lists its parts, their attributes, and their interrelationships. Structural descriptions can be used to represent object models or to represent information extracted from an image of an object. Structural matching is the process of comparing two structural descriptions to determine how similar they are. Matching a structural description extracted from an image to one representing an object model can tell us whether the object in the image is an instance of the object being modeled. Matching two structural descriptions, one from each of the images of a stereo pair can determine the correspondences between the images. Matching two structural descriptions of object models can determine the similarities between the models for purposes of organizing a database of models.

Most structural matching has been done by some variation of a backtracking tree search with discrete or continuous relaxation (Rosenfeld, Hummel, and Zucker, 1976). Such a procedure has exponential complexity and is infeasible when the objects to be matched have large numbers of parts and impractical in industrial vision scenarios where near real time speeds are often required. In such cases, a polynomial time algorithm is desirable.

Although the general structural matching problem is an NP complete problem, vision researchers have, for a number of years, found ways to severely prune the tree to be searched. Researchers who have developed such procedures include Waltz (1975), who was the first to use discrete relaxation to solve a computer vision problem and Gaschnig (1974), Freuder (1978), Haralick and Elliot (1980), Mackworth (1977), Montanari (1974), Haralick and Shapiro (1979), and Nudel (1983), who have all published related work on constraint satisfaction problems. Freuder (1976) used a form of what we are calling ordered structural matching to compare line drawings. He converted each line drawing to a list of vertex lists; each vertex list represented the bounding vertices of a region in counterclockwise order and the vertex lists themselves were arranged in order according to the surrounds relationship. Once he had a list representation of each line drawing, he performed a simple string matching algorithm to determine isomorphism. He did not extend the technique to inexact matching.

The pruning procedures are based on the fact that in most vision problems, spatial relationships severely constrain the possible mappings from the parts of one structural description to the parts of a second structural description. Thus for some classes of

NATO ASI Series, Vol. F45
Syntactic and Structural Pattern Recognition
Edited by G. Ferraté et al.
© Springer-Verlag Berlin Heidelberg 1988

problems, the tree search time is greatly reduced. We now consider a class of problems where the tree search can entirely eliminated. In order to discuss this class of problems, we first review the notation we use for structural matching problems.

Let O_A and O_B be two objects that we wish to compare. Set A is the set of *primitives* or parts of object O_A, and B is the set of primitives of O_B. A *relational description* D_A of object O_A is a sequence of relations $< R_1, \ldots, R_I >$ where each relation R_i is a subset of A^{n_i} for some positive integer n_i. Similarly, a relational description D_B of object O_B is a sequence of relations $< S_1, \ldots, S_I >$ where each relation S_I is a subset of B^{n_i}. Thus, a relational description consists of a sequence of relations, each relation consisting of n-tuples of primitives of the object being described and each such n-tuple describes the relationship among the n primitives involved in that n-tuple. This can be extended to the concept of an *attributed relational description* where there is a set Q of attributes and each relation R_i is a subset of $A^{n_i} \times Q^{m_i}$ for non-negative integers n_i and m_i. In this case, a tuple consists of n primitives and the m attributes of that relationship.

We have published several definitions of structural matching using the above terminology and several algorithms for determining the similarity between two structural descriptions (Shapiro and Haralick, 1985; Shapiro and Haralick, 1981; Haralick and Shapiro, 1979). All of these share the basic goal of finding a mapping from A to B that preserves the relationships among primitives and the attributes of those relationships. In these definitions and algorithms, the primitives don't have any inherent ordering, and any primitive of A can potentially map to any primitive of B.

However, in many computer vision problems, the spatial arrangement of the primitives allows the definition of an *ordering* on the primitives. Suppose that we wish to compare D_A to D_B. Suppose the ordering for A is $< a_1, a_2, \ldots, a_s >$ and the ordering for B is $< b_1, b_2, \ldots, b_t >$. Suppose we hypothesize that primitive a_i maps to primitive b_j. The ordering tells us that either

1) a_{i+1} maps to b_{j+1},
2) a_{i+1} maps to $b_{i+k}, k > 1$, and no primitive of D_A maps to any of $b_{i+1}, b_{i+2}, \ldots, b_{i+k-1}$, or
3) a_{i+1} maps to no primitive of D_B,

where a_{i+1} is the next primitive (circularly) after a_i in the ordering. Thus, once a_i is mapped to b_j, the ordering can be used to find the correspondences between all the other primitives in polynomial time. If k can be bounded, then the algorithm is particularly efficient.

Ordered structural matching requires 1) a method of ordering the primitives of an object and 2) a method of deciding, once a correspondence is hypothesized between a primitive of D_A and a primitive of D_B, whether the next primitives in the ordering can correspond.

We have used ordered structural matching in a two-dimensional shape matching problem (Shapiro, et al., 1986). In this paper, we describe the technique used in that application and discuss possible extensions to other two-dimensional applications and to three-dimensional problems.

2. Two-Dimensional Shape Matching

A two-dimensional shape is an entity that has a definite structure; thus structural matching is an obvious method for determining how similar one shape is to another. In most of the previous work on structural shape matching, the shape itself was regarded either as a sequence of boundary points or as an area of the plane. In the former case, the extracted primitives were generally line or arc segments as in Davis (1977,1979) and Bjorklund and Pavlidis (1981). In the second case, the primitives were usually pieces of this area. Pavlidis (1972) decomposed shapes into convex pieces, Feng and Pavlidis (1975) decomposed them into convex parts, T-shaped parts and spirals, and Maruyama (1972) segmented them into angularly simple regions where each such region had at least one interior point that could "see" its entire boundary. Avis and Toussaint (1981) presented a fast algorithm for the Maruyama type of decomposition. Shapiro and Haralick(1979) decomposed shapes into "near convex pieces" which were the clusters of a binary relation that associated pairs of boundary points which could "see" each other. More recently, Fairfield (1983) segmented a shape into "intuitively pleasing" pieces using the Voronoi diagram of the dot pattern of the shape, and Nevins (1982) decomposed shapes into convex, spiral, and biconcave regions using knowledge of smoothness and symmetry criteria.

All of these decomposition procedures begin with a set of points, usually a polygonal approximation to the boundary of the shape, which must be extracted from the image of the shape before the decomposition, description, and matching can take place. Furthermore, the decomposition algorithms take at least $n \log n$ time for n points, and the structural matching itself is usually an exponential tree search, as discussed above.

In this section, a new approach to both decomposition and matching is described. This approach uses the operations of mathematical morphology (Serra,1982) to decompose the shape and a polynomial time ordered structural matching algorithm. Thus, it constitutes a feasible method for industrial applications which require very short computational times.

2.1 Primitive Extraction Using Mathematical Morphology

Extraction of primitives using image processing followed by segmentation or decomposition techniques can be too slow a process to enable its use in real time machine vision applications. It is desireable, instead, to be able to extract the primitives using only extremely simple image processing operations and no separate decomposition procedure. The operations of mathematical morphology fit this description and are available on fast pipeline machines.

A binary image can be represented by a set of points, one point for each pixel with value 1 in the image. The morphologic operators work with two binary images: the original data to be analyzed and a structuring element. Each structuring element has a shape which can be thought of as a parameter to the operators. Typical shapes for structuring elements are circles and rectangles, but any arbitrary shape may be used. A structuring element also has one point denoted as its origin.

Let X be the set of points representing the original binary image and B be the set of points representing a structuring element. The *dilation* of X by B, denoted $X \oplus B$, is defined by

$$X \oplus B = \bigcup_{x \epsilon X} B_x$$

where

$$B_x = \{b + x \mid b \epsilon B\}.$$

The *erosion* of X by B, denoted $X \ominus B$, is defined by

$$X \ominus B = \{y \mid B_y \subset X\}.$$

More intuitively, the dilation of X by B is the set of all points that are covered by B when the origin of B is placed at every point of X. The erosion of X by B is the set of all points where the origin of the structuring element B can be placed so that B is completely contained in X.

Dilations and erosions frequently occur in pairs. A dilation of X by B followed by an erosion of the result by B is called a *closing* and is denoted $X \bullet B$.

$$X \bullet B = (X \oplus B) \ominus B.$$

The closing of X by B is the set of all points that are contained in some translation of B whose intersection with X is non-empty. An erosion of X by B followed by a dilation of the result by B is called an *opening* and is denoted $X \circ B$.

$$X \circ B = (X \ominus B) \oplus B.$$

The opening of X by B is the set of all points of X that are contained in some translation of B which is entirely contained in X. The operations of mathematical morphology, along with a set of logical and arithmetic operations on images form a powerful image processing system.

A two-dimensional shape can be digitized to produce a binary image containing a single connected set of pixels having value 1. A *morphological shape transformation* is a sequence of morphological and logical operators. The result of applying a morphological shape transformation to the original binary image of a shape is a new image with zero or more connected components. These components become the primitives of the shape. Consider the shapes shown in Figure 1. For these simple shapes, the sequence of operators consisting of

1) an opening by a disk whose radius depends on the size of the shape and

2) a logical operator that subtracts the result of the opening from the original image leaving a residue

yields a set of primitives that can be used by a structural matching algorithm to distinguish the shapes. Figure 2 shows the primitives extracted for each of the shapes of Figure 1. The time to apply the sequence of operators to the 512 X 512 binary image was 1.4 seconds on a GENESIS 2000 development system; this system was described in Sternberg (1985).

After the primitives have been isolated, a connected components procedure is run on the residue image to determine the features of each individual primitive. The features measured for each primitive and for the original shape are centroid, area, and ratio of minor to major axis length of the best fitting ellipse. We define the models created from this data in the next section.

2.2 Structural Shape Models

An ordered structural matching algorithm, when comparing two structural descriptions and when primitive a_i of the first description maps to primitive b_j of the second, must have a way of deciding the mappings for all the other primitives without a combinatorial search. This can be achieved if there is a way to order the primitives of each shape so that when a_i maps to b_j, a_{i+1} can only map to b_{j+1} or another primitive that falls (circularly) soon after b_{j+1} in the ordering of the primitives of the second shape. We use the centroid of the shape to order the primitives as follows.

Let S be a shape. Denote its centroid by centroid(S). Let C be the set of primitives extracted from $S, \mid C \mid = N$. For each primitive $c \epsilon C$, consider the vector from centroid(S) to centroid(c). The vector has a direction D_c and a length L_c. Then the primitives of C can be placed in the ordering $c_1, \ldots, c_N$, where $D_n < D_{n+1}, n = 1, \ldots, N-1$. Using this ordering concept, a *centroid based model* captures the spatial relationships among the primitives in their relationships to the centroid. Such a model M is a 6-tuple M = (model_name, centroid, area, minor_major_ratio, number_of_primitives, sequence_of_primitives) where sequence_of_primitives is an ordered list of primitive descriptions. Each primitive description P is a 4-tuple P = (centroid, area, direction, length) where direction and length refer to the vector from the centroid of the shape to the centroid of the primitive. Figure 3 gives the model produced for a sample shape.

Given two models M_c (the candidate model) and M_u (the unknown model), and a mapping

$$f : \text{primitives}(M_c) \Rightarrow \text{primitives}(M_u) \bigcup \{\text{nil}\},$$

the error of the mapping f is computed as the normalized sum of the area error, the length error, and the direction error. The area error is defined by

area_error (f) =

$$\sum_{f(i)=j} \mid \text{area}(i) - \text{scaled\_area}(j) \mid$$

$$+ \sum_{f(i)=nil} \mid \text{area}(i) \mid + \sum_{\substack{j \\ \text{there is no } i \\ \text{with } f(i)=j}} \mid \text{scaled\_area}(j) \mid .$$

The area of each primitive of the unknown model M_u is scaled according to the ratio of area(c) to area(u), resulting in a scaled area for each primitive of M_u. Thus,

Figure 1. Three simple shapes.

Figure 2. The primitives extracted from the shapes of Figure 1.

the area error is independent of scale. Note that the first term of area_error refers to differences in pairs of primitives associated by mapping f, the second term pertains to primitives of M_c that are missing in M_u, and the third term pertains to primitives of M_u that are missing in M_c. Similarly, the length error is defined by

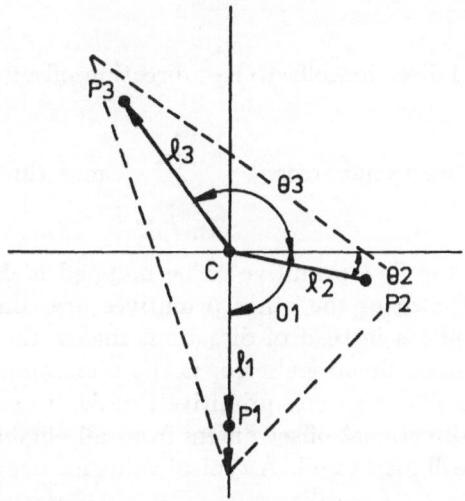

Model: right_tri

centroid = 360 270   area = 15463

major length = 105   minor length = 55

minor/major = 0.523809

Primitives:

| | centroid | | | best fitting ellipse axes | | | vector from main centroid to centroid | |
	x	y	area	maj	min	min/maj	direction	length
P1	360	176	680	18	16	.88889	-90.0°	84.1
P2	422	264	206	13	6	.46154	-4.4°	69.5
P3	317	355	1508	32	20	.62500	122.3°	90.0

Figure 3. A triangular shape and the centroid based model constructed from the shape.

length_error(f) =

$$\sum_{f(i)=j} | \text{length}(i) - \text{scaled\_length}(j) |$$

$$+ \sum_{f(i)=nil} | \text{length}(i) | + \sum_{\substack{j \\ \text{there is no } i \\ \text{with } f(i)=j}} | \text{scaled\_length}(j) |.$$

Finally, the direction error is defined by

direction_error(f) =

$$\sum_{f(i)=j} \mid \text{direction\_offset}(i) - \text{direction\_offset}(j) \mid$$

$$+ \sum_{f(i)=nil} \text{max\_directional\_error} + \sum_{\substack{j \\ \text{there is no } i \\ \text{with } f(i)=j}} \text{max\_directional\_error}.$$

The directional offset of the first primitive to be mapped is defined to be zero, and the directional offsets of each of the other primitives are computed relative to the first. Using directional offsets instead of directions makes the procedure orientation invariant. The constant max_directional_error is the maximum amount of directional error allowed in order for f to map any primitive i of M_c to some primitive j of M_u. Primitives of M_c whose directional offset differs from all eligible primitives of M_u by more than this constant will map to nil. A typical value for max_directional_error is 15 degrees.

The area, length, and direction errors can be normalized as follows. Let n_c be the number of primitives of candidate model M_c and n_u the number of primitives of unknown model M_u. Then the area normalization factor ANM is given by

$$\text{ANM} = \sum_{i \in M_c} \text{area}(i) + \sum_{j \in M_u} \text{scaled\_area}(j),$$

the length normalization factor LNM is given by

$$\text{LNM} = \sum_{i \in M_c} \text{length}(i) + \sum_{j \in M_u} \text{scaled\_length}(j),$$

and the directional normalized factor DNM is given by

$$\text{DNM} = (n_c + n_u) * \text{max\_directional\_error}.$$

Using the above definitions, the mapping error of f can be given by
mapping_error(f) =

$$\frac{ka * \text{area\_error}(f)}{\text{ANM}} + \frac{kl * \text{length\_error}(f)}{\text{LNM}} + \frac{kd * \text{direction\_error}(f)}{\text{DNM}}.$$

where ka, kl, and kd are weighting constants.

The mapping error only deals with the attributes and spatial relationships of the primitives. In order to also be able to take into account the attributes of the entire shape, we define

$$\text{main\_error} = \mid \text{minor\_major\_ratio}(M_c) - \text{minor\_major\_ratio}(M_u) \mid .$$

Then the total error of a match with mapping f is given by

$$\text{total\_error}(f) = \text{mapping\_error}(f) + \text{main\_error}.$$

2.3 The Matching Algorithm

Each primitive of M_c can possibly map to any of the primitives of M_u or may map to nil. Once $f(cfirst) = ufirst$ is fixed the remainder of the mapping may be deterministically constructed. The procedure MAP that constructs the rest of the mapping is summarized below.

procedure MAP(cfirst, ufirst, f, tried)
*
integer cfirst, ufirst, cpart, upart, n_c, n_u;
real cangle, uangle, max_directional_error;
boolean array tried [*, *];
mapping f;
*
* *Map cfirst in M_c to ufirst in M_u.*
*
f (cfirst) = ufirst;
*
* *Begin with cpart, the part after cfirst*
* *and upart, the part after ufirst.*
*
  cpart = successor(cfirst, n_c),;
  upart = successor(ufirst, n_u);
  while (cpart $\wedge = $ cfirst)
*
  {
*
* *Retrieve cangle, the directional offset*
* *of cpart and uangle, the directional offset*
* *of upart.*
*
  cangle = dir_offset(cpart);
  uangle = dir_offset(upart);
*
* *Initially cpart has nothing to map to.*
*
  f(cpart) = nil;
*
* *Try upart and, if necessary,*
* *those of its successors*
* *that are not too far away in directional*
* *offset for a match to cpart.*
*
  while (uangle < cangle + max_directional_error
      and upart $\neq$ ufirst)
*
* *If the directional offsets of cpart and upart*
* *are close enough, then map cpart to upart.*
*
      if closenuf (uangle, cangle)
      {
      f(cpart) = upart;
      tried [cpart,upart] = true;
      upart = successor (upart,n_u);
      break;

```
        }
*
* If the directional offsets are not
* close enough then move on to the part
* following cpart in M_u.
*
     else
      {
      upart = successor(upart,n_u);
      uangle = dir_offset(upart);
      }
*
* Move on to the next part following cpart in M_c.
*
   cpart = successor (cpart, n_c);
   }
end MAP
```

In this procedure, the *successor* function, when given a primitive index and number of primitives, is assumed to return the index of the next primitive, circularly, like a modulo function. The function *closenuf* decides if the directional offsets of two primitives are close enough to map the first to the second. The boolean array *tried* marks which pairs have been instantiated as part of a mapping, so they will not be tried again as the first pair of a mapping.

MAP goes through each of the n_c parts of M_c trying a few parts of M_u that are close enough in directional offset to match. If we assume that the number of parts close enough is bounded by a constant, then MAP has complexity $O(n_c)$.

Given this procedure MAP, it is possible to determine and calculate the error for each mapping f that maps a fixed primitive i of M_c to some starting primitive of M_u. If it were known that primitive i of M_c must map to some primitive of M_u and not to *nil*, then the lowest error mapping of these n_u mappings would be the best mapping and no more work would need to be done. However, since it is possible in the general case for a primitive i of M_c to map to no primitive of M_u in the true best mapping, it is necessary to try each primitive of M_c as the first. Thus the matching procedure is as follows.

procedure MATCH
```
*
integer n_c,n_u;
boolean tried [n_c, n_u];
real besterr, err;
mapping bestmap,amap;
*
besterr = 9999.;
*
* Try each part of M_c as the first part.
*
for i = 1 to n_c
*
* Try each part of M_u as the match of the first part.
*
  for j = 1 to n_u
*
* Don't bother to try any pair (i,j) that has
```

```
* already been instantiated as part of a previous
* mapping
*
    if (∧ tried [i,j])
      {
*
* Map returns the constructed mapping amap.
*
      map(i, j, amap, tried);
*
* Err is the mapping error of mapping amap.
*
      err = map_err(amap,i);
*
* If this is a lower error mapping than any so far, keep it.
*
        if (err < best_err)
          {
          besterr = err;
          bestmap = amap;
          }
        }
*
end MATCH
```

Within procedure MATCH, the function *maperr* is called to compute the error associated with a given mapping and first primitive mapped. There is no directional error for the first primitive mapped since its offset and the offset of the primitive it maps to are both zero.

Since MATCH has an $O(n_c * n_u)$ loop in which it calls MAP, which has complexity $O(n_c)$ and map_err, which has complexity $O(n_c)$, the entire procedure has worst case complexity $O(n_c^2 * n_u)$. The use of the *tried* array makes it even better in practice. The important thing is that it is of polynomial, not exponential, complexity. A discussion of experimental results of this shape matching system is given in Shapiro, MacDonald, and Sternberg (1987).

3. Proposed Extensions

In the previous section, we described a two-dimensional shape matching system that utilizes ordered structural matching. We believe that the ordered structural matching concept can be used in more complex realms. We discuss two possible extensions here: two-dimensional matching of more complex entities and three-dimensional matching.

Our simple two-dimensional shapes could be decomposed into primitives arranged circularly around the border of the shape, each at a different angle from the horizontal with respect to the centroid of the shape. The more general two-dimensional matching problem can involve more complex entities where the features that become primitives are located in arbitrary positions on the shape. Such features may include edges, areal features, holes, and symbols printed on the shape. The main difference between comparing two of these more general shapes and the simple ones of the previous section

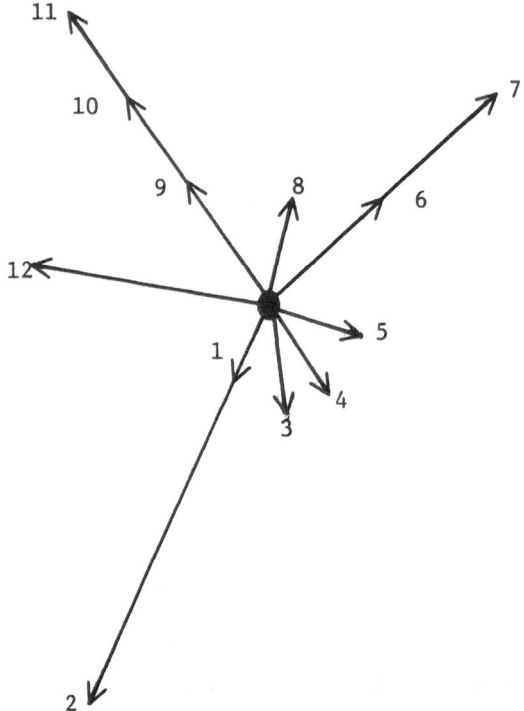

Figure 4. The arrangement of the primitives of a complex two-dimensional object.

is that two or more primitives may lie on or close to the same vector from the centroid of the shape. This is illustrated in Figure 4.

Even with this more complex arrangement of primitives, we should not have to resort to general graph matching. We can order the primitives according to the angles of the vectors from the centroid of the object to the centroids of the primitives and within these angles by the lengths of the vectors. The numbers in Figure 4 show the ordering of this hypothetical object. Furthermore, to avoid problems due to similar, but not identical angles, we can group together primitives that are close in angle and distinguish the primitives in that group only by the differences in vector length. Primitives that are very close together in both angle and length are not suitable for matching unless there is another obvious way to distinguish them.

Three-dimensional object matching is much more difficult than two-dimensional object matching, whether the data being analyzed are gray tone images or range images. Consider the case of a single gray tone image of the three-dimensional object. We would like to map the primitive features detected in the image to the three-dimensional features from which they came. One way to do this is to work with representative two-dimensional views of the object instead of a truly three-dimensional object model. Certain features are detectable in each of the infinite number of views on the viewing sphere. A clustering procedure can group this infinite set of views into a finite set of view clusters where the views belonging to each cluster share the property of having the same features present in the same topological arrangement. A two-dimensional model is constructed for each viewing cluster. The two-dimensional structural description extracted from the image can now be mapped to any of the two-dimensional models representing the viewing clusters using ordered structural matching techniques. And by using effective indexing techniques, it should be possible to reduce the number of possible views that need be checked to a very small number of views.

4. Conclusions

We have discussed a technique called ordered structural matching, described its use in matching simple two-dimensional shapes, and proposed its use for more complex two-dimensional objects and for three-dimensional objects. This is a new technique, but it is also an old one. Researchers in syntactic pattern recognition convert a two-dimensional object to a one-dimensional string and use string parsing techniques to recognize the object. The strings they work with must be generatable with a grammar, preferably a regular grammar or a context free grammar that can be parsed with little or no backtracking. Ordered structural matching depends on rules that specify the spatial arrangement of the primitives instead of requiring a grammar. If these rules can, in the future, be automatically derived from examples of or CAD models of the objects, then ordered structural matching will become a very useful tool.

References

Avis, D. and G.T. Toussaint, "An Efficient Algorithm for Decomposing a Polygon into Star Shaped Polygons", *Pattern Recognition*, Volume 13, Number 6, 1981, pp. 395-398.

Bjorklund, C. and T. Pavlidis, "Global Shape Analysis by k-Syntactic Similarity", *IEEE Transactions on Pattern Analysis and Machine Intelligence*, Volume PAMI-3, Number 2, March 1981, pp. 144-154.

Davis, L.S., "Shape Matching Using Relaxation Techniques", *IEEE Transactions on Pattern Analysis and Machine Intelligence*, Volume PAMI-1, Number 1, January 1979, pp. 60-72.

Davis, L.S., "Understanding Shape: Angles and Sides", *IEEE Transactions on Computing*, Volume C-26, March 1977, pp. 236-242.

Fairfield, J., "Segmenting Dot Patterns by Voronoi Diagram Concavity", *IEEE Transactions on Pattern Analysis and Machine Intelligence*, Volume PAMI-5, Number 1, January 1983, pp. 104-110.

Feng, H.F. and T. Pavlidis, "Decomposition of Polygons into Simpler Components: Feature Generation for Syntactic Pattern Recognition", *IEEE Transactions on Computers*, Volume C-24, June 1975, pp. 636-650.

Freuder, E.C., "Synthesizing Constraint Expressions", *Communications of the ACM*, Vol. 21, No. 11, 1978.

Freuder, E.C., "Structural Isomorphism of Picture Graphs", in *Pattern Recognition and Artificial Intelligence*, ed. C.H. Chen, Academic Press, New York, 1976, pp. 248-257.

Haralick, R.M. and L.G. Shapiro, "The Consistent Labeling Problem - Part I", *IEEE Transactions on Pattern Analysis and Machine Intelligence*, Vol. PAMI-1, No. 2, 1979, pp. 173-194.

Haralick, R.M. and G. Elliot, "Increasing Tree Search Efficiency for Constraint Satisfaction Problems", *Artificial Intelligence*, Vol. 14, 1980, pp. 263-313.

Mackworth, A., "Consistency in Networks of Relations", *Artificial Intelligence*, Vol. 8, 1977, pp. 95-118.

Maruyama, K., *A Study of Visual Shape Perception*, Department of Computer Science, University of Illinois, Urbana, October 1972.

Montanari, U., ""Networks of Constraints: Fundamental Properties and Applications to Picture Processing", *Information Science*, Vol. 7, 1974, pp. 95-132.

Nevins, J.A., "Region Extraction from Complex Shapes", *IEEE Transactions on Pattern Analysis and Machine Intelligence*, Volume PAMI-4, Number 5, September 1982, pp. 500-510.

Nudel, B., "Solving the General Consistent Labeling (or Constraint Satisfaction) Problem: Two Algorithms and their Expected Complexities", *Proceedings of AAAI-83*, Washington, D.C., 1983, pp. 292-296.

Pavlidis, T., "Representation of Figures by Labelled Graphs", *Pattern Recognition*, Volume 4, 1972, pp. 5-17.

Serra, J., *Image Analysing and Mathematical Morphology*, Academic Press, London, 1982.

Shapiro, L.G., "A Structural Model of Shape", *IEEE Transactions on Pattern Analysis and Machine Intelligence*, Volume PAMI-2, Number 2, March 1980, pp. 111-126.

Shapiro, L.G. and R.M. Haralick, "Decomposition of Two Dimensional Shapes by Graph Theoretic Clustering", *IEEE Transactions on Pattern Analysis and Machine Intelligence*, Volume PAMI-1, Number 1, January, 1979. pp. 10-19.

Shapiro, L.G. and R.M. Haralick, "A Metric for Comparing Relational Descriptions", *IEEE Transactions on Pattern Analysis and Machine Intelligence*, Vol. PAMI-7, No. 1, Jan. 1985, pp. 90-94.

Shapiro, L.G. and R.M. Haralick, "Structural Descriptions and Inexact Matching", *IEEE Transactions on Pattern Analysis and Machine Intelligence*, Vol. PAMI-3, No. 5, Sept., 1981, pp. 504-519.

Shapiro, L.G., R. MacDonald, and S.R. Sternberg, "Ordered Structural Shape Matching with Primitive Extraction by Mathematical Morphology", *Pattern Recognition*, January, 1987.

Sternberg, S.R., "An Overview of Image Algebra and Related Architectures", in *Integrated Technology for Parallel Image Processing*, S. Levialdi, ed., Academic Press, London, 1985, pp. 79-100.

Waltz, D., "Understanding Line Drawings of Scenes with Shadows", in *The Psychology of Computer Vision*, P.H. Winston, ed., McGraw-Hill, New York, 1975, pp. 19-113.

II.   MATCHING AND PARSING II

A PARSING ALGORITHM FOR WEIGHTED GRAMMARS AND SUBSTRING RECOGNITION

F. Casacuberta, E. Vidal

Univ. Politecnica de Valencia
Dept. Sistemas Informaticos y
Computacion
46071 Valencia
Spain

ABSTRACT

In Syntactic Pattern Recognition, the weighted grammars are very useful in the representation of the inexact nature of many problems. Basically, such grammars are defined as a conventional grammar and a function from the set of rules into some weighting space. Some algorithms are used for particular grammars and particular spaces. In this paper, a general parsing algorithm for weighted context-free grammars is presented. This algorithm parses a given string and supplies the corresponding weight. An extension of this algorithm whose main goal is to achieve, in parallel with the parsing, a segmentation of the input string into substrings which are recognized by the weighted grammars used is also presented.

1.- INTRODUCTION.

Context-free languages and grammars are of considerable interest in many areas of Computer Science, being powerful tools in Programming Languages and Syntactic Pattern Recognition. One of the main problems of this type of grammar is the complexity of parsers; nevertheless, linear-time algorithms are available for many subclasses of context-free grammars, and cubic-time algorithms seem to be those of least parsing complexity for general context-free grammars (AHO,72a).

There are two well-known ("tabular") algorithms for general context-free grammars, which are based on Dynamic

NATO ASI Series, Vol. F45
Syntactic and Structural Pattern Recognition
Edited by G. Ferraté et al.
© Springer-Verlag Berlin Heidelberg 1988

Programming techniques. The first is the Cocke-Younger-Kasami (CYK) algorithm (AHO,72a), which requires the grammar to be in Chomsky Normal Form (CNF). Based on this algorithm, an improved version was presented in (VALIANT,75), which used multiplications of Boolean Matrices, and an error-correcting parser was developed in (TANAKA,78). The CYK algorithm presents a worst case time complexity of $O(n^3)$ and can be used for substring recognition (AHO,72a).

The second method is Earley's parsing algorithm, which does not require the grammar to be in any special form (AHO,72a) and presents a time complexity $O(n^3)$, where n is the length of the parsed string, and $O(n^2)$ if the grammar is not ambiguous. The concepts involved in this algorithm have some properties which facilitate its extension to other applications. This is the case of the method proposed in (DON,85) for image segmentation, some ideas of which follow in this paper.

In (GRAHAM,80) an improved algorithm is presented with a less than cubic asymptotical complexity, and it is shown that the two mentioned tabular methods are "almost" identical.

A parallel Earley's algorithm aimed at VLSI implementations, has been introduced in (CHIANG,84) for both string recognition and parse extraction.

All the parsing algorithms mentioned above make a left to right string analysis. Recently, a non-left-to-right and non-directional parser has been introduced which makes an "island" driven strategy easy, working in $O(n^3)$ worst case time complexity (GIORDANA,85).

Earley's algorithm has also been used with some weighted grammars for Error-Correcting Parsing, and for stochastic Error-Correcting Parsing (FU,82). In both cases the weights are associated with the error rules, for computing their contributions in the analysis of a string. However, the first case takes into account some distance measures between strings, and the second case uses a stochactic model for Syntax Errors.

In Syntactic Pattern Recognition, the weighted grammars are used to try to account for the inexact nature of many problems. The main development is produced with regular stochactic grammars or equivalently with Markov model sources, for example, in Automatic Speech Recognition (CASACUBERTA,86b). In some others areas, the theoretic interest in these types of grammars (or some specific cases) has increased (KIM,75)(WETHERELL,80).

In this paper, we will present an extended version of Earley's algorithm for generalizing the parsing of strings for any weighted context-free grammars. This version can be conveniently modified to take into account the problem of substring recognition with such context-free grammars.

2. <u>PARSING WITH CONTEXT-FREE WEIGHTED GRAMMARS.</u>

This section deals with parse algorithms for some classes of grammars with "weighting spaces", in the sense defined in (MIZUMOTO,75) for automata.

Let W be a weighting space, with two defined operations $W=(W,\square,o)$, such that $(W,\square)$ and (W,o) are monoids, and "o" verifies the distributive property respect "$\square$". We denote the unit elements for "$\square$" and "o" as O and I respectively. A context-free W-grammar is a pair (G,f), where G is a context-free grammar $G=(N,E,P,S)$ defined as: N and E are finite (non-empty) sets of non-terminals and terminal symbols, respectively; $S \in N$ is the starting symbol; and $P \subset N \times (N \cup E)^*$ is a finite set of rules, where $(N \cup E)^*$ is the free monoid over N U E. And f is a function $f:P \longrightarrow W$.

Let Di be a left derivation of $x \in E^*$ from S; i.e. Di is a sequence of rules from P $(r_1^i, r_2^i,..,r_{n_i}^i)$ such that $S \overset{*}{\underset{D_i}{\Rightarrow}} x$ (AHO,72a). We can define a W-language as $\mathcal{L}:E^* \longrightarrow W$:

$$(2.1) \quad x \in E^* \quad \mathcal{L}(x) = \underset{D_i}{\square} \ (f(r_1^i) \ o \ f(r_2^i) \ o \ ... \ o \ f(r_{n_i}^i))$$

We will use an extended version of Earley's parsing algorithm to parse a string $x \in E^*$ ($x = a_1 .. a_n$) with a given W-grammar (G,f). The modifications are carried out in a similar way as in (FU,82) for stochastic error correcting parsers, i.e., by introducing a new definition of an item in a list $I(j)$

(2.2) $(A \longrightarrow \alpha . \beta , i, p)$

where $p \in W$, and

1.) $A \longrightarrow \alpha\beta$ is a rule in P, "." is a meta-symbol not in E U N, and $i \in Z^{\geq 0}$.

2.) for some $\gamma, \delta \in (N \cup E)^*$, we have:

$\quad$ 2.a) $S \overset{*}{\Rightarrow} \gamma A \delta$.

$\quad$ 2.b) $\gamma \overset{*}{\Rightarrow} a_1 .. a_i$.

$\quad$ 2.c) $\delta \overset{*}{\Rightarrow} a_{i+1} .. a_i$.

To handle these types of items, it is necessary to introduce some simple modifications in the steps of Earley's algorithm.

The first modification is the initialization, where some value in the new items must be assigned to p, for example $f(A \longrightarrow \alpha)$.

The second important definition concerns step (B) in order to take into account the computation of p.

The third modification is introduced to account for the case where G is ambiguous, and some items of the form $(S \longrightarrow \alpha ., 0, p)$ can appear in $I(n)$. Further discussion about the problem of the ambiguity of G is presented later. At this point we will only introduce this modification on a grammatical level, that is, an expanded grammar (G', f') is built from (G,f) as follows: $G' = (N', E, P', S')$, where:

$\quad$ 1.) $N' = N \cup \{S'\}$: S' is a new non-terminal.

$\quad$ 2.) $P' = P \cup \{S' \longrightarrow S\}$.

$\quad$ 3.) $f'(S' \longrightarrow S) = I$, and $f'(r) = f(r)$ $\forall r \in P$.

The modified Earley's parsing algorithm for W-grammars is as follows:

ALGORITHM 1 MODIFIED EARLEY'S PARSING ALGORITHM FOR W-GRAMMARS

INPUT: An expanded context-free grammar $G'=(N',E,P',S')$, and $f':P' \longrightarrow W$. And a string $x \in E^*$ ($x=a_1..a_n$).

METHOD

Initialization

   I(0):=NIL;

   add (S' $\longrightarrow$ S.,0,I) to I(0);

Iteration

   for j:=0 to n do

     repeat

      (A) for (A $\longrightarrow \alpha.B\beta$,i,p) in I(j)

         for (B $\longrightarrow \gamma$) $\in$ P

           add (B $\longrightarrow \cdot \gamma$,j,f'(B $\longrightarrow \gamma$)) to I(j);

         endfor;

       endfor;

      (B) for (A $\longrightarrow \alpha.$,i,p) in I(j)

         for (B $\longrightarrow \beta.A\gamma$,k,q) in I(i)

         if (B $\longrightarrow \beta A.\gamma$,k,r) not in I(j)

         then add (B $\longrightarrow \beta A.\gamma$,k,p o q) to I(j);

         else substitute r of (B $\longrightarrow \beta A.\gamma$,k,r)

                   by (r □ (p o q));

         endif;

         endfor;

       endfor;

     until no new items can be added or modified to I(j);

     if j < n

     then I(j+1):=NIL;

      (C) for (A $\longrightarrow \alpha.a_{j+1}\beta$,i,p) to I(j)

         add (A $\longrightarrow \alpha a_{j+1}. \beta$,i,p) to I(j+1);

       endfor

     endif;

   endfor;

OUTPUT: The parse list I(0), I(1),.., I(n) of x in L(G)

END OF ALGORITHM 1

    The following theorem relates the weights which appear in the items of a parse list with the weights of the W-language generated by (G,f).

Theorem 1. Given $x=a_1..a_n \in E^*$, and a parse list from algorithm 1, if $(S' \longrightarrow S.,0,p)$ in $I(n)$, then $\mathcal{L}(x)=p$.

Proof. In general, if there is an item in $I(j)$ such as

$$(2.3) \quad (A \longrightarrow \alpha_0 A_1 \alpha_1 A_2 .. A_m \alpha_m .,i,p) \quad \alpha_k \in E^* \quad 0 \leq k \leq m; \quad A_k \in N \quad 1 \leq k \leq m$$

then for $1 \leq k \leq m$ and for $1 \leq l \leq n_k$ there are the following items:

$$(2.4) \quad (A_k \longrightarrow \beta_{kl} .., h_{kl}, p_{kl}) \quad \text{in } I(g_l) \quad \text{for some } g_l : h_{kl} \leq g_l \leq j$$

where by $A \longrightarrow \beta_{kl}$ $(1 \leq l \leq n_k)$ we denote the n_k different rules with the same left hand side, which have succeeded in the analysis of $a_{h_{kl}} .. a_{g_l}$. And

$$(2.5) \quad p = f_A \circ (\overset{n_1}{\underset{l=1}{\square}} p_{1l}) \circ (\overset{n_2}{\underset{l=1}{\square}} p_{2l}) \circ .. \circ (\overset{n_m}{\underset{l=1}{\square}} p_{ml})$$

where $f_A = f(A \longrightarrow \alpha_0 A_1 .. A_m \alpha_m)$.

If there are terminal symbols only on the right hand side of (2.3), then $p=f_A$.

From the properties of W, we can rewrite (2.5) as:

$$(2.6) \quad p = \overset{n_1}{\underset{l_1=1}{\square}} \overset{n_2}{\underset{l_2=1}{\square}} .. \overset{n_m}{\underset{l_m=1}{\square}} (f_A \circ p_{1l_1} \circ p_{2l_2} .. \circ p_{ml_m})$$

Each p_{kl_k} is computed in a way similar to that in (2.6), from (2.4), and so on, until only items without non-terminals on the right hand side have been considered. If we substitute each p_{kl_k} by its computational form, we have:

$$(2.7) \quad p = \underset{s}{\square} (f(r_1^s) \circ .. \circ f(r_{n_s}^s))$$

where by s we denote all different sequences of rules $(r_1^s, r_2^s, .., r_{n_s}^s)$, such that each one defines a derivation of $a_{i+1} .. a_j$ from A.

In particular, if $x \in L(G)$ there is only one item $(S' \longrightarrow S.,0,p)$ in $I(n)$, and p is computed from (2.7) for $A=S'$, obtaining the value $\mathcal{L}(x)$ defined in (2.1) ///.

I(0) (initialization)

```
(S' ———→ .S,  0,   1.00)
(S ———→ .SS,  0,   1.00)
(S ———→ .A,   0,   5.00)
(A ———→ .aA,  0,   2.00)
(A ———→ .a,   0,   5.00)
```

I(1) (analyzed symbol="a")

```
(A ———→ a.A,  0,   2.00)
(A ———→ a.,   0,   5.00)
(S ———→ A.,   0,  25.00)
(S' ———→ S.,  0,  25.00)
(S ———→ S.S,  0,  25.00)
(A ———→ .aA,  1,   2.00)
(A ———→ .a,   1,   5.00)
(S ———→ .SS,  1,   1.00)
(S ———→ .A,   1,   5.00)
```

I(2) (analyzed symbol="a")

```
(A ———→ a.A,  1,   2.00)
(A ———→ a.,   1,   5.00)
(A ———→ aA.,  0,  10.00)
(S ———→ A.,   1,  25.00)
(S ———→ A.,   0,  50.00)
(S ———→ SS.,  0, 625.00)
(S ———→ S.S,  1,  25.00)
(S' ———→ S.,  0, 625.00)
(S ———→ S.S,  0, 625.00)
(A ———→ .aA,  2,   2.00)
(A ———→ .a,   2,   5.00)
(S ———→ .SS,  2,   1.00)
(S ———→ .A,   2,   5.00)
```

I(3) (analyzed symbol="a")

```
(A ———→ a.A,  2,    2.00)
(A ———→ a.,   2,    5.00)
(A ———→ aA.,  1,   10.00)
(S ———→ A.,   2,   25.00)
(A ———→ aA.,  0,   20.00)
(S ———→ A.,   1,   50.00)
(S ———→ SS.,  1,  625.00)
(S ———→ SS.,  0,15625.00)
(S ———→ S.S,  2,   25.00)
(S ———→ A.,   0,  100.00)
(S ———→ S.S,  1,  625.00)
(S' ———→ S.,  0,15625.00)
(S ———→ S.S,  0,15625.00)
(A ———→ .aA,  3,    2.00)
(A ———→ .a,   3,    5.00)
(S ———→ .SS,  3,    1.00)
(S ———→ .A,   3,    5.00)
```

FIG. 1. A parse list of /aaa/ using the max-weighted grammar of example 1. Since (S'→S.,15625.00) is in I(3), $\mathcal{L}$(aaa)=15625.00.

```
I(0) (initialization)
```

```
(S' ──→ .S, 0, 1.00000)
(S  ──→ .SS, 0, 0.75000)
(S  ──→ .A, 0, 0.25000)
(A  ──→ .aA, 0, 0.66660)
(A  ──→ .a, 0, 0.33340)
```

```
I(1) (analyzed symbol="a")
```

```
(A  ──→ a.A, 0, 0.66660)
(A  ──→ a., 0, 0.33340)
(S  ──→ A., 0, 0.08335)
(S' ──→ S., 0, 0.08335)
(S  ──→ S.S, 0, 0.06251)
(A  ──→ .aA, 1, 0.66660)
(A  ──→ .a, 1, 0.33340)
(S  ──→ .SS, 1, 0.75000)
(S  ──→ .A, 1, 0.25000)
```

```
I(2) (analyzed symbol="a")
```

```
(A  ──→ a.A, 1, 0.66660)
(A  ──→ a., 1, 0.33340)
(A  ──→ aA., 0, 0.22224)
(S  ──→ A., 1, 0.08335)
(S  ──→ A., 0, 0.05556)
(S  ──→ SS., 0, 0.00521)
(S  ──→ S.S, 1, 0.06251)
(S' ──→ S., 0, 0.06077)
(S  ──→ S.S, 0, 0.04558)
(A  ──→ .aA, 2, 0.66660)
(A  ──→ .a, 2, 0.33340)
(S  ──→ .SS, 2, 0.75000)
(S  ──→ .A, 2, 0.25000)
```

```
I(3) (analyzed symbol="a")
```

```
(A  ──→ a.A, 2, 0.66660)
(A  ──→ a., 2, 0.33340)
(A  ──→ aA., 1, 0.22224)
(S  ──→ A., 2, 0.08335)
(A  ──→ aA., 0, 0.14815)
(S  ──→ A., 1, 0.05556)
(S  ──→ SS., 1, 0.00521)
(S  ──→ SS., 0, 0.00760)
(S  ──→ S.S, 2, 0.06251)
(S  ──→ A., 0, 0.03704)
(S  ──→ S.S, 1, 0.04558)
(S' ──→ S., 0, 0.04463)
(S  ──→ S.S, 0, 0.03348)
(A  ──→ .aA, 3, 0.66660)
(A  ──→ .a, 3, 0.33340)
(S  ──→ .SS, 3, 0.75000)
(S  ──→ .A, 3, 0.25000)
```

FIG. 2. A parse list of /aaa/ using the probabilistic grammar of example 2. Since $(S' \to S., .04463)$ is in I(3), $\mathcal{L}(aaa)=0.04463$.

In this section, we present three examples of the application of the algorithm 1, which use the following grammar:

rules

(r1)   S ⟶ SS

(r2)   S ⟶ A

(r3)   A ⟶ aA

(r4)   A ⟶ a

the string "aaa" has four left derivations:

D1:S ⟹ A ⟹ aA ⟹ aaA ⟹ aaa;

D2:S ⟹ SS ⟹ AS ⟹ aS ⟹ aA ⟹ aaA ⟹ aaa;

D3:S ⟹ SS ⟹ AS ⟹ aAS ⟹ aaS ⟹ aaA ⟹ aaa;

D4:S ⟹ SS ⟹ SSS ⟹ ASS ⟹ aSS ⟹ aAS ⟹ aaS ⟹ aaA ⟹ aaa;

Example 1: In this example we use the following weighting space $W=(\,[0,\infty[\,,MAX,X)$, for the Max-weighted grammars (MIZUMOTO,75), and the corresponding weights of the rules are:

f(r1) = 1.0; f(r2) = 5.0; f(r3) = 2.0; f(r4) = 5.0

The weight associated with each derivation is:

p(D1)=100.0; p(D2)=1250.0; p(D3)=1250.0; p(D4)=15625.0;

therefore, $\mathcal{L}$(aaa) = 15625.0. In figure 1, the parse list of "aaa" is presented.

Example 2: In this case a probabilistic grammar $(W=(\,[0,1],+,X))$ is used; and the corresponding weights of the rules are:

f(r1) = 0.75; f(r2) = 0.25; f(r3) = 0.66; f(r4) = 0.34

and the probabilities associated with each derivations are:

p(D1)=0.037; p(D2)=0.0035; p(D3)=0.0035; p(D4)=0.00033.

And $\mathcal{L}$(aaa) ≃ 0.044. In fig.2 we show the parse list of "aaa", we must observe that in I(3) there is an item (S' ⟶ S.,0,044) //.

Example 3: In this example we use the following weighting space $W=(\,[0,1],MAX,MIN)$, and the same weights for the rules of example 2, therefore (G,f) is a special case of an L-grammar, known as a Fuzzy grammmar. The parse list for the string "aaa" is presented in fig.3, and the evidence associated with the

I(0) (initialization)

(S'⟶ .S, 0,1.00000)
(S ⟶ .SS, 0, 0.75000)
(S ⟶ .A, 0, 0.25000)
(A ⟶ .aA, 0, 0.66660)
(A ⟶ .a, 0, 0.33340)

I(1) (analyzed symbol="a")

(A ⟶ a.A, 0, 0.66660)
(A ⟶ a., 0, 0.33340)
(S ⟶ A., 0, 0.25000)
(S'⟶ S., 0, 0.25000)
(S ⟶ S.S, 0, 0.25000)
(A ⟶ .aA, 1, 0.66660)
(A ⟶ .a, 1, 0.33340)
(S ⟶ .SS, 1, 0.75000)
(S ⟶ .A, 1, 0.25000)

I(2) (analyzed symbol="a")

(A ⟶ a.A, 1, 0.66660)
(A ⟶ a., 1, 0.33340)
(A ⟶ aA., 0, 0.33340)
(S ⟶ A., 1, 0.25000)
(S ⟶ A., 0, 0.25000)
(S ⟶ SS., 0, 0.25000)
(S ⟶ S.S, 1, 0.25000)
(S'⟶ S., 0, 0.25000)
(S ⟶ S.S, 0, 0.25000)
(A ⟶ .aA, 2, 0.66660)
(A ⟶ .a, 2, 0.33340)
(S ⟶ .SS, 2, 0.75000)
(S ⟶ .A, 2, 0.25000)

I(3) (analyzed symbol="a")

(A ⟶ a.A, 2, 0.66660)
(A ⟶ a., 2, 0.33340)
(A ⟶ aA., 1, 0.33340)
(S ⟶ A., 2, 0.25000)
(A ⟶ aA., 0, 0.33340)
(S ⟶ A., 1, 0.25000)
(S ⟶ SS., 1, 0.25000)
(S ⟶ SS., 0, 0.25000)
(S ⟶ S.S, 2, 0.25000)
(S ⟶ A., 0, 0.25000)
(S ⟶ S.S, 1, 0.25000)
(S'⟶ S., 0, 0.25000)
(S ⟶ S.S, 0, 0.25000)
(A ⟶ .aA, 3, 0.66660)
(A ⟶ .a, 3, 0.33340)
(S ⟶ .SS, 3, 0.75000)
(S ⟶ .A, 3, 0.25000)

FIG. 3. A parse list of /aaa/ using the fuzzy grammar of example 3. Since (S'→S.,,.25) is in I(3), $\mathscr{L}$(aaa)=0.25.

four derivations are:

$$p(D1) = p(D2) = p(D3) = p(D4) = 0.25,$$

and $\mathcal{L}(aaa) = 0.25$ //.

3.- SUBSTRING SEGMENTATION WITH WEIGHTED CONTEXT-FREE GRAMMARS.

In the parsing of a string $a_1 .. a_n$ by algorithm 1, the following obvious property is verified by some $p \in W$:

(3.1) If $(S \longrightarrow \alpha., i, p)$ in $I(j)$ then $S \overset{*}{\Longrightarrow} a_{i+1} .. a_j$

Nevertheless, the inverse of this statement is not always true, as it can be derived from the definition of an item.

If we could build an algorithm which verifies (3.1) and its inverse, the problem of substring recognition through the parsing of a given string x is solved.

Another problem, related with substring recognition, appears in step (C) of algorithm 1 if there is not any item $(A \longrightarrow \alpha.a_j\beta, i, p)$ in $I(j-1)$. In this case, no new item can be added to $I(j)$ using steps (C) and, consequently, (A) and (B), and the parsing process is stopped.

In this section we propose some modifications to algorithm 1, in order to overcome the problems mentioned above. The basic idea is to introduce a _reinitialization_, each time the construction of a list is concluded. This process can be seen as n parsings which share items. And the substring recognition consists of verifying whether a new item is of the form $(S \longrightarrow \alpha., i, p)$. Therefore, at the end of the parsing we know whether either x, or any of its substrings, have or have not been generated by the grammar.

ALGORITHM 2 MODIFIED EARLEY'S PARSING ALGORITHM FOR SUBSTRING SEGMENTATION AND FOR W-GRAMMARS

INPUT: An expanded context-free W-grammar $G'=(N',E,P',S')$, and $f':P' \longrightarrow W$. And a string $x \in E'*$ $(x=a_1..a_n)$, with $E \subseteq E'$.

METHOD

Initialization

    $I(0):=NIL; RECOG:=NIL; \underline{add}$ $(S' \longrightarrow .S,0,I)$ $\underline{to}$ $I(0)$;

Iteration

    $\underline{for}$ $j:=0$ $\underline{to}$ n $\underline{do}$

      $\underline{repeat}$

        (A) $\underline{for}$ $(A \longrightarrow \alpha.B\beta ,i,p)$ $\underline{in}$ $I(j)$

              $\underline{for}$ $(B \longrightarrow \gamma) \in P$

                $\underline{add}$ $(B \longrightarrow .\gamma ,j,f'(B \longrightarrow \gamma))$ $\underline{to}$ $I(j)$;

              $\underline{endfor}$;

            $\underline{endfor}$;

        (B) $\underline{for}$ $(A \longrightarrow \alpha. ,i,p)$ $\underline{in}$ $I(j)$

              $\underline{for}$ $(B \longrightarrow \beta.A\gamma ,k,q)$ $\underline{in}$ $I(i)$

                $\underline{if}$ $(B \longrightarrow \beta A.\gamma,k,r)$ $\underline{not~in}$ $I(j)$

                $\underline{then}$ $\underline{add}$ $(B \longrightarrow \beta A.\gamma,k,p \circ q)$ $\underline{to}$ $I(j)$;

                $\underline{else}$ $\underline{substitute}$ r $\underline{of}$ $(B \longrightarrow \beta A.\gamma,k,r)$

                                   $\underline{by}$ $(r \square (p \circ q))$;

              $\underline{endif}$;

            $\underline{endfor}$;

          $\underline{endfor}$;

      $\underline{until}$ no new items can be added or modified to $I(j)$;

    (B') $\underline{for}$ $(S \longrightarrow \alpha. ,i,p)$ $\underline{in}$ $I(j)$

            $\underline{add}$ $(i+1,j,p)$ $\underline{to}$ RECOG;

        $\underline{endfor}$;

    (D) $\underline{for}$ $(A \longrightarrow .\alpha ,0,q)$ $\underline{in}$ $I(0)$

          $\underline{add}$ $(A \longrightarrow .\alpha ,j,q)$ $\underline{to}$ $I(j)$;

        $\underline{endfor}$;

        $\underline{if}$ $j<n$

        $\underline{then}$ $I(j+1):=NIL$;

          (C) $\underline{for}$ $(A \longrightarrow \alpha.a_{j+1}\beta ,i,p)$ $\underline{to}$ $I(j)$

                $\underline{add}$ $(A \longrightarrow \alpha a_{j+1}.\beta ,i,p)$ $\underline{to}$ $I(j+1)$;

              $\underline{endfor}$;

        $\underline{endif}$;

      $\underline{endfor}$;

OUTPUT: 1. A parse list I(0), I(1), ..., I(n).

2. A list RECOG of triplets of starting points, ending points, and weights of corresponding substrings accepted by the grammar G

<u>END OF ALGORITHM 2</u>

The differences between algorithm 1 and 2 are first step (D), which represents the reinitialization of parsing; and second substep (B') in which a list of recognized substrings is built.

Following theorem 2, algorithm 2 aims at solving the problem of substring recognition using weighted grammars.

<u>Theorem 2.</u> Given a string $x=(a_1..a_n)$ from E^*, we have $(S' \longrightarrow S., i, p)$ in $I(j)$ if and only if $S' \overset{*}{\Rightarrow} a_{i+1}..a_j$, and $\mathscr{L}(a_{i+1}..a_j)=p$.

<u>Proof.</u> a) "only if": If $(S' \longrightarrow S., i, p)$ in $I(j)$ then $S' \overset{*}{\Rightarrow} a_{i+1}..a_j$. The proof of this statement is trivial from the definition of the item itself, and the theorem 1.

b) "if": If $S' \overset{*}{\Rightarrow} a_{i+1}..a_j$ and $\mathscr{L}(a_{i+1}...a_j)=p$ then $(S' \longrightarrow S., i, p)$ in $I(j)$. The first part of this statement means that $a_{i+1}..a_j$ is a string generated by the grammar used, therefore, by step (D), with the sequence of lists which begins with $I(i)$, this substring must be recognized, and so $(S \longrightarrow \alpha., i, q)$ in $I(j)$ for some q, but if this substring is recognized by the weighted grammar, following the theorem 1, $q=p$ ///.

Worst case time complexity of algorithm 2 is achieved when no item is shared by two or more sequences of lists. In this case, this algorithm presents a time complexity of $O(n^4)$. On the other hand, if all items are shared by the sequence of lists which begins with $I(0)$, the time complexity is the same as Earley's algorithm, that is $O(n^3)$. If the grammar is not ambiguous, the time complexity is a function between $O(n^3)$ and $O(n^2)$.

```
I(0) (initialization)
─────────────────────────────
(S ──→ .A,  0,   1.0)
(S ──→ .B,  0,   1.0)
(A ──→ .aAb, 0,   3.0)
(A ──→ .0,  0,   2.0)
(B ──→ .bbBa, 0,  4.0)
(B ──→ .1,  0,   2.0)
```

```
I(2) (analyzed symbol="0")
─────────────────────────────
(A ──→ 0.,  1,   2.0)
(A ──→ aA.b, 0,   5.0)
(S ──→ A.,  1,   3.0)
(S ──→ .A,  2,   1.0)
(S ──→ .B,  2,   1.0)
(A ──→ .aAb, 2,   3.0)
(A ──→ .0,  2,   2.0)
(B ──→ .bbBa, 2,  4.0)
(B ──→ .1,  2,   2.0)
```

```
I(1) (analyzed symbol="a")
─────────────────────────────
(A ──→ a.Ab, 0,   3.0)
(A ──→ .aAb, 1,   3.0)
(A ──→ .0,  1,   2.0)
(S ──→ .A,  1,   1.0)
(S ──→ .B,  1,   1.0)
(A ──→ .aAb, 1,   3.0)
(A ──→ .0,  1,   2.0)
(B ──→ .bbBa, 1,  4.0)
(B ──→ .1,  1,   2.0)
```

```
I(4) (analyzed symbol="b")
─────────────────────────────
(B ──→ bb.Ba, 2,  4.0)
(B ──→ b.bBa, 3,  4.0)
(B ──→ .bbBa, 4,  4.0)
(B ──→ .1,  4,   2.0)
(S ──→ .A,  4,   1.0)
(S ──→ .B,  4,   1.0)
(A ──→ .aAb, 4,   3.0)
(A ──→ .0,  4,   2.0)
(B ──→ .bbBa, 4,  4.0)
(B ──→ .1,  4,   2.0)
```

```
I(3) (analyzed symbol="b")
─────────────────────────────
(A ──→ aAb.,  0,   5.0)
(B ──→ b.bBa, 2,   4.0)
(S ──→ A.,  0,   6.0)
(S ──→ .A,  3,   1.0)
(S ──→ .B,  3,   1.0)
(A ──→ .aAb, 3,   3.0)
(A ──→ .0,  3,   2.0)
(B ──→ .bbBa, 3,  4.0)
(B ──→ .1,  3,   2.0)
```

```
I(6) (analyzed symbol="a")
─────────────────────────────
(B ──→ bbBa.,  2,   6.0)
(A ──→ a.Ab, 5,   3.0)
(S ──→ B.,  2,   7.0)
(A ──→ .aAb, 6,   3.0)
(A ──→ .0,  6,   2.0)
(S ──→ .A,  6,   1.0)
(S ──→ .B,  6,   1.0)
(A ──→ .aAb, 6,   3.0)
(A ──→ .0,  6,   2.0)
(B ──→ .bbBa, 6,  4.0)
(B ──→ .1,  6,   2.0)
```

```
I(5) (analyzed symbol="1")
─────────────────────────────
(B ──→ 1.,  4,   2.0)
(B ──→ bbB.a, 2,   6.0)
(S ──→ B.,  4,   3.0)
(S ──→ .A,  5,   1.0)
(S ──→ .B,  5,   1.0)
(A ──→ .aAb, 5,   3.0)
(A ──→ .0,  5,   2.0)
(B ──→ .bbBa, 5,  4.0)
(B ──→ .1,  5,   2.0)
```

FIG. 4.   Example of recognition of two concatenated strings generated by the discriminant grammar of example 5. Two concatenations are possibles: /a0b/-/1/ and /0/-/bb1a/. The recognition of /a0b/ can be seen in I(3) (S──→A.,0,6.0); that of /1/, in I(5) with (S──→B.,4,3.0); that of /0/, in I(2) (S──→A.,1,3), and that of /bb1a/, in I(6) with (S ──→ B.,2,7).

In the following example we consider substring segmentation with a discriminant grammar (FILIPSKI,80).

Example 4. Let (G,f) be a discriminant grammar defined as:
$$N = \{S, A, B\}; \quad E = \{a, b, 0, 1\}$$
and the rules with their corresponding weights are:

weights	rules
1.0	S $\longrightarrow$ A
1.0	S $\longrightarrow$ B
3.0	A $\longrightarrow$ aAb
2.0	A $\longrightarrow$ 0
4.0	B $\longrightarrow$ bbBa
2.0	B $\longrightarrow$ 1

In this grammar the weights are related to the number of terminals generated by a rule, and $W=(\mathbb{R},MAX,+)$. In fig. 5 we present a parse list of "a0bb1a", and the substring recognized by the grammar. From this string two possible concatenations are possible: "a0b"-"1" and "0"-"bb1a", if we use the operator "+" to compute the evidence of the concatenation of two or more strings, we have $\mathscr{L}(a0b)+\mathscr{L}(1)=9.0$, and in the second case $\mathscr{L}(0)+\mathscr{L}(bb1a)=10.0$. The latter seems to be the most evident, since the number of terminals is greater than that in the former and, therefore, a greater accumulated weight is obtained

3. CONCLUSIONS AND FUTURE WORKS.

In this paper, an extension of Earley's parsing algorithm for general weighted context-free grammars is presented. This algorithm allows us to obtain the weight associated to a string which is generated by the grammar. A modification of this algorithm is also introduced in order to obtain a procedure for substring recognition using weighted grammars.

Other parsers can be used to achieve some performances of the algorithm presented in this paper, at the expense of the loss of generality. For example, the problem of using the CYK algorithm for substring recognition, if we can modify it in order to use weighted grammars, is that the grammar must be in CNF. Furthermore, if the grammar is not ambiguous, worst time complexity of the algorithm presented here is a function between $O(n^3)$ and $O(n^2)$, while the CYK algorithm is always $O(n^3)$. Finally, there are L-fuzzy-grammars which have not an equivalent L-grammar in CNF (KIM,75), and therefore, the algorithm presented here is more general since there are not any constraints on the form of the grammar.

Two main works are in progress in our laboratory. One is concerned with the use of substring segmentation procedures in Error-correcting parsing, the aim being to obtain all substrings of a given string which are "close" to a given grammar. On the other hand, we are developing an extension of these algorithms to parse strings of fuzzy-symbols (VIDAL,85) (CASACUBERTA,86a) (CASACUBERTA,86b) or more general L-symbols (CASACUBERTA,86c).

An application of the combination of substring segmentation and parsing with some kind of L-grammars is now under research for Connected (and Continuous) Word Recognition. Within this framework, we will represent lexical knowledge as a context-free L-grammar, whose terminal symbols are some kind of sub-lexical categories (pseudo-phonemes, diphones, etc..). Parsing with substring segmentation could supply, simultaneously, both the _limits_ and the _identification_ (with some degree of evidence) of the lexical categories in a phrase of connected words.

4. REFERENCES.

(AHO,72a) A.V. Aho and J.D. Ullman: "The Theory of Parsing, Translation and Compiling". Vol.1. Prentice-Hall. Englewood Cliffs.

(AHO,72b) A.V. Aho and T.G. Peterson: "A Minimum Distance Error-Correcting Parse for Context-Free Languages". SIAM J.Comput. Vol.1 No.4.

(CASACUBERTA,86a) F.Casacuberta, E.Vidal, J.M.Benedí: "Interpretation of Fuzzy Data by means of Fuzzy Rules with Applications to Speech Recognition". To be published in International J. on FUZZY Sets and Systems.

(CASACUBERTA,86b) F.Casacuberta, E.Vidal: "Reconocimiento Automático del Habla". In press. Marcombo. Barcelona.

(CASACUBERTA,86c) F.Casacuberta: "Some Preliminary Notes about Languages, Grammars and Automata of Fuzzy Symbols". Internal Report.

(CHIANG,84) Y.T. Chiang and K.S. Fu: "Parallel Parsing Algorithm and VLSI Implementations for Syntactic Patterns Recognition": IEEE Trans. Pattern. Anal. and Mach. Intell. Vol. PAMI-6, No.3 pp. 302-312.

(DON,85) H.D. Don and K.S. Fu: "A Syntactic Method for Image Segmentation and Object Recognition". Pattern Recognition. Vol.18, No.1 pp. 73-87.

(FILIPSKI,80 A.J.Filipski: "A least Mean-Squared Error Approach to Syntactic Classification". IEEE Trans. Pattern Anal. and Machine Intel., Vol. PAMI-2, No. 3, pp. 252-255.

(FU,82) K.S.Fu: "Syntactic Pattern Recognition and Applications": Prentice-Hall, Englewood Cliffs.

(GIORDANA,85) A. Giordana and L. Saitta: "A non-left-to-right, non-directional Parser for Ambiguous Lattices". Pattern Recognition Letters Vol.3, No.2 pp.105-111.

(GRAHAM,80) S.L.Graham, M.A.Harrison and W.L.Ruzzo: "An Improved Context-Free Recognizer": ACM Trans. on Program. Lang. and Systems. Vol. 2, No.3, pp 415-462.

(KIM,75) H.H.Kim, M.Mizumoto, J.Toyoda and K.Tanaka: "L-Fuzzy Grammars". Information Sciences. Vol. 8, pp. 123-140.

(MIZUMOTO,75) M.Mizumoto, J.Toyoda, K.Tanaka: "Various kinds of Automata with Weights". J. of Comp. and Syst. Sci. Vol. 10, pp. 219-236.

(TANAKA,78) E.Tanaka and K.S. Fu: "Error-Correcting Parsers for Formal Languages": IEEE Trans. on Comp. Vol.C-27, No.7, pp 605-616.

(VALIANT,75) L.G. Valiant: "General Context-Free Recognition in Less than Cubic Time": J. of Comp. and Sys. Sci. Vol.10, pp 308-315.

(VIDAL,85) E.Vidal, F.Casacuberta, E.Sanchís, J.M.Benedí: "A General Fuzzy-Parsing Scheme for Speech Recognition". In "New Systems and Architectures for Automatic Speech Recognition and Synthesis". Eds.: R.deMori and C.Y.Suen. Springer-Verlag.

(WETHERELL,80) G.S.Wetherell: "Probabilistic Languages: A Review and Some Open Questions". Comp. Surveys. Vol. 12, No. 4, pp. 361-379.

COMPUTING THE MINIMUM ERROR DISTANCE OF GRAPHS IN O(n³) TIME WITH PRECEDENCE GRAPH GRAMMARS

M. Kaul

Universität Passau, Fak. für Informatik
8390 Passau, Germany

Summary. A major part in structural pattern recognition is *inexact graph matching*. Typically some node and edge labelled graph has to be matched against a possibly infinite graph language, which represents the set of all correct patterns. The task is to identify that correct pattern, that is most similar to the input graph. Similarity is defined by weighted editing operations, yielding an error distance. Computing the minimum error distance even for two graphs only is NP-complete. In order to gain efficient procedures *application dependent knowledge* has to be involved.

In this paper a graph parser generator is presented, that can be adapted to a wide range of applications easily. A precedence graph grammar is used to describe the *structural* knowledge about the class of all correct patterns. The weights for editing operations on graphs provide the *statistical* knowledge. By restricting backtracking to subgraphs of constant size, the minimum error distance is computed in $O(n^3)$ time, n the number of nodes of the whole input graph. Furthermore the parse tree of the most similar graph is computed, thus providing further processing steps with an efficient hierarchical decomposition.

1. Introduction

Structural descriptions of patterns consists of their primitives and interrelationships, which are best expressed as *node and edge labelled graphs*. Pattern recognition systems have to match input patterns against a set of prototypes. Usually the input graph is distorted, but only undistorted prototypes are specified. So the task is to identify that prototype that is most similar to the input graph, which is done by *inexact graph matching*. Similarity is defined by weighted editing operations on graphs, yielding an error distance. Computing the minimum error distance even for two graphs only is NP-complete. In order to make inexact graph matching tractable, in literature a lot of approaches are discussed for tuning brute-force backtracking methods to special applications, e.g. [ShHa 81], [BuAl 83] to similarity of silhouettes. The problems with these approaches are: The algorithms remain exponential in the worst case. The user has to capture his intuition of similarity by specifying the weights for the editing operations. No structural knowledge about the application domain can be incorporated into the system. Only a finite set of prototypes can be specified.

CR Categories and Subject Descriptors: I.5.1 [Pattern Recognition]: Models- *statistical; structural* ; F.4.2 [Mathematical Logic and Formal Languages]: Grammars and Other Rewriting Systems-*parsing*, G.2.2 [Discrete Mathematics]: Graph Theory- *graph algorithms; trees*

Additional Key Words and Phrases: inexact graph matching, similarity of graphs, error distance between graphs, graph grammar, graph parser, precedence relations, parallel parsing, hierarchical graph model

Author´s address: Dr.Manfred Kaul, Fakultät für Mathematik und Informatik, Universität Passau, Postfach 2540, D-8390 Passau, Fed. Rep. Germany, Tel. 0851-509344, West-German Electronic Mail Address: unido!unipas.uucp!kaul

By *error correcting precedence graph parsing* the problems mentioned above can be avoided. We first present our method, list some important results, show the advantages and finally give as example a precedence graph grammar specifying all staircases with an unbounded number of steps. For a distorted picture of a stair the most similar graph generated by the precedence graph grammar is recognized. The method is demonstrated step by step in this example. Proofs are omitted in this paper. They are included in the author's dissertation [Ka 86], where precedence graph grammars and parsers are described in depth. The parser has been implemented in Pascal and is available for interested readers.

First let us introduce node and edge labelled graphs. All notions *not* introduced explicitly are assumed to be standard and can be taken e.g. from [Har 69]. Let Σ_V, Σ_E be two alphabets of vertex and edge labels resp. $G = (V_G, E_G, L_G)$ is a graph (over Σ_V, Σ_E) if $V_G \neq \varnothing$ is a finite set (the vertex set of G), $E_G \subseteq \{(v,x,w) \in V_G \times E_G \times V_G |\ v \neq w\}$ (the edge set of G) and $L_G : V_G \to \Sigma_V$ is a total mapping (the node labelling of G). $E_G(v,w) =_{def} \{x \in \Sigma_E |\ (v,x,w) \in E_G\}$ is the set of all labels from v towards w. $H \lhd G$ means that H is a full subgraph of G. $G|V$ is the restriction of G to a full subgraph $H \lhd G$ iff $V_H = V$. $R_G(H)$ is a full subgraph H' of $G \backslash H$ iff every node from H' is adjacent to some node in H. If two graphs G,H are isomorphic, we write $G \cong H$. The label tuple of two nodes $v,w \in V_G$ is $lab_G(v,w) =_{def} (L_G(v), E_G(v,w), E_G(w,v), L_G(w))$.

Next let us define graph grammars briefly. Introduction and survey can be found e.g. in [ClEhRo 79]. $GG = (\Sigma_V, \Sigma_T, \Sigma_E, P, S)$ is a graph grammar, short gg, if (i) $\Sigma_V, \Sigma_T, \Sigma_E$ are finite alphabets (for node, terminal node and edge labels resp.), $\Sigma_T \subseteq \Sigma_V$; (ii) P is a finite set of productions (explained below); and (iii) $S \in \Sigma_V \backslash \Sigma_T$ is the start symbol. The start graph is a single node labelled by S. Node labels from $\Sigma_N =_{def} \Sigma_V \backslash \Sigma_T$ are called nonterminals. A production $p = (L,R,I,O) \in P$ consists of two graphs L,R (the left and right hand side of p, short lhs and rhs) over Σ_V, Σ_E, each of them weakly connected, and the embedding rules $I,O \subseteq V_R \times \Sigma_V \times \Sigma_E \times \Sigma_E$ for incoming and outgoing edges resp. If L is one node only, p is called contextfree. In this paper only contextfree productions are considered. $s = (p, \tilde{L}, \tilde{R}, \tilde{b})$ is a derivation specification of GG if $p = (L,R,I,O) \in P$, $\tilde{L} \cong$ L, $\tilde{R} \cong R$, $\tilde{b} : V_{\tilde{R}} \to V_R$ is an isomorphism. s is applicable to a graph G if $\tilde{L} \lhd G$ and $V_G \cap V_{\tilde{R}} = \varnothing$. $G \underset{s}{\Rightarrow} G'$ is a derivation step if s is applicable to G, $\tilde{R} \lhd G'$, $G|(V_G \backslash V_{\tilde{L}}) = G'|(V_G \backslash V_{\tilde{R}})$ and edges between $\tilde{R}$ and the restgraph are as follows: For all $v \in V_{\tilde{R}}$, any node w from $R_G(\tilde{L})$ we have $E_{G'}(w,v) = \{x' \in \Sigma_E |\ \exists\ x \in E_G(w,\tilde{L}) : (\tilde{b}(v), L_G(w), x, x') \in I\}$ and $E_{G'}(v,w) = \{x' \in \Sigma_E |\ \exists\ x \in E_G(\tilde{L},w) : (\tilde{b}(v), L_G(w), x, x') \in O\}$.

Our first example is $GG_1 = (\{S,a\}, \{a\}, \{x,y\}, \{p_1, p_2\}, S)$ with two productions $p_1 = (S, R_1, I_1, O_1)$, $p_2 = (S, R_2, I_2, O_2)$, graphically described below. R_1 is the graph with vertices 1,2, R_2 is the graph with vertex 3, 4. The embedding relations are

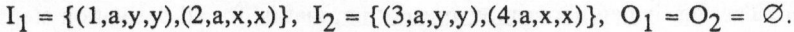

$I_1 = \{(1,a,y,y),(2,a,x,x)\}$, $I_2 = \{(3,a,y,y),(4,a,x,x)\}$, $O_1 = O_2 = \varnothing$.

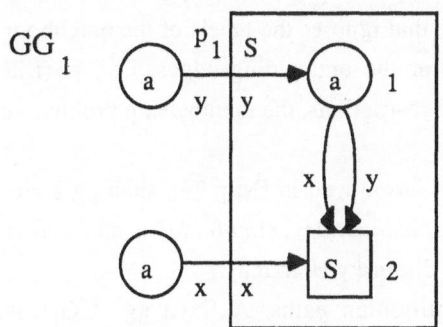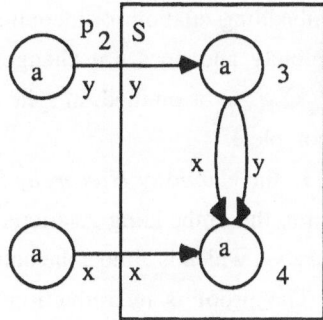

A sequence of derivation steps $D = (G_{i-1} \overrightarrow{s_i} G_i \mid 1 \le i \le n)$, $n \in \mathbb{N}$, $s_i = (p_i, \tilde{L}_i, \tilde{R}_i, \tilde{b}_i)$, is called <u>derivation sequence</u>, if G_0 is the start graph and in every step only new nodes are introduced, that is $V_{\tilde{R}_i} \cap V_{\tilde{R}_j} \ne \varnothing \Rightarrow i = j$. G_n is called <u>derivable</u> by GG. $S(GG)$ is the <u>class of all derivable graphs</u>. The <u>language</u> of GG, short $L(GG)$, is the class of all graphs from $S(GG)$, that are labelled with terminals only.

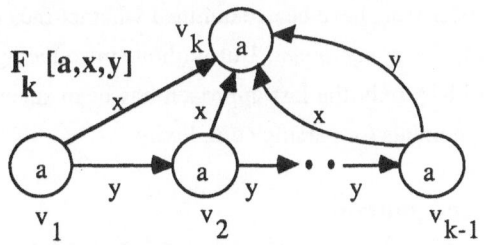

$L(GG_1) = \{G \cong F_k[a,x,y] \mid k \ge 2\}$ is the graph language in our example, which is an infinite set of <u>k-fans</u> $F_k[a,x,y]$, where a is a vertex label, x,y the edge labels. Note that v_1 is the first node, v_k is the last node generated during a derivation. The degree of nonterminals in derivable graphs is unbounded. But there is an equivalent gg GG'_1 s.t. the degree is at most 3.

2. Complexity of the membership problem

The membership problem for the class of gg´s introduced in chapter 1 is PSPACE-complete [GaJo 79]. The proof is by reduction from the membership problem for context-sensitive string languages. Even if we add the following restrictions on the productions, especially the embedding rules, the membership problem remains intractable: (1.) There is only one nonterminal on every rhs, that has incoming edges only. Every rhs has at most three nodes and two edges. (2.) Every nonterminal in every derivable graph has no outgoing edges, that is outdegree zero. (3.) The gg is <u>neighbourhood preserving</u>, that is, for all derivation steps $G \underset{s}{\Rightarrow} G'$, $s = (p,\tilde{L},\tilde{R},\tilde{b})$, $G,G' \in S(GG)$ we have $R_G(\tilde{L}) = R_{G'}(\tilde{R})$. (4.) There is only an embedding relation for incoming edges I, that ignores the labels of the neighbour nodes completely and does not change the label of the embedding edges, i.e. $p=(L,R,I)$, I $\subseteq V_R \times \Sigma_E$, O is omitted. In spite of all these restrictions, the membership problem remains NP-complete.

In the *Chomsky Hierarchy* for gg's, as introduced in [Nag 79], such gg's are called *regular*, the embedding relations *monotone, elementary, simple, label and orientation preserving*, which is almost the bottom of the Chomsky Hierarchy.

The proof is by reduction from Hamiltonian paths. A fixed gg GG_0 with all restrictions mentioned above can be given, that has NP-complete membership problem. That is, even if the gg is not part of the input, the membership problem remains intractable. In fact, it does not matter whether the gg is <u>node label controlled</u> (NLC, cf. [JaRo 80]), or <u>edge label controlled</u>(ELC). If there is a way to divide the neighborhood into three different classes, the proof referred to above, will work. Obviously graph parsing is much harder than string parsing. *Cocke-Younger-Kasami*-type and *Earley*-type parsers, which are commonly used in syntactic pattern recognition [Fu 82], cannot be extended to the full class of contextfree gg's. Therefore other types of parsers have been examined whether they can be transferred to graph parsing: *LL(k)-, LR(k)-* and *precedence*. But without introducing unnatural orderings into graphs, as e.g. in [Lud 81], only the last approach has been successful [Fra 78], [Ka 86] in gaining low-order polynomials for parsing complexity.

3.   Derivation Sequences

Let $D = (G_{i-1} \underset{s_i}{\Rightarrow} G_i \mid 1 \leq i \leq n)$, $n \in \mathbb{N}$, $s_i = (p_i,\tilde{L}_i,\tilde{R}_i,\tilde{b}_i)$, be a derivation sequence. s_i <u>precedes</u> s_j if $\tilde{L}_j \lhd \tilde{R}_i$, $1 \leq i,j \leq n$. The reflexive and transitive closure of this relation is denoted $\leq_D$. s_i , s_j are <u>incomparable</u> if neither $s_i \leq_D s_j$ nor $s_j \leq_D s_i$. $\leq_D$ is a partial order, called <u>derivation order</u> of D. The *Hasse-diagram* is a tree, called <u>derivation tree</u> of D. Let $s_D(v) =_{def} s_i$ if $v \in V_{\tilde{R}_i}$, $1 \leq i \leq n$. The derivation order imposes an ordering of the nodes in G_n as follows: $v \Theta w \Leftrightarrow s_D(v) \Theta_D s_D(w)$, $v,w \in V_{G_n}$ adjacent, $\Theta \in \{=,<,>\}$. Instead of $=,<,>$ we write $\doteq,\langle,\rangle$. If $s_D(v), s_D(w)$ are incomparable we write $v \bowtie w$. $\doteq,\langle,\rangle,\bowtie$ are the <u>precedence relations between nodes</u>. <u>The precedence relations between labels</u> R_Θ , $\Theta \in \{\doteq,\langle,\rangle,\bowtie\}$ is the set of all $lab_G(v,w)$ s.t. there is a derivable graph G, $v,w \in V_G$, $v\Theta w$. E.g. the precedence relations of GG_1 are $R_{\doteq} = \{(a,\{x,y\},\varnothing,S),(S,\varnothing,\{x,y\},a),(a,\{x,y\},\varnothing,a),(a,\varnothing,\{x,y\},a)\}$, $R_{\langle} = \{(a,\{x\},\varnothing,S)\} \cup \{ (a,\xi,\varnothing,a) \mid \xi \in \{\{x\}, \{y\}\}\}$, $R_{\rangle} = \{(S,\varnothing,\{x\},a)\} \cup \{ (a,\varnothing,\xi,a) \mid \xi \in \{\{x\}, \{y\}\}\}$, $R_{\bowtie} = \varnothing$.

If the precedence relations between labels are pairwise disjoint, the precedence relations between nodes can easily be inferred from them. A _precedence conflict_ is a label tuple t, that occurs in more than one precedence relation.

Let $n \in \mathbb{N}$, $D = (G_{i-1} \overset{\rightarrow}{s_i} G_i \mid 1 \le i \le n)$, $F = (H_{i-1} \overset{\rightarrow}{t_i} H_i \mid 1 \le i \le n)$ be two derivation sequences of the same length n, $s_i = (p_i, \tilde{L}_i, \tilde{R}_i, \tilde{\delta}_i)$, $p_i = (L_i, R_i, I_i, O_i)$, $t_i = (q_i, \tilde{L}_i', \tilde{R}_i', \tilde{\delta}_i')$, $1 \le i \le n$. Let $\alpha = (a_i : V_{R_i} \to V_{R_i} \mid 1 \le i \le n)$ be a family of automorphisms a_i on the rhs's R_i and β be a permutation on $\{1,...,n\}$ with $p_i = q_{\beta(i)}$ and $s_i \le_D s_j \Leftrightarrow t_{\beta(i)} \le_F t_{\beta(j)}$, $1 \le i,j \le n$. D,F are called _symmetrically equivalent_ under (α, β) if $\tilde{L}_j \triangleleft \tilde{R}_i \Rightarrow a_i(\tilde{\delta}_i(\tilde{L}_j)) = \tilde{\delta}'_{\beta(i)}(\tilde{L}'_{\beta(j)})$, $1 \le i,j \le n$. (α,β) is called _symmetric derivation permutation_. Note that derivation steps as well as symmetric nodes with rhs's may be interchanged. A gg GG is _unambiguous_ if any two derivation sequences of GG yielding isomorphic graphs are symmetrically equivalent. GG is _confluent_ if for all derivable graphs G_1, G_2, G_3 and incomparable derivation specifications s_1, s_2 with $G_1 \overset{\rightarrow}{s_1} G_2$, $G_1 \overset{\rightarrow}{s_2} G_3$ there is a derivable graph G_4 s.t. $G_2 \overset{\rightarrow}{s_2} G_4$ and $G_3 \overset{\rightarrow}{s_1} G_4$ are derivation steps. Let $p = (L,R,I,O)$ be a production, $v \in V_R$, $I_v =_{def} \{(w,A,x,x') \in I \mid w = v\}$, $O_v =_{def} \{(w,A,x,x') \in O \mid w = v\}$. p is _symmetric_ if $I_v = I_{a(v)}$, $O_v = O_{a(v)}$ for all automorphisms $a : V_R \to V_R$, $v \in V_R$. GG is _symmetric_ if all productions are. If GG is both symmetric and confluent, then all symmetric equivalent derivation sequences yield isomorphic results. GG is _uniquely invertible_ if every derivation step can be inverted uniquely only by inspection of the substituted subgraph and its direct neighbourhood. A nonterminal B is called _reflexive_ if B can be derived from B in at least one step.

4. Precedence Graph Grammars

A gg that is confluent, symmetric, uniquely invertible, has no reflexive nonterminals and no precedence conflicts, is called _precedence graph grammar_, short _pgg_. Main properties of pgg's are:

Pgg's are neighbourhood preserving and unambiguous. For any contextfree gg, it can easily be decided, whether it is a pgg. In particular precedence relations can efficiently be computed by taking into account subgraphs H of derivable graphs with $|V_H| \le 2$. The same technique applies to the test of confluence. On the other hand it is undecidable, whether an equivalent pgg can be found. The membership of a graph G in L(GG) is decidable in $O(|V_G|^2)$ time and $O(|V_G|+|E_G|)$ space. If one of both properties, (1.) unique invertibility and (2.) no precedence conflicts, is omitted, the membership problem becomes NP-complete. The proof is by reduction from Hamiltonian paths.

There is a _trade-off_ between unique invertibility and precedence conflicts. E.g. consider $L_{star} = \{S_k[a,x] \mid k \ge 1\}$, where $S_k[a,x]$, called _k-star_ with node label a and edge label x, as shown below, consists of a node v_k adjacent to k-1 nodes $v_1,...,v_{k-1}$, which are pairwise nonadjacent. The following gg's generate L_{star}: GG_2 is a uniquely invertible gg , GG_3 is a gg without precedence conflicts. But there is no pgg for k-stars.

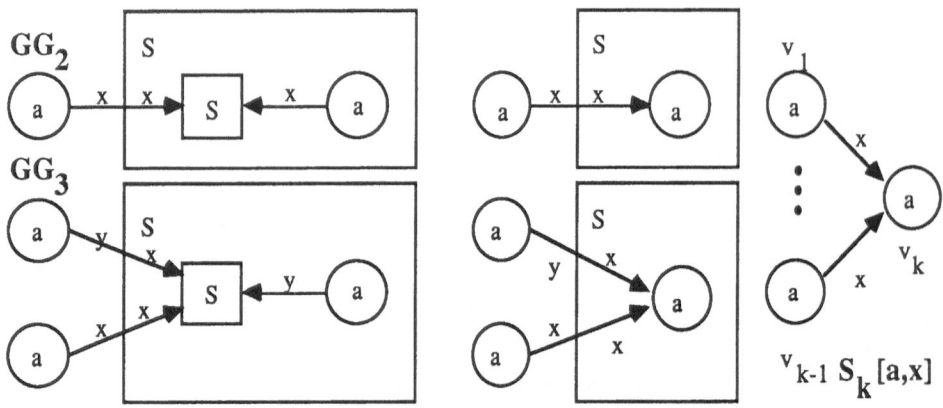

In some sense, precedence conflicts can always be eliminated: Let $\Sigma_E^{\Diamond} =_{def}$ $\Sigma_E \times \{\doteq, \langle, \rangle, \bowtie\}$ is the <u>extended edge label alphabet</u>. $\Diamond$ maps every graph G over $\Sigma_E^{\Diamond}$ to a graph $G^{\Diamond}$ over Σ_E as follows: $V_G^{\Diamond} = V_G$, $L_G^{\Diamond} = L_G$, $E_G^{\Diamond} = \{ (v,x,w) \mid v,w \in V_G , x \in \Sigma_E , (v,(x,\mu),w) \in E_G , \mu \in \{\doteq, \langle, \rangle, \bowtie\} \}$. Then for every gg GG a gg $GG^{\Diamond}$ without precedence conflicts can effectively be constructed, s.t. $[L(GG^{\Diamond})]^{\Diamond} = L(GG)$. $GG^{\Diamond}$ is the <u>conflictfree normal form</u> of GG. But this construction is at the expense of unique invertibility, which is often lost during the construction of $GG^{\Diamond}$.

Some results are known about the generative power of pgg's. If a graph language L contains too many nonisomorphic symmetric graphs of certain types, (e.g. k-stars $S_k[a,x]$ as shown above, or symmetric k-chains $K_k[a,x]$, or symmetric k-cycles $C_k[a,x]$ as shown below), then there is no pgg GG with $L = L(GG)$.

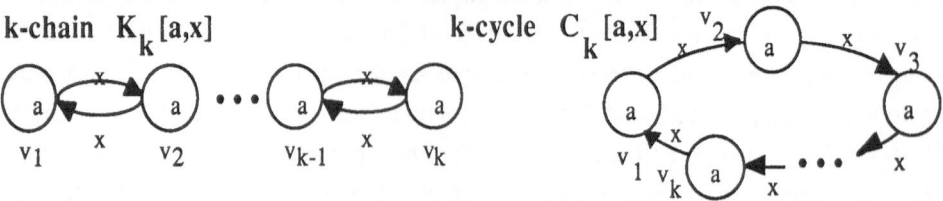

Introducing auxiliary edges that destroy symmetry often helps to construct a pgg. E.g. the graph $F_k[a,x,y]$ of $L(GG_1)$ may be considered as a k-star $S_k[a,x]$, to which auxiliary y-edges $v_i \to v_{i+1}$, $1 \le i \le k-1$, have been added. Furthermore there is an information theoretic argument, that limits the number of nonisomorphic graphs with n nodes within

L(GG) to 2^{c*n}, c some constant depending on GG, if GG is a pgg. The reason is that the length of derivation sequences yielding graphs with n nodes is $O(n)$.

5. Precedence Graph Parser

The membership problem for gg´s as introduced in chapter 3 is PSPACE-complete. So the parsing problem is at least as hard. A parser that constructs the parse tree efficiently, has obiously to be provided with more information about the syntactical structure of the input graph. The precedence relations describe parts of the parse tree indirectly and thus supply such helpful information. A parse tree is a hierarchical decomposition of the input graph, that may be described by putting the graph into hierarchically nested boxes according to the following rules: (1.) Nodes of equal precedence are derived from the same node and therefore lie within the same box. (2.) Along ascending precedence no box is leaved, but at least one box entered. (3.) Along ⋈ at least one box is left, and then at least one box is entered, as shown below:

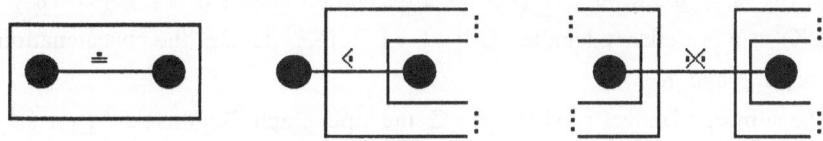

Although there are infinitely many ways to put a graph into nested boxes according to these rules, there is only one parse tree T for G up to symmetric equivalence. (Recall Unambiguity.) Two observations show that T can be constructed efficiently: Let $n \in \mathbb{N}$, D = $(G_{i-1} \overrightarrow{s_i} G_i \mid 1 \leq i \leq n)$, be a derivation sequence of length n, $s_i = (p_i, \tilde{L}_i, \tilde{R}_i, \mathfrak{b}_i)$, $1 \leq i \leq n$. A full subgraph $H \triangleleft G_n$ is a <u>handle</u> w.r.t. D if there is some $\tilde{R}_i$. Of course, when parsing G_n bottom-up, a handle must be reduced first. On the other hand let G be some derivable graph. A full subgraph $H \triangleleft G$ is called a <u>precedence handle</u> in G if $lab_G(v,w) \in R_{\pm}$ for all adjacent $v,w \in V_H$ and $lab_G(v´,w´) \notin R_{\pm} \cup R_{\rangle}$ for all adjacent nodes v´ from $R_G(H)$, $w´ \in V_H$. In syntactically correct graphs a precedence handle is always a handle and vice versa.

The second observation is that if the parser proceeds along ascending precedence, a precedence handle has to be found after at most $|V_G|$ steps, G being the input graph. Precedence parsers are well known as shift-reduce parsers [AhUl 72], that store intermediate results into a stack. String parsers store symbols, graph parsers store node addresses into the stack. *Precedence graph parsing* is performed as follows: (1.) Start at an arbitrary node. (2.) Go along equal or ascending precedence as long as possible. Shift traversed nodes into the stack. (3.) If no further ascending is possible, a precedence handle is reached, which is isomorphic to a rhs of a production. Reduce the precedence handle to the lhs using the production in the opposite direction, in particular the inverse embedding rules, and proceed with (2.). Of course all precedence handles available at the same time can be reduced in parallel (recall confluence), which may reduce parsing time to $O(\log n)$, if the derivation tree is balanced.

A graph parser that does not visit any node more than two times is described in the following: First we denote an <u>instantaneous description</u>, short ID, by (G,K,Ψ), where G is the instantaneous graph, K an ordered list of nodes in G, Ψ a set of derivation specifications constructed so far. The initial ID is $(G_0,K_0,\varnothing)$, where G_0 is the input graph, K_0 some node of G_0. Let (G,K,Ψ) be some ID, $K = \langle v_1, \dots ,v_k\rangle$, $k \geq 1$. Let j be the minimum index $1 \leq j \leq k$, s.t. there is some path in G from v_j to v_k along equal precedence. Then $\underline{TOP(G,K)} =_{def} G|\{v_j,\dots,v_k\}$. There are two types of <u>moves</u>, called shift and reduce: $(G,K,\Psi) \vdash_S (G,Kw, \Psi)$ is a <u>shift</u> if (i) $w \in V_G$ does not occur in K already; (ii) $lab_G (v,w) \in R_{\doteq} \cup R_{\lessdot}$ for some v in $TOP(G,K)$. $(G,K_1K_2,\Psi) \vdash_R (G',K_1w,\Psi\cup\{s\})$ is a <u>reduce</u> if (i) w is a node not used in G or Ψ already; (ii) $s = (p,\tilde{L},\tilde{R},\tilde{b})$, $G' \underset{s}{\Rightarrow} G$ is a derivation step, $s \notin \Psi$, $K_2 = V_{\tilde{R}}$, $G|K_2 = TOP(G,K)$ is a precedence handle, $G'|\{w\} = \tilde{L}$, K_1K_2 denotes the concatenation of both sequences K_1 and K_2.

The number of moves is $O(|V_G|)$, G the input graph. But there are gg's s.t. a reduce may need $O(|V_G|)$ steps as well, e.g. our example GG_1. But there is an equivalent pgg $GG_1{}'$ s.t. every reduce can be performed by $O(1)$ steps only. The idea is to generate v_{n+1} first, then the chain $v_1,\dots,v_n$ within the graph G_n. It is still an open problem whether always such a "better" gg can be found.

Comparing precedence graph parsers with other approaches such as LL(k), LR(k), it turns out to be superior concerning the following *advantages*: (1.) No starting node has to be distinguished. There is no fixed linear order in which the graph has to be traversed. All traversals, even in parallel, yield the same parse tree up to equivalence, because pgg's are unambiguous. (2.) In [Ka 86] an abstract concept of precedence relations is introduced, that is general enough to describe every bottom-up parser as a precedence graph parser. Furthermore the schema for computing precedence relations is independent of the special type of embedding rules, and is therefore applicable to a wide range of gg-classes. (3.) Attributing grammars [Knu 68] is a well known technique to incorporate additional semantic information into a grammar, which may be applied to pgg's as well. [Fu 82] applies precedence string grammars as well as attributed string grammars to pattern recognition. In [Sch 85] algorithms for attribute evaluation on attributed pgg's are given. There exactly the pgg-class given in this paper is used.(4.) Introducing application conditions, an interface to knowledge based production systems [Nag 82] can easily be introduced. Pattern descriptions that could be expressed in logical formulas more comfortably, can be incorporated into the precedence parser easily. (5.) Precedence string grammars have been subject to a wide range of publications since 1966. Several techniques are already known for precedence string grammars and can be transferred to pgg's easily. E.g. [Has 74], [Bab 79] give efficient parsers for context-sensitive precedence languages, e.g. $\{a^n b^n c^n d^n \mid n \geq 1\}$ and $\{$ wcw $\mid$ w $\in \{a,b\}^*\}$. [Fu 82] describes stochastic precedence parsers.

6. Error Handling

Error handling consists of error detection and error correction. Of course, only local errors can be detected without parsing the whole graph completely. Error correction modifies the graph in a minimal way s.t. a shift/reduce can be performed again.

Let (G,K,Ψ) be some ID. A vertex w in G is called a shift error if w can be shifted onto the stack, but (i) there are vertices v,v´ in TOP(G,K) s.t. $lab_G(v,w) \in R_{\doteq}$ and $lab_G(v´,w) \in R_{\lessdot}$; or (ii) there is a vertex $v'' \in K \backslash V_{TOP(G,K)}$ s.t. $lab_G(v'',w) \notin R_{\lessdot}$. TOP(G,K) is a reduce error if neither a shift nor a reduce can be applied to (G,K,Ψ). Clearly, TOP(G,K) is a precedence handle, that matches with no rhs, or there is some embedding edge between TOP(G,K) and the host graph with illegal precedence. Both shift and reduce error can be detected by regarding the local context only.

In order to discuss error correction, first minimal modification of graphs has to be specified. For this purpose the difference of two graphs G,G´, short $\Delta(G,G´)$ is introduced as follows: $\Delta(G,G´)$ is the union of $\Delta_j(G,G´)$, $1 \leq j \leq 4$, where $\Delta_1(G,G´) =_{def} \{(1,v,A) \mid v \in V_G \backslash V_{G´}$, $A = L_G(v)\}$, $\Delta_2(G,G´) =_{def} \{(2,v,A,B) \mid v \in V_G \backslash V_{G´}$, $A = L_G(v)$, $B = L_{G´}(v)$, $A \neq B\}$, $\Delta_3(G,G´) =_{def} \{(3,v,A) \mid v \in V_{G´} \backslash V_G$, $A = L_{G´}(v)\}$, $\Delta_4(G,G´) =_{def} \{(4,v,w,X,Y) \mid v,w \in V_G \cup V_{G´}$, $X = \{x \mid (v,x,w) \in E_G \backslash E_{G´}\}$, $Y = \{y \mid (v,y,w) \in E_{G´} \backslash E_G\}\}$. An element from $\Delta_j(G,G´)$, $1 \leq j \leq 3$, is an elementary vertex transformation, in particular a (1) vertex deletion, (2) vertex relabelling, (3) vertex insertion. An element from $\Delta_4(G,G´)$ is an elemenatry edge transformation, called edge relabelling. $LAB_j[d]$, $d \in \Delta_j(G,G´)$, $1 \leq j \leq 4$ denotes the tuple of those components of d, that are labels, e.g. $LAB_4[(4,v,w,X,Y)] = (X,Y)$. Let $COST_j$ assign costs to elementary transformations from $\Delta_j(G,G´)$, dependent on labels only, that is $COST_{1,3} : \Sigma_V \to \Re^+ \cup \{\infty\}$, $COST_2 : \Sigma_V \times \Sigma_V \to \Re^+ \cup \{\infty\}$, $COST_4 : 2^{\Sigma_E} \times 2^{\Sigma_E} \to \Re^+ \cup \{\infty\}$ are mappings, $\Re^+$ denotes the set of positive real numbers. The error distance between G,G´ is $\delta(G,G´) =_{def} \sum_{d \in \Delta_i(G,G´)} COST_j[LAB_j[d]]$, $1 \leq j \leq 4$.

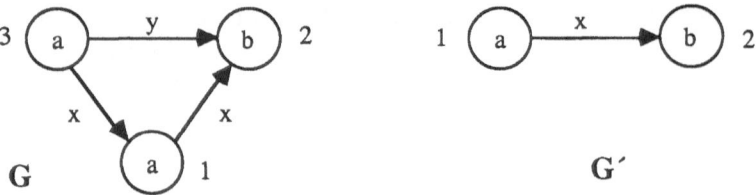

$\Delta_1(G,G') = \{(1,3,A)\}$, $\Delta_2(G,G') = \Delta_3(G,G') = \varnothing$, $\Delta_4(G,G') = \{(4,1,2,\{y\},\{x\})\}$.

δ should be a <u>metric</u>, that is $\delta(G,G') = 0$ iff $G = G'$, $\delta(G,G') = \delta(G',G)$, $\delta(G,G') \leq \delta(G,G'') + \delta(G'',G')$. In particular $COST_1[A] = COST_3[A]$, $COST_2[A,B] = COST_2[B,A]$, $COST_4[X,Y] = COST_4[Y,X]$, $A,B \in \Sigma_V$, $X,Y \in 2^{\Sigma_E}$

Furthermore δ should be <u>compatible</u> with the pgg. This means e.g. infinite loops of vertex relabellings and reduce moves have to be avoided by a proper choice of $COST_2$, that is, the relation $\vdash^{RC} =_{\text{def}} \{(A,B) \in \Sigma_N \times \Sigma_N \mid B \Rightarrow A$ or $\forall C \in \Sigma_N : COST_2[A,B] \leq COST_2[A,C]\}$ has to be cyclefree: $\forall A \in \Sigma_N : (A,A) \notin (\vdash^{RC})^+$, the transitive closure of $\vdash^{RC}$.

We are now ready for specifying shift and error correction: Let (G,K,Ψ) be an ID, w a shiftable vertex. G' is a <u>shift correction</u> of G if G' if G' is identical to G up to edge labels between w and K, w remains shiftable within G', and w is no shift error any more. G' is a <u>minimal shift correction</u> if for all shift corrections G'' $\delta(G',G) \leq \delta(G'',G)$. (G',K',Ψ) is a <u>reduce correction</u> if (i) G',K' are identical to G,K up to $TOP(G,K)$ and its embedding edges, i.e. $G|(V_G \backslash V_{TOP(G,K)}) = G'|(V_G \backslash V_{TOP(G',K')})$; and (ii) to (G',K',Ψ) a reduce can be applied; and(iii) if $TOP(G,K)$ is a single vertex only, then $TOP(G',K')$ is a single vertex, too. (The last condition guarantees decreasing size of the parsed graph.) (G',K',Ψ) is a <u>minimal reduce correction</u> if for all reduce corrections (G'',K'',Ψ) $\delta(G',G) \leq \delta(G'',G)$.

Error correcting moves first detect some error, compute some minimal error correction and finally proceed with an ordinary move. But there are several startegies to do so in the case of shift errors: <u>Strategy A ("Eliminate errors as early as possible"):</u> Traverse all shiftable vertices and apply minimal shift correction to all of them. Shift all traversed vertices onto the stack. <u>Strategy B ("Modify graph in a minimal way"):</u> Search for a shiftable vertex, such that its minimal shift correction is less than those of all other shiftable vertices. Clearly, if shiftable vertices with no shift error exist, then one of them is shifted onto the stack. <u>Startegy C ("Shift at random"):</u> Take any shiftable vertex w. If w is a shift error, apply minimal shift correction. Shift w onto the stack. The choice is application dependent and not anticipated here.

In the case of reduce errors, a minimal reduce correction is applied, and finally a reduce performed. This is simply an application of inexact graph matching [ShHa 81], [BuAl 83] to subgraphs of constant size (the constant given by the pgg). That is, the rhs of some production, that is most similar to the instantaneous precedence handle is found by a backtracking method. Embedding edges with illegal precedence are relabelled with minimal costs s.t. TOP(G,K) remains a precedence handle. For both shift and reduce correction the deletion of an edge may be the cheapest modification. In this case the instantaneous graph G may fall apart into several connected components $G_1, \ldots, G_k$, $k \geq 2$. The parser continues to process the current connected component G_i in which TOP(G,K) is a subgraph, until G_i is reduced to a single vertex only. then the parser proceeds with the other components G_j, $j \neq i$. Switching from one component to another one os called a <u>next move</u>, denoted $(G,K,\Psi) \vdash_{\overline{n}} (G,K',\Psi)$.

Contrary to the exponential complexity of the algorithms given in [ShHa 81], [BuAl 83], we arrive at $O(n)$ time complexity for a single reduce move, n the number of vertices of the input graph. For a single error correcting shift, $O(n^2)$ time in the case of strategies A,B and $O(n)$ time in the case of strategy C is needed. Since the number of shift and reduce moves is $O(n)$, we finally end up with $O(n^2)$, $O(n^3)$ time complexity resp. for the entire parsing process, which essentially is a divide-and-conquer strategy for inexact graph matching. Note that this approach is not intended to yield $O(n^3)$ time complexity for a NP-complete problem. Instead we changed the problem specification in a profitable way: The user of the parser has to specify his notion of similarity of graphs in a different way. Instead of specifying weights for graph editing operations only, as in [ShHa 81], [BuAl 83], helpful structural knowledge about the family of graphs is added to the input. The additional information reduces complexity.

Let $D =_{def} ((G_{i-1}, K_{i-1}, \Psi_{i-1}) \vdash_i (G_i, K_i, \Psi_i) \mid 1 \leq i \leq n)$, $n \geq 0$, be a sequence of error correcting and ordinary moves, that is a shift or reduce. To (G_n, K_n, Ψ_n) no further move should be applicable. Let $\delta_D(G_0)$ be the sum of $\delta(G_n, S)$, $\delta(G_{i-1}, G_i)$, for all error correcting moves $\vdash_i$ $\delta_D(G_0)$ is called <u>error distance</u> of G_0 w.r.t. D. Of course, D as well as $\delta_D(G_0)$ are not uniquely determined by G_0. Syntactically correct graphs, and only those, have error distance 0: Let GG be a pgg, G_0 some input graph, δ a metric. There is a sequence of error correcting and ordinary moves D of GG with $\delta_D(G_0) = 0$ iff $G_0 \in S(GG)$.

Fortunately the number of error corrections can be limited by $O(n^3)$ in the case of strategies A,B and $O(n^2)$ in the case of strategy C. This is especially important, if some error report has to be processed by further processing steps.

7. Example

A simple example of line drawings describing stairs is presented in this chapter to demonstrate the basic ideas of our method. First, line drawings have to be encoded as node and edge labelled directed graphs, thereby mapping classes of geometrical features to topological representations.

The edge direction is obtained from the undirected lines by orientation from left to right, from the bottom to the top. The edge labels describe approximate direction and length of a basic line. Our notation is based on the well-known quarters of the heavens, that is $\Sigma_E = \{N,E,NE,NW\}$. Node labels $\Sigma_T = \{a,b,c,d,e,f,g,h\}$ describe the shape of intersections of lines, that is their number and the angles between them approximately as follows:

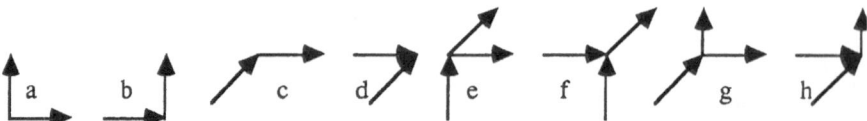

All line drawings describing stairs viewed from the right with an arbitrary number of steps are generated by the following pgg GG_4:

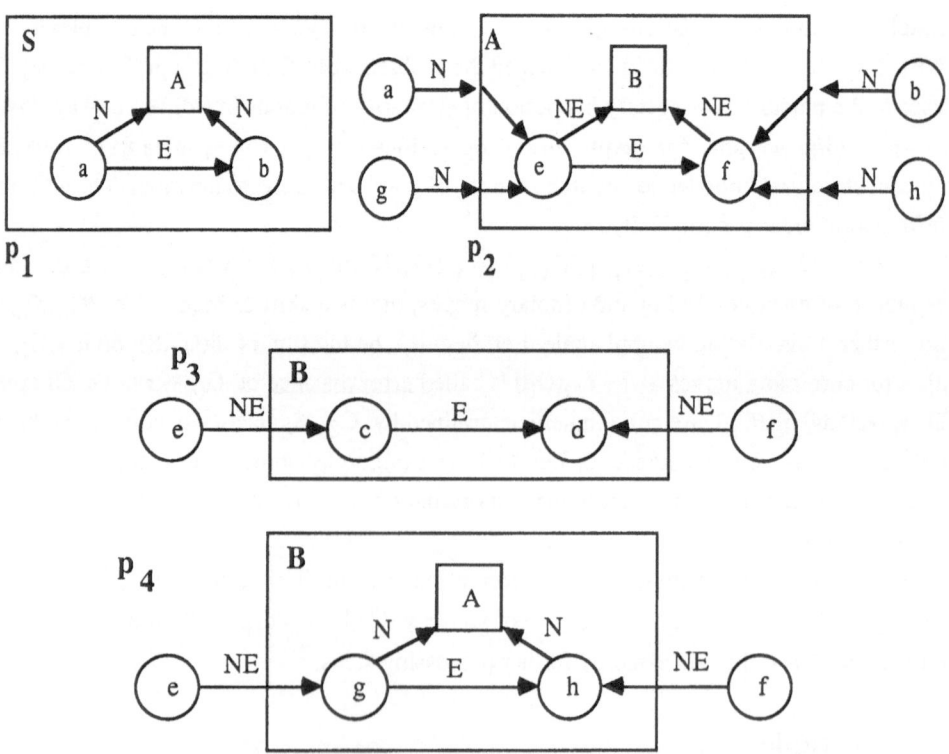

Precedence relations are as follows: In $R_{\pm}$ all label tuples are included from rhs´s of

GG$_4$. In R$_4$ we have edges between terminals labelled by N or NE with ascending precedence in the direction of the edge. R$_>$ consists of the same edges as R$_4$, but with opposite direction. R$_{><}$ = ∅. By GG$_4$ only undistorted stairs are described, such as a stair with one step, see G$_1$ below. Now let D$_2$ as shown below be a distorted line drawing of a stair with two steps. First we assume D$_2$ to be encoded as the graph G$_2$, shown below. Note that there are several intersections of lines in D$_2$, for which no appropriate vertex label of GG$_4$ exist. At these places the most similar vertex labels are used instead. Let G$_2$ be the input of our error correcting precedence graph parser.

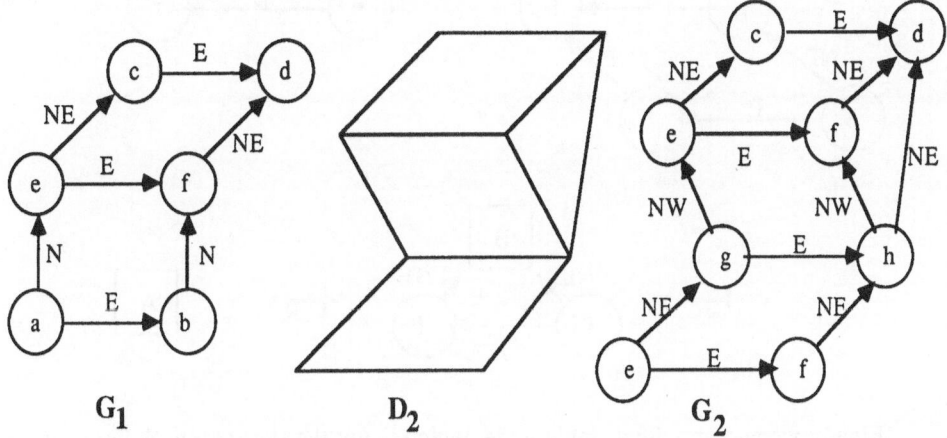

First the parser chooses an arbitrary node to start with, e.g. the c-node at the top. The d-node at the top has equal precedence, and there is no adjacent node with higher precedence. So a precedence handle is found, that is isomorphic to the rhs of p$_3$. But p$_3$ has no embedding relation for the adjacent h-node. Therefore a reduce error occurs and a reduce correction is applied, which simply deletes the annoying edge. Next a reduce can be performed:

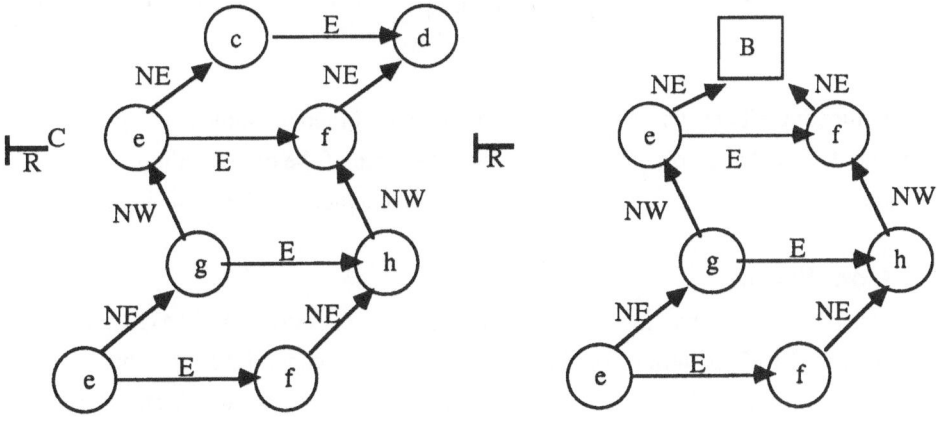

Next we arrive at a precedence handle isomorphic to the rhs of p_2. There are no embedding relation for NW-edges, but for N-edges, which are most similar in this case:

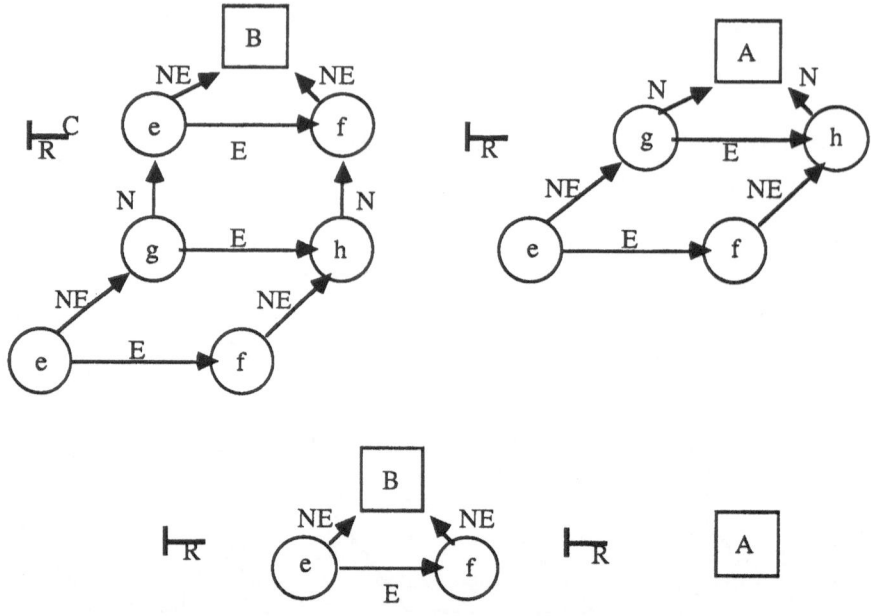

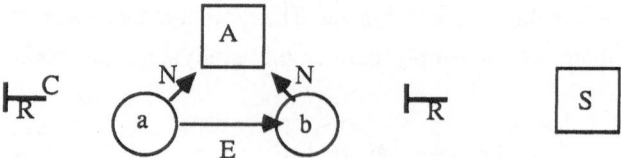

Finally we end up with a single node, which is not the start graph S, but subgraph of the rhs of p_1, that has S as its lhs. The easiest way to end up with the start graph is to insert the missing nodes of the rhs of p_1.

Collecting all derivation specifications involved in reduce moves, we get a derivation sequence D. The Hasse-diagram of D is the parse tree of the corrected input. The result of D is the syntactically correct stair with two steps, that is most similar to the input.

8. Conclusion

Compiler construction has much gained from parser generators like YACC under UNIX. According to [Rf 82] pattern recognition research lacks of general theories, that enable us to build such general tools for pattern recognition as well. Our contribution is an error correcting precedence graph parser, which allow to incorporate *application dependent knowledge* systematically: The *structural* knowledge about the pattern class is described by the pgg, the *statistical* knowledge is given by the weights for the editing operations. By exploiting application dependent knowledge inexact graph matching, which in general is intractable, becomes solvable in $O(n^3)$ time. (It should be emphasized, that no polynomial time solution to an NP-complete problem is attempted, but that the problem specification is changed in such a way, that the problem becomes tractable.) Furthermore the parse tree of the most similar graph is computed, thus providing further processing steps with an efficient hierarchical decomposition. The method is described in depth in the author´s dissertation [Ka 86].

9.   References

[AhUl 72]   A.V.Aho/J.D.Ullman: The Theory of Parsing, Translation, and Compiling; I,II, Prentice-Hall, Englewood Cliffs, NJ (1972)

[Bab 79]   J.P.Babinov: Class of generalized context-sensitive prcedence languages; Progr.Comput. Software 5 (1979) 117-126

[BuAl 83]   H.Bunke/G.Allermann: Inexact Graph Matching for Structural Pattern Recognition; Pat.Rec.Let. 1 (1983) 245-253

[ClEhRo 79]   V.Claus/H.Ehrig/G.Rozenberg: Graph-Grammars and Their Application to Computer Science and Biology; 1st Int. Workshop, LNCS 73, Springer (1979)

[EhNaRo 83]   H.Ehrig/M.Nagl/G.Rozenberg(Eds.): Graph-Grammars and Their Application to Computer Science, 2nd Int. Workshop, LNCS 153, Springer (1983)

[Fra 78]   R.Franck: A Class of Linearly Parsable Graph Grammars, Acta Inform. 10(1978)175-201

[Fu 82]   K.S.Fu: Syntactic Pattern Recognition; Prentice-Hall, Englewood Cliffs, NJ (1982)

[GaJo 79]   M.R.Garey/D.S.Johnson: Computers and Intractability; A Guide to the Theory of NP-Completeness; Freeman, San Francisco(1979)

[Har 69]   F.Harary: Graph Theory; Addison-Wesley Publ. Comp., Reading Mass. (1969)

[Has 74]   R.Haskell: Symmetrical precedence relations on general phrase structure grammars; Comp. Journ. 17 (1974) 234-241

[JaRo 80]   D.Janssens/G.Rozenberg: On the structure of Node Label Contolled Graph Languages; Inform.Sci. 20 (1980) 191-216

[Ka 86]   M.Kaul: Syntaxanalyse von Graphen bei Präzedenz-Graph-Grammatiken; Techn. Report MIP-8610, Uni. Passau, FRG

[Knu 68]   D.E.Knuth: Semantic of Context-free Languages; Math. Syst. Theo. (1968)

[Lud 81]   H.Ludwigs: Properties of Ordered Graph Grammars; in: H.Noltemeier(Ed.): Graphtheoretic Concepts in Comp. Science; LNCS 100, Springer (1981) 70-79

[Nag 79]   M.Nagl: Graph-Grammatiken - Theorie, Implementierung, Anwendung; Vieweg, Braunschweig (1979)

[Nag 82]   M.Nagao: Control Strategies in Pattern Analysis; Proc. Pat. Rec. Vol. I, 6th Int. Conf., Munich 1982 (1982) 996-1006

[Rf 82]   A.Rosenfeld: Image Analysis: Progress, Problems, and Prospects; Proc. Pat. Rec.Vol. I, 6th Int. Conf., Munich 1982 (1982) 7-15

[Sch 85]   A.Schütte: Einführung in die Theorie und Konzepte von attributierten Zeichenketten- und Graphgrammatiken; TR1/85, EWH Koblenz, FRG (1985)

[ShHa 81]   L.G.Shapiro/R.M.Haralick: Structural Descriptions and Inexact Matching;IEEE Trans. Pat. Ana. PAMI-3, No. 5 (1981)

A UNIFIED VIEW ON TREE METRICS

K. Ohmori, E. Tanaka

Utsunomiya University
Dept. of Information Science
Faculty of Engineering
Utsunomiya 321-31
Japan

1. Introduction

A considerable amount of work has been done on sequence comparison for problems such as string correction, molecular biology, human speech, codes and error control, and so on[1]. Tree metrics have been also studied[3-14]. Potential applications of tree metric include the areas of behavioral science[3], data base[5], clustering[7], waveform correlation[9], and so on. Among various tree metrics, Tai´s metric[6] seems to be the most fundamental one. Selkow´s metric[4] is a strictly restricted Tai´s metric. Between these two metrics, several metrics have been defined[10-12]. However, the interrelation between these metrics is not known clearly. In this paper, by introducing a concept "the nearest ancestor determined by a mapping", we give a unified point of view for tree metrics. Furthermore, we propose a new similarity between two trees.

2. Definitions

In this paper all trees we discuss are rooted, ordered, and labeled.

[Definition 1] Numbering in preorder.
A tree T is numbered from one in preorder for nodes of T. A positive integer represents a node.

NATO ASI Series, Vol. F45
Syntactic and Structural Pattern Recognition
Edited by G. Ferraté et al.
© Springer-Verlag Berlin Heidelberg 1988

[Definition 2] Notations.

T(k) denotes a subtree of a tree T whose root is k. Ch(k) and An(k) denote the set of children of k and that of ancestors of k, respectively. N(k) denotes the number of nodes of T(k). Let N mean N(1). t(k) denotes the label of node k. The right-most leaf of T(k) is called the end leaf of T(k) and denoted by el(k).

[Definition 3] Separation of nodes and subtrees.

For any nodes k_1 and k_2 ($k_1 \neq k_2$), k_1 and k_2 are said to be separated if k_1 is neither an ancestor of k_2 nor a descendant of k_2. Furthermore, if k_1 and k_2 are separated, $T(k_1)$ and $T(k_2)$ are said to be separated.

[Definition 4] Forest.

A sequence of separate subtrees $T(k_1), T(k_2), \cdots, T(k_n)$ ($k_1 < k_2 < \cdots < k_n$) is called a forest of T. A subtree T(k) is also a forest. If the forest is composed of all the nodes from k to m($k \leq m$), it is written F(k,m).

In Fig.1, Ch(4)={5,7}, An(4)={1,2}, N(4)=4, t(4)="E", el(4)=7.

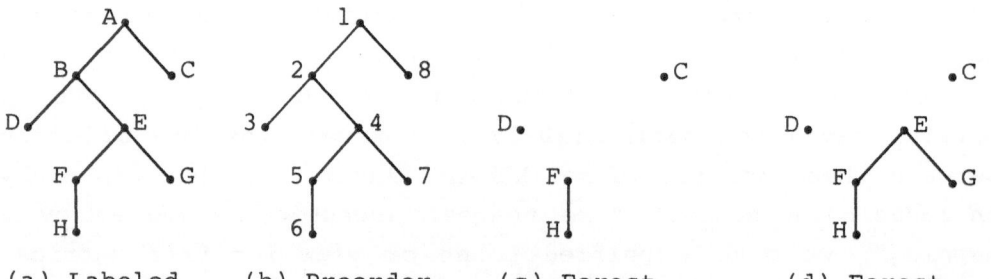

(a) Labeled tree T. (b) Preorder numbering for T. (c) Forest T(3),T(5),T(8). (d) Forest F(3,8).

Fig.1 Tree and forest.

Before discussing tree metrics, let us review briefly the weighted Levenshtein distance[2].

The following three operations to transform one string into the other and their weights(or costs) are considered: (1)sub-stitute another symbol for a symbol(cost p); (2)insert an extra symbol(cost q); and (3)delete a symbol(cost r). In general, p<q+r. Because if p>q+r, a substitution is always regarded as a pair of an insertion and a deletion.

[Definition 5] The weighted Levenshtein distance.

Let $\alpha = a_1 a_2 \cdots a_m$ and $\beta = b_1 b_2 \cdots b_n$ be two finite strings of symbols. A mapping M from α to β is a set of ordered pairs (i,j) $(1 \leq i \leq m, 1 \leq j \leq n)$. Let $I = \{i \mid (i,j) \in M\}$ and $J = \{j \mid (i,j) \in M\}$. Then, M represents a transformation from α to β under the following interpretation :

(1) For $(i,j) \in M$, if $a_i \neq b_j$, b_j is substituted for a_i;

(2) If $j \notin J$, b_j is inserted;

(3) If $i \notin I$, a_i is deleted.

Let $\widetilde{M}$ be the set of all possible these mappings from α to β. Then, the minimum cost from α to β, denoted by $D(\alpha, \beta)$, can be defined as follows :

$$D(\alpha,\beta) = \min_{M \in \widetilde{M}} \{ \sum_{(i,j) \in M} p(i,j) + (n - |J|)*q + (m - |I|)*r \}. \qquad (2\text{-}1)$$

where $|I|$ denotes the number of elements in I, and

$$p(i,j) = \begin{cases} 0 & : \text{if } a_i = b_j; \\ p & : \text{if } a_i \neq b_j. \end{cases}$$

$D(\alpha,\beta)$ is called the weighted Levenshtein distance(WLD) from α to β if the mapping M satisfies the following conditions.

For any pairs $(i_1, j_1), (i_2, j_2) \in M$,

(1) $i_1 = i_2$ iff $j_1 = j_2$;

(2) $i_1 < i_2$ iff $j_1 < j_2$.

$D(\alpha,\beta)$ can be computed by applying the following formula, iteratively :

$$d[i,j] = \min \begin{cases} d[i-1, j-1] + p(i,j), \\ d[\,i\,, j-1] + q, \\ d[i-1,\,j\,] + r, \end{cases} \qquad (2\text{-}2)$$

where,

$d[0,j] = j*q$ ($0 \leq j \leq n$),

$d[i,0] = i*r$ ($0 \leq i \leq m$).

Then, $D(\alpha,\beta) = d[m,n]$.

The time and space complexities to compute $D(\alpha,\beta)$ are proportional to mn.

In the mathematical literature, the word "distance" is ordinarily used to indicate a function ´d´ which satisfies the metric axioms:

For all α, β and γ,

(1) Nonnegative property : $d(α,β) \geq 0$;

(2) Zero property   : $d(α,β)=0$ iff $α=β$;

(3) Symmetry   : $d(α,β)=d(β,α)$;

(4) Triangle inequality : $d(α,γ) \leq d(α,β)+d(β,γ)$.

WLD satisfies the metric axioms if the insertion cost equals to the deletion cost, that is, q=r.

Let us turn to tree metrics. We will use a similar approach to define transformation between trees and tree-to-tree distances. The three edit operations on a labeled node, that is, substitution (cost p), insertion (cost q) and deletion (cost r) are considered. A mapping between trees is regarded as a transformation between trees. (i,j) M means that a labeled node i mapped to a labeled node $j(1 \leq i \leq N_α, 1 \leq j \leq N_β)$. Since the mapping conditions of WLD have no information on tree structures, a mapping M from $T_α$ to $T_β$ must satisfy conditions about tree structures. Tai proposed the following mapping.

[Definition 6] The mapping conditions of Tai´s distance[6].

For any pairs $(i_1,j_1),(i_2,j_2) \in M$,

(1) $i_1=i_2$ iff $j_1=j_2$;

(2) $i_1<i_2$ iff $j_1<j_2$;

(3) $i_1 \in An_α(i_2)$ iff $j_1 \in An_β(j_2)$.

Here we call it the Tai mapping. (1) means one-to-one correspondence, (2) means that the preorder of tree nodes does not change and (3) means that the ancestor-descendant relation does not change. If $i_1 \in An_α(i_2)$ and $j_1 \in An_β(j_2)$, then it is obvious from preorder numbering that $i_1<i_2$ and $j_1<j_2$. That is, (2) states that the order relation between separate nodes is preserved.

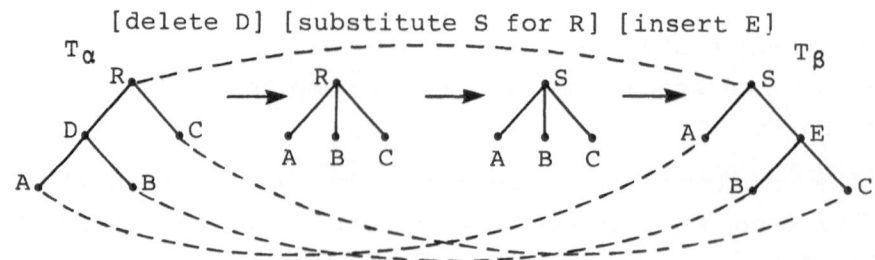

Fig.2 Transformation and mapping from $T_α$ to $T_β$.

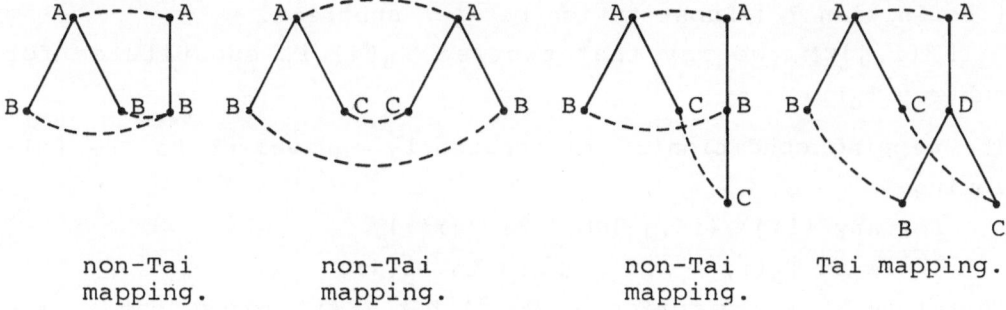

non-Tai   non-Tai   non-Tai   Tai mapping.
mapping.   mapping.   mapping.

Fig.3   Examples of non-Tai mappings and a Tai mapping.

The minimum cost under the Tai mapping is called the Tai distance. The Tai distance $D(T_\alpha, T_\beta)$ satisfies the metric axioms in
case of q=r.

3. Computation of the Tai distance

In this section, we propose a simple algorithm for computing
the Tai distance.

The word "mapping" and M mean a Tai mapping in this section.
The sets I and J, defined in definition 5, are again used
hereafter.

[Lemma 1]

Let R_α and R_β be the roots of T_α and T_β, respectively. Suppose M is the minimum cost mapping from T_α to T_β. Then $R_\alpha \in I$
and/or $R_\beta \in J$.

(Proof) Assume, for the sake of contradiction, that $R_\alpha \not\in I, R_\beta \not\in J$
and M is the minimum cost mapping. Let $An(I) = \{ \cap An_\alpha(i) | i \in I \}$.
Then $An(I) \neq \{\}$ because $R_\alpha \in An(I)$. Similarly, $An(J) \neq \{\}$. Let $M' =$
$M \cup \{(i', j')\}$ for $i' \in An(I)$ and $j' \in An(J)$. Since M is a Tai mapping, M' is also a Tai mapping. Assume that the cost of the
transformation represented by M be c+q+r, that is, i' is
deleted and j' is inserted. Then the cost of the transformation by M' is c+p, since j' is substituted for i'. Apparently,
c+p<c+q+r. This contradicts our assumption that M is the minimum cost mapping.

[Definition 7] Substitution between subtrees.

If $(i,j) \in M$, we say that subtree $T_\beta(j)$ is substituted for subtree $T_\alpha(i)$.

The mapping condition(3) is apparently equivalent to the following.

For any $(i,j),(i',j') \in M$ $(i \neq i', j \neq j')$,

i' is in $T_\alpha(i)$ iff j' is in $T_\beta(j)$.

Therefore, a substitution of $T_\beta(j)$ for $T_\alpha(i)$ means that a mapping from $T_\alpha(i)$ to $T_\beta(j)$ meets the mapping condition (3).

From lemma 1 and the above definition, one of the three cases (a),(b) and (c) gives the minimum cost mapping from $T_\alpha(x)$ to $T_\beta(y)$:

(a) $T_\beta(y)$ is substituted for $T_\alpha(x)$;

(b) One subtree of $T_\beta(y)$ is substituted for $T_\alpha(x)$;

(c) $T_\beta(y)$ is substituted for one subtree of $T_\alpha(x)$.

Let $\Delta a(x,y)$, $\Delta b(x,y)$ and $\Delta c(x,y)$ be the minimum costs in case (a),(b) and (c), respectively. Then the Tai distance $D(T_\alpha(x), T_\beta(y))$, which is stored in $D[x,y]$, is the minimum value of $\Delta a(x,y)$, $\Delta b(x,y)$ and $\Delta c(x,y)$.

The main algorithm for computing the Tai distance is as follows:

[Main algorithm]

```
    for  x:=Nα  downto  1  do
    for  y:=Nβ  downto  1  do
    begin
      if { x is a leaf }  then                            (3-1-1)
```

$$D[x,y]:=\begin{cases}(N_\beta(y)-1)*q & : \text{if } t_\alpha(x) \in Lab_\beta(y); \\ (N_\beta(y)-1)*q + p & : \text{if } t_\alpha(x) \notin Lab_\beta(y);\end{cases}$$

```
      if { y is a leaf }  then                            (3-1-2)
```

$$D[x,y]:=\begin{cases}(N_\alpha(x)-1)*r & : \text{if } t_\beta(y) \in Lab_\alpha(x); \\ (N_\alpha(x)-1)*r + p & : \text{if } t_\beta(y) \notin Lab_\alpha(x);\end{cases}$$

```
      if { Neither x nor y is a leaf }  then              (3-1-3)
        D[x,y]:= min { Δa(x,y), Δb(x,y), Δc(x,y) };
    end;
    D(Tα,Tβ):= D[1,1];

    where, Lab(k)={t(k')|k' in T(k)}.
```

If x is a leaf, an arbitrary node of $T_\beta(y)$ can be substituted for x. The remaining nodes of $T_\beta(y)$ are considered to be inserted. Hence, we get the formula (3-1-1). If y is a leaf, we have the formula(3-1-2).

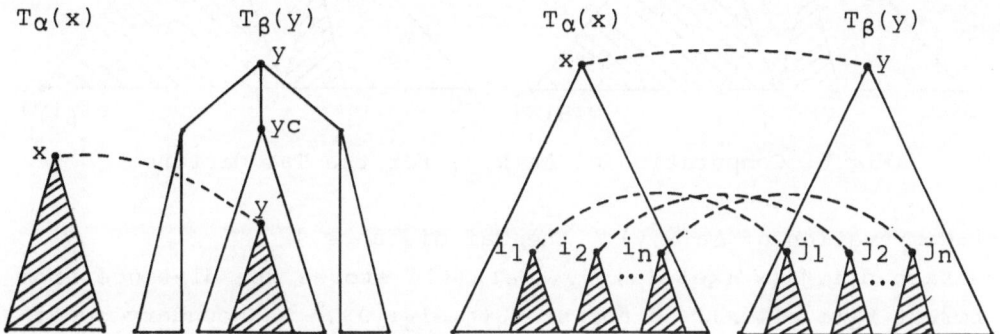

Fig.4   Substitution of $T_\beta(y')$
         for $T_\alpha(x)$.

Fig.5   Substitution of $T_\beta(y)$
         for $T_\alpha(x)$.

Let us consider the case(b) (See Fig.4). Let $(x,y')\epsilon M$, y' in $T_\beta(yc)$ and $yc\epsilon Ch_\beta(y)$. Then the remaining subgraph by removing $T_\beta(yc)$ from $T_\beta(y)$ are inserted. Furthermore, the minimum cost mapping from $T_\alpha(x)$ to $T_\beta(y')$ is identical with that from $T_\alpha(x)$ to $T_\beta(yc)$. Therefore, $\Delta b(x,y)$ can be computed by the following formula :

$$\Delta b(x,y)= \min_{y_j\epsilon Ch_\beta(y)} \{ D[x,y_j] + (N_\beta(y)-N_\beta(y_j))*q \}. \qquad (3-2)$$

Similarly,

$$\Delta c(x,y)= \min_{x_i\epsilon Ch_\alpha(x)} \{ D[x_i,y] + (N_\alpha(x)-N_\alpha(x_i))*r \}. \qquad (3-3)$$

Let us investigate the case (a). As in Fig.5, using the mapping conditions and definition 7, substitution of $T_\beta(y)$ for $T_\alpha(x)$ can be decomposed into that of $T_\beta(j_1)$ for $T_\alpha(i_1)$, that of $T_\beta(j_2)$ for $T_\alpha(i_2),\cdots$, that of $T_\beta(j_n)$ for $T_\alpha(i_n)$ such that both $T_\alpha(i_1),T_\alpha(i_2),\cdots,T_\alpha(i_n)$ and $T_\beta(j_1),T_\beta(j_2),\cdots,T_\beta(j_n)$ are forests , and any node outside these forests is inserted or deleted. Hence, to compute $\Delta a(x,y)$, we must check up the all possible pairs of forests. However, we can use the technique of dynamic programming.

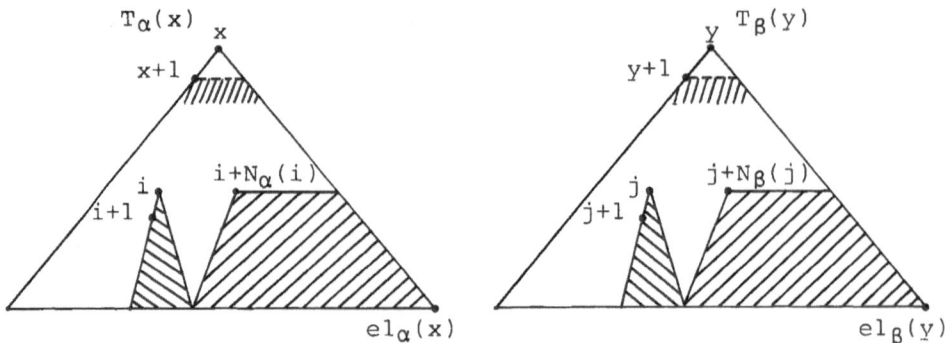

Fig.6   Computation of Δa(x,y) for the Tai distance.

[Computation of Δa(x,y) : the Tai distance]

Each d[i,j](x<i≤el$_α$(x), y<j≤el$_β$(y)) stores the distance from forest F$_α$(i,el$_α$(x)) to forest F$_β$(j,el$_β$(y)). As boundary conditions, d[el$_α$(x)+1,j] is the cost of insertions of nodes j,j+1, ·· ,el$_β$(y), and d[i,el$_β$(y)+1] is the cost of deletions of nodes i,i+1,··,el$_α$(x). Δa(x,y) can be computed by applying the following formula, iteratively:

$$δ1= d[i+N_α(i),j+N_β(j)] + D[i,j]; \qquad (3-4-1)$$
$$δ2= d[\ i\ ,j+1] + q; \qquad (3-4-2)$$
$$δ3= d[i+1,\ j\ ] + r; \qquad (3-4-3)$$
$$d[i,j]= min \{\ δ1,\ δ2,\ δ3\ \}, \qquad (3-4)$$

where, the boundary conditions are

$$d[el_α(x)+1,j]= \{el_β(y)+1-j\}*q \quad (\ y<j≤el_β(y)+1\ ),$$
$$d[i,el_β(y)+1]= \{el_α(x)+1-i\}*r \quad (\ x<i≤el_α(x)+1\ ).$$

Let $p(x,y)= \begin{cases} 0 : if\ t_α(x)=t_β(y); \\ p : if\ t_α(x)≠t_β(y). \end{cases}$

Then,

$$Δa(x,y)= d[x+1,y+1] + p(x,y).$$

The formula of Δa(x,y) is a straightforward extension of that of WLD.

4. Subclasses of the Tai mappimg

There are some problems that the Tai mapping can not be applied to. Consider a Tai mapping between classification trees

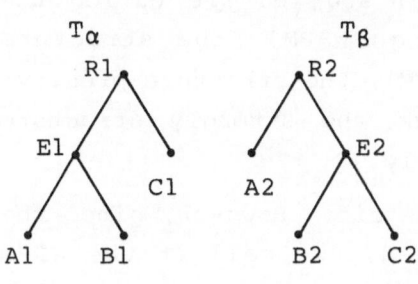

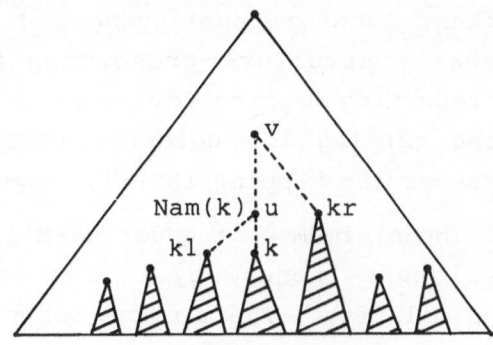

Fig.7 Mapping from T_α to T_β.　　　Fig.8　Nam(k).

T_α and T_β in Fig.7. In T_α, B1 is closer to A1 than to C1. On
the other hand, B2 is closer to C2 than to A2 in T_β. Therefore,
the Tai mapping is not appropriate to classification trees.

To introduce other mappings, we define a special ancestor,
called "the nearest ancestor determined by a mapping".
[Definition 8] The nearest ancestor determined by a mapping.

Let M be a Tai mapping. For some i∈I, let i′∈I be any
separate node of i. The nearest common ancestor of i and i′ is
called the nearest ancestor of i determined by mapping M, and
denoted by $Nam_\alpha(i)$. If i can not be determined, $Nam_\alpha(i)$ can
not be also determined. $Nam_\beta(j)$ is defined in the same way.

Let us illustrate Nam using Fig.8. Let kl and kr be left and
right neighbor separate nodes of k such that k,kl,kr∈I (or ∈J),
respectively. Let u and v be the nearest common ancestors of k
& kl and k & kr, respectively. Since u and v have a common descendant k, u and v are not separated. Then, Nam(k) is the
nearest node between u and v.

By putting restrictions on insertion and deletion of Nam, we
propose the following mappings.
[Definition 9] WSPM, ISPM, DSPM, SSPM.
Let M be a Tai mapping. For any (i,j)∈M, if $Nam_\alpha(i)$ and $Nam_\beta(j)$
can be determined,
   (a) $Nam_\alpha(i)∈I$　 or 　$Nam_\beta(j)∈J$;
   (b) $Nam_\alpha(i)∈I$;
   (c) $Nam_\beta(j)∈J$;
   (d) $Nam_\alpha(i)∈I$ 　and 　$Nam_\beta(j)∈J$.

Then, mappings satisfying (a),(b),(c) and (d) are called the weakly structure preserving mapping (WSPM), the structure preserving mapping for insertion (ISPM), the structure preserving mapping for deletion (DSPM), and the strongly structure preserving mapping (SSPM), respectively.

The minimum cost under WSPM is symmetric. However, since the triangle inequality is not satisfied, we call it the WSPM "similarity". The minimum costs under other mappings are called the "distance". Neither the ISPM distance nor the DSPM distance is symmetric. If q=r, the SSPM distance satisfies the metric axioms.

Since these four mappings belong to the Tai mapping, $\Delta b(x,y)$ and $\Delta c(x,y)$ for the Tai distance are available to compute $D[x,y]$. We will explain $\Delta a(x,y)$.

As illustrated in the previous section, substitution of $T_\beta(y)$ for $T_\alpha(x)$ is decomposed into that of forest $T_\beta(j_1), T_\beta(j_2), \cdots$, $T_\beta(j_n)$ for forest $T_\alpha(i_1), T_\alpha(i_2), \cdots, T_\alpha(i_n)$. In the case of WSPM, by definition 9(a), the following holds:

$\quad Nam_\alpha(i_1) = x \quad$ or $\quad Nam_\beta(j_1) = y, \cdots,$

$\quad Nam_\alpha(i_k) = x \quad$ or $\quad Nam_\beta(j_k) = y, \cdots,$

$\quad Nam_\alpha(i_n) = x \quad$ or $\quad Nam_\beta(j_n) = y.$

Without loss of generality, assume $Nam_\alpha(i_k)=x$. Let xc be a child of x such that i_k in $T_\alpha(xc)$. Then, $i_h(h \neq k)$ is not in $T_\alpha(xc)$. Because if i_h is in $T_\alpha(xc)$, xc is a common ancestor of i_k and i_h, and x is not $Nam_\alpha(i_k)$. In order to ensure that only i_k is in $T_\alpha(xc)$, we must use $D[xc,y_k]$, not $D[i_k,j_k]$ in computing $\Delta a(x,y)$. Therefore, we can get the formula of $\Delta a(x,y)$ by only replacing (3-4-1) with the following formula.

[$\Delta a(x,y)$ for the WSPM similarity]

$$\delta 1 = \begin{cases} d[i+N_\alpha(i), j+N_\beta(j)] + D[i,j] & : \text{if } i \epsilon Ch_\alpha(x) \text{ or } j \epsilon Ch_\beta(y); \\ \text{infinite} & : \text{otherwise.} \end{cases} \quad (4\text{-}1\text{-}1)$$

Similarly, we can compute other $a(x,y)$ by replacing (3-4-1) with the following formulae.

[$\Delta a(x,y)$ for the ISPM distance]

$$\delta 1 = \begin{cases} d[i+N_\alpha(i), j+N_\beta(j)] + D[i,j] & : \text{if } i \epsilon Ch_\alpha(x); \\ \text{infinite} & : \text{otherwise.} \end{cases} \quad (4\text{-}2\text{-}1)$$

[$\Delta a(x,y)$ for the DSPM distance]

$$\delta 1= \begin{cases} d[i+N_\alpha(i),j+N_\beta(j)] + D[i,j] & : \text{if } j\epsilon Ch_\beta(y); \\ \text{infinite} & : \text{otherwise.} \end{cases} \quad (4-3-1)$$

[$\Delta a(x,y)$ for the SSPM distance]

$$\delta 1= \begin{cases} d[i+N_\alpha(i),j+N_\beta(j)] + D[i,j] & : \text{if } i\epsilon Ch_\alpha(x) \text{ and } j\epsilon Ch_\beta(y); \\ \text{infinite} & : \text{otherwise.} \end{cases} \quad (4-4-1)$$

However, except the WSPM similarity, we can improve the algorithms. We will explain how to improve the algorithm for computing the ISPM distance.

[Computation of $\Delta a(x,y)$: the ISPM distance]

The children of x are named $x_1, x_2, \cdots, x_m$ from left to right.

$$d[i,j]= \min \begin{cases} d[i+1,j+N(j)] + D[x_i,j], \\ d[i,j+1] + q, \\ d[i+1,j] + N_\alpha(x_i)*r. \end{cases} \quad (4-5)$$

The boundary conditions are

$$d[m+1,j] = \{el_\beta(y)+1-j\}*q \qquad (y<j\leq el_\beta(y)+1),$$
$$d[i,el_\beta(y)+1]= \{N_\alpha(x_i)+ \cdots +N_\alpha(x_m)\}*r \qquad (1\leq i\leq m).$$

Then,

$$\Delta a(x,y)= d[1,y+1] + p(x,y).$$

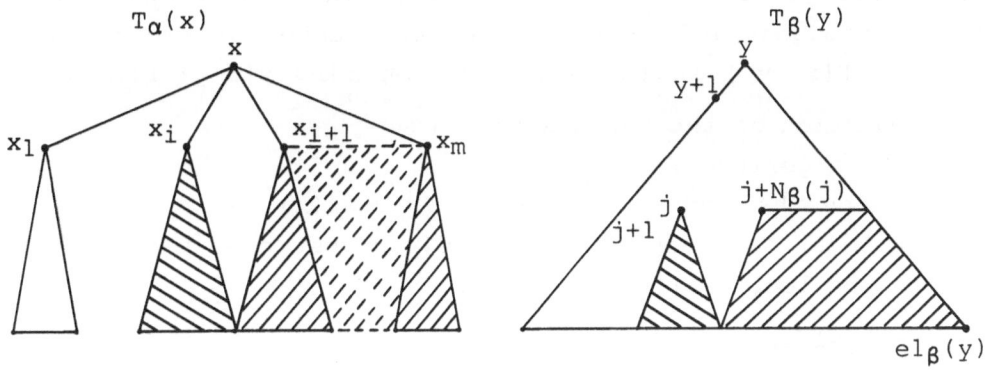

Fig.9 Computation of $\Delta a(x,y)$ for the ISPM distance.

5. Improved algorithms

We can further improve the algorithms for computing the Tai, ISPM and DSPM distances. We take the Tai distance as an example.

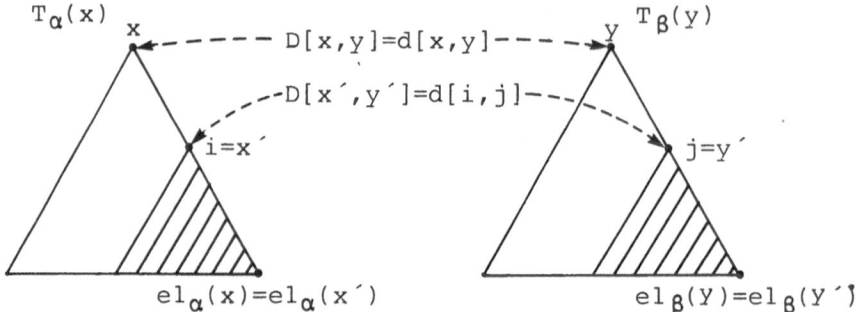

Fig.10   Computation of D[x,y] and D[x´,y´].

Note that $\Delta b(x,y)$ and $\Delta c(x,y)$ can be computed by the formula (3-4). That is, although this formula is applied to d[i,j] such that $x<i\leq el_\alpha(x)$ and $y<j\leq el_\beta(y)$, we can compute d[x,j] and d[i,y] using it. Then, D[x,y] is as follows :

$$D[x,y]= d[x,y]= \min \begin{cases} d[x+1,y+1] + p(x,y) \\ d[\ x\ ,y+1] + q \\ d[x+1,\ y\ ] + r \end{cases} \qquad (5-1)$$

Consider the case shown in Fig.10. The end leaf of $T_\alpha(x)$ ($T_\beta(y)$) is also that of $T_\alpha(x´)$($T_\beta(y´)$). In the algorithm shown in section 3, we compute d[i,j] for getting D[x,y] separately from d[i,j] for D[x´,y´]. However, since the computation of d[i,j] proceeds leftward from the end leaves of two trees, we can get D[x´,y´] in the midst of computing d[i,j] for D[x,y].

[Computation of the Tai distance]

(1) Main algorithm :

Let L be the number of leaf in T, leaf(k) be k-th leaf from the leftmost leaf, and R(u) be the youngest node v such that u is the end leaf of T(v).

```
for h:=1 to Lα  do
for k:=1 to Lβ  do
begin
  x:=Rα(leafα(h));
  y:=Rβ(leafβ(k));                              (5-2)
  "compute d[x+Nα(x),y+Nβ(y)]  d[x,y]";
  (*  where, z+N(z)=el(z)+1 (z=x or z=y), *)
end;
D(Tα,Tβ):=D[1,1].
```

(2) Computation of d[i,j] :

If $el_\alpha(x)=el_\alpha(i)$ and $el_\beta(y)=el_\beta(j)$, we get D[i,j] by formula (5-1). Otherwise, we compute d[i,j] by formula (3-4).

$$\delta1=\begin{cases} d[i+1,j+1] + p(i,j) & : \text{ if } el_\alpha(x)=el_\alpha(i) \text{ and } \\ & \quad el_\beta(y)=el_\beta(j); \\ d[i+N_\alpha(i),j+N_\beta(j)] + D[i,j] & : \text{ otherwise;} \end{cases}$$

$\delta2 = d[\ i\ ,j+1] + q;$

$\delta3 = d[i+1,\ j\ ] + r;$

$d[i,j]= \min \{\ \delta1,\ \delta2,\ \delta3\ \};$ (5-3)

If $el_\alpha(x)=el_\alpha(i)$ and $el_\beta(y)=el_\beta(j)$
  then $D[i,j]= d[i,j];$

where, the boundary conditions are

$d[el_\alpha(x)+1,j]= \{el_\beta(y)+1-j\}*q$ ($y\leq j\leq el_\beta(y)+1$),
$d[i,el_\beta(y)+1]= \{el_\alpha(x)+1-i\}*r$ ($x\leq i\leq el_\alpha(x)+1$).

The time and space complexities of the above algorithm are $O(L_\alpha L_\beta N_\alpha N_\beta)$ and $O(N_\alpha N_\beta)$, respectively. The expected computational complexities of the algorithms for computing other distances and similarity are shown in Table 1.

Table 1. Time and space complexities.

	Time complexity	Memory requirement	
		D	d
Tai	$O(L_\alpha L_\beta N_\alpha N_\beta)$	$N_\alpha N_\beta$	$N_\alpha N_\beta$
WSPM	$O(N_\alpha^2 N_\beta^2)$	$N_\alpha N_\beta$	$N_\alpha N_\beta$
ISPM	$O(C_\alpha L_\beta N_\alpha N_\beta)$	$N_\alpha N_\beta$	$C_\alpha N_\beta$
DSPM	$O(C_\beta L_\alpha N_\alpha N_\beta)$	$N_\alpha N_\beta$	$C_\beta N_\alpha$
SSPM	$O(C_\alpha C_\beta N_\alpha N_\beta)$	$N_\alpha N_\beta$	$C_\alpha C_\beta$

where, C_α denotes the maximum number of children in T_α.

6. Relation of metrics already proposed.

We clarify the relation between metrics in this paper and metrics in the previous work. The WSPM similarity, as a special case, becomes the metric between binary trees by Nakabayashi and Kamata[12]. Although the definition of ISPM is different from that of SPM by Tanaka[10], they are equivalent in meaning. In this paper SPM is called ISPM to express the meaning of SPM clearly. If the inverse mapping of a given mapping is SPM, this is called DSPM. SSPM in this paper is the same as SSPM by Tanaka[11]. The formula given by Lu[7] is the same as the formula of the SSPM metric. However, her definition of a tree metric is equivalent to the Tai metric[11]. Selkow's metric is the restricted SSPM metric. We can compute Selkow's metric by letting $D[x,y] := \Delta a(x,y)$ in the formula (3-1-3), where $\Delta a(x,y)$ is that of the SSPM metric. Therefore, it can be defined by the following mapping.

[Definition 10] The mapping conditions of Selkow's metric.

Let M be a SSPM. For any pair $(i,j) \in M$, $d_\alpha(i) = d_\beta(j)$, where $d(k)$ denotes the depth of node k, that is, the number of branches from root to k.

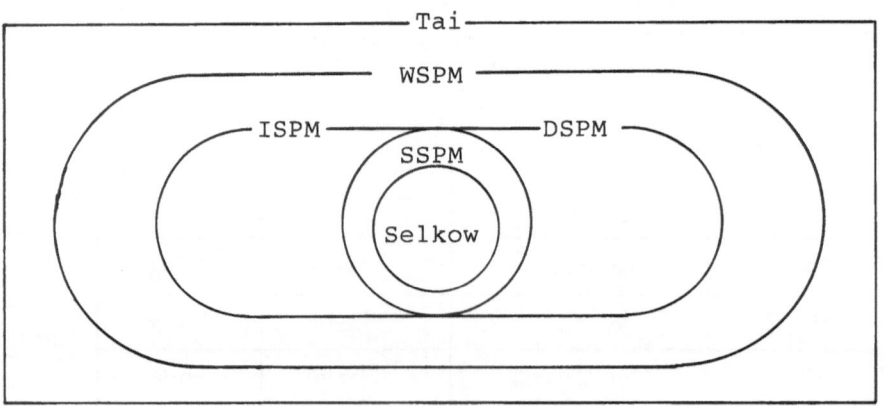

Fig.11 Hierarchy of mappings.

7. Concluding Remarks

We have introduced "the nearest ancestor determined by a mapping" and defined the four mappings, WSPM, ISPM, DSPM and SSPM. Using these mappings, we have succeeded in classifying the tree metrics already proposed, and proposed the new similarity, WSPM similarity. Furthermore, judging from the definitions of the four mappings, it seems that all possible tree metrics and similarities that satisfy the Tai mapping conditions have already proposed.

Among the Tai mapping conditions, the ancestor-descendant relation is necessary to rooted trees. If the order relation is disregarded, the problem is generally NP-complete. If one-to-one correspondence is excluded, node splitting and merging operations must be introduced instead of substitution. With respect to skeletal trees, this approach was done by Lu[8].

In addition, we have proposed the simple algorithm for computing the Tai metric. This is not at all inferior to other algorithms in time and space complexities. However, it is not known whether the time complexity for the WSPM similarity can be reduced.

References

[1] J.B.Kruskal, "An overview of sequence comparison: Time warps, string edits, and macromolecules," SIAM Review vol.25, no.2, pp.201-237,(1983)

[2] T.Okuda, E.Tanaka and T.Kasai, "A method for the correction of garbled words based on the Levenshtein metric," IEEE Trans. Com., vol.C-25, no.2, pp.172-178,(1976)

[3] S.A.Boorman and D.Oliver, "Metrics on spaces of finite trees," J.Math.Psychology, vol.10, pp.26-59,(1973)

[4] S.M.Selkow, "The tree-to-tree editing problem," Inf.Proc. Letters, vol.6, no.6, pp.184-186,(1977)

[5] S.Yamasaki and T.Nishida, "On the metric between tree structures," (in Japanese) Annual meeting of IECE Japan, pp.1107,(1977)

[6] K.C.Tai, "The tree-to-tree correction problem," JACM, vol.26, no.3, pp.422-433,(1979)

[7] S.Y.Lu, "A tree-to-tree distance and its application to cluster analysis," IEEE Trans.PAMI, vol.PAMI-1, no.2, pp.219-224,(1979)

[8] S.Y.Lu, "A tree-matching algorithm based on node splitting and Merging," IEEE Trans.PAMI, vol.PAMI-6, no.2, pp.249-256,(1984)

[9] Y.C.Cheng and S.Y.Lu, "Waveform correlation by tree matching," IEEE Trans.PAMI, vol.PAMI-7, no.3, pp.299-305,(1985)

[10] E.Tanaka and K.Tanaka, "A metric on trees and its computing method,"(in Japanese) Trans.IECE, vol.J65-D, no.5, pp.511-518,(1982)

[11] E.Tanaka, "The metric between trees based on the strongly structure preserving mapping and its computing method,"(in Japanese) Trans.IECE, vol.J67-D, no.6, pp.722-723,(1984)

[12] K.Nakabayashi and K.Kamata, "A distance between two binary trees and its computing algorithm,"(in Japanese) Trans. IECE, vol.J66-D, no.4, pp.445-451,(1983)

[13] E.Tanaka, "A computing algorithm for the tree metric based on the structure preserving mapping," Trans.IECE, vol.E68, no.5, pp.317-324,(1985)

[14] E.Tanaka, "Efficient computing algorithm for the Tai metric," Proc. International Computer Symposium 1986, Tainan, Taiwan.

III. APPLICATIONS I

PROBLEMS IN RECOGNITION OF DRAWINGS

T. Pavlidis

Dept. of Electrical Engineering
SUNY
Stony Brook
NY 11794
U.S.A.

ABSTRACT

The paper examines the problems that need be solved in order to convert automatically technical drawings into computer readable form. There are three major applications for such a process. Archives of publications containing both text and line illustrations; human-machine interface; and integration of electronic CAD systems with the paper archives. The automation of this process involves the following steps: (1) Digitization (2) Conversions into binary form (3) Discrimination between solid, halftone, and line area and extraction of line information. (4) Discrimination between symbols (text) and graphics. (5) Recognition of text. (6) Normalization of graphics. (7) Recognition of graphics.

The major challenges are the use of gray scale images to extract line information (relevant to steps 2 and 3) and the recognition of schematics and graphics (relevent to steps 6 and 7).

Introduction

Today there are numerous machines in the market that are able to read printed text. While some of the manufacturers claims may be exaggerated, it is likely that within the next couple of years the performance of such machines would improve sufficiently so that most printed text could be entered reliably and economically into computers. Such machines are unable to handle illustrations. At best they can identify them and skip the part of the page where they are printed. For many applications (e.g. patent data bases) it is desirable to include diagrams together with the text. Therefore there is a need for the *recognition of drawings,* in order to expand the class of documents that can be stored in computer readable form and made available to later searches as part of data bases. (For example, a user might want to list all documents that contain diagrams with at least a certain number of transistor symbols.)

There are two other major applications for the recognition of drawings. One is *human-machine interface.* It is often easier to draw something on paper rather than electronically while electronic media are more convenient for subsequent editing. The other is *integration of electronic CAD systems with the paper archives.* One factor that has slowed down the introduction of Computer Aided Design in some industries is the existence of a large set of paper documents that must be integrated into any new system (for example diagrams of the telephone network). Currently such integration is done manually (through tablet digitizers) at considerable cost and low reliability. The challenge there is not only to con-

NATO ASI Series, Vol. F45
Syntactic and Structural Pattern Recognition
Edited by G. Ferraté et al.
© Springer-Verlag Berlin Heidelberg 1988

vert such documents into a raster or vector file (many products do that now) but also to describe their contents in a high level language compatible with the CAD system. The Human-Machine Interface problem appears at first unrelated to the CAD problem but this is not the case. Current CAD systems present large menu list for drawing specific symbols. An alternative is to let the user draw informally and have the machine infer the symbols.

While recognition of diagrams may be viewed as an extension of recognition of text, it is much more difficult because it must produce both names and descriptions. For example, "A circular sector of radius 1.2 inches and angle of 36°." In some cases we may require even higher level recognition such as the symbol for a telephone switch. This usually comes as the last step of a series of processes. When we deal with paper input we must pay attention to the following sequence of processes.

1. Digitization

2. Conversions into binary form

3. Discrimination between solid, halftone, and line area and extraction of line information.

4. Discrimination between symbols (text) and graphics.

5. Recognition of text.

6. Normalization of graphics.

7. Recognition of graphics.

When we deal with electronic input we need worry only about steps 6 and 7. (Lines are entered directly through a tablet or a mouse and text through a keyboard.)

This paper will survey the challenges that must be overcome in order to build a drawing recogntition machine.

Binarization (Step 2)

This a topic which is often considered unimportant and the literature discusses it only in the context of thresholding. Our experience with text recognition suggests that quantization noise introduced at this step creates serious difficulties in subsequent steps. It appears that considerable improvement in performance may be achieved by focusing on the binarization process and if possible integrating it with vectorization. The reason is that while the input may be a binary image, the digitized image is gray scale. There are two causes for that. One is the "cross-talk" amongst sensors of the digitizer, usually referred to as the *point spread function (psf)*. The signal just before the A/D converter is given by an expression of the form

$$g(t,s) = \int_R h(t-x,s-y)u(x,y)dtds \tag{1}$$

where $u(x,y)$ is the brightness at point (x,y) of the (analog) input image and $g(t,s)$ the output of the tth sensor when the sth line is scanned. $h()$ is a weighting function describing the contribution of the various points of the analog image. In an ideal digitizer $h()$ is a δ-function so that $g(t,s) = u(t,s)$. In practice, it has a bell-shaped form and the faster it drops, the better is the digitizer. The other cause for converting binary into gray scale images is nonuniform light reflection from the paper. In that case Equ. (1) is modified into

$$g(t,s) = \int_R h(t-x,s-y)L(x,y)u(x,y)dtds \qquad (2)$$

where $L(x,y)$ denotes the variation of illumination.

It can be shown that if $u(x,y)$ is a binary function and $L(x,y)$ is linear, then $g(x,y)$ is convex in points corresponding to light areas and concave to points corresponding to dark areas. A method for recovering the binary image based on this result is described in [PW86].

An alternate approach is to omit binarization altogether, and in particular use deliberate defocusing in order to apply multiresolution techniques. A major problem in the analysis of engineering drawings is the identification of groups of elements as a single entity. A good example is offered by shading (Figure 1).

Figure 1

Sampling that image at a high resolution will yield a large number of lines and require a fair amount of computation to recover the structure present. It is possible to apply a low pass filter and sample at a coarser resolution so that shaded areas appear within a range of gray scale values. The structure of the coarser image can be overlayed on the image of higher resolution and guide the analysis there. In particular one can eliminate empty areas from the coarser resolution if the gray value there is nearly constant and close to the maximum brightness. This method will fail if there are significant variations in illumination (i.e. if $L(x,y)$ in equ. (2) is not a constant) but if it is shown to be useful under constant illumination, then it will be worthwhile to make sure that the scanning device is designed so that $L(x,y)$ is indeed constant.

Vectorization (Steps 3 and 4)

Vectorization is the term used in the industry to describe the conversation of a binary image into a set of lines. This is typically preceded by binarization which is often performed by some form of adaptive thresholding by the digitizer. There is a considerable literature on various forms of thinning, including papers by this author [Pa82, Pa86].

If we consider the gray scale image as a surface, then we may search for features corresponding to the centers of dark lines. It is easily shown that under the transformation of Equ. (1) such centers correspond to minima of $g(x,y)$. If

$L(x,y)$ varies slowly compared to the line thickness, then the result will be approximately valid for the transformation of Equ. (2) as well. Therefore line finding is reduced to valley tracking. While this problem is relatively simple for isolated lines, many interesting cases of surface geometry occur at line intersections. Such aspects of surface geometry have been investigated in the context of robot vision [MN84, BPYA85, BJ86] and investigations along this line may offer an interesting unified methodology applicable both to document analysis and 3D-vision. In particular some form of a *surface primal sketch* may correspond to the thin lines of the drawing. The classification of critical surface points by Nackman [Na84] provides a basis for following the surface form. The major difficulty in the application of such approaches to document analysis is the presence of large flat areas in the surfaces when large solid areas appear in the input document. Such areas could be best described by their contour and plan to include provisions for switch from the surface tracing to flat region analysis. Figure 2 illustrates this problem. (The figure has been constructed on the basis of actual data from scanned text.)

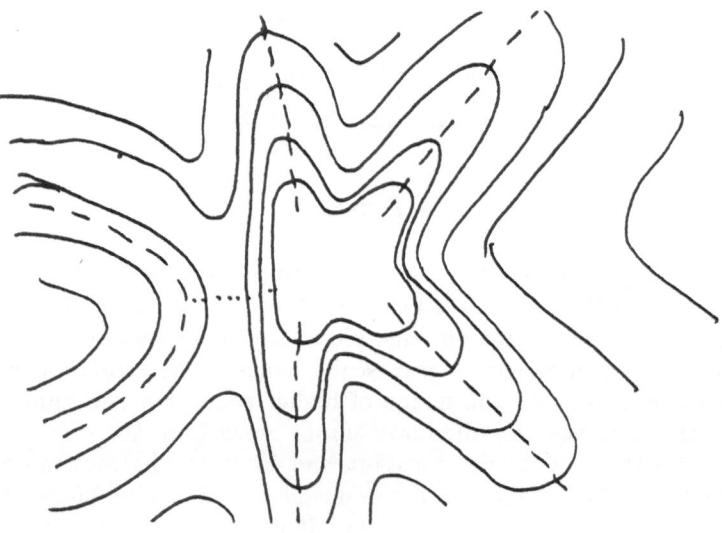

Figure 2

There the solid lines denote areas of equal brightness and the dashed and dotted lines valleys. The central area of the letter K appears flat and valley tracing fails there. It is interesting to note that such areas present problems for thinning algorithms of thresholded images. However the valley tracing method identifies such areas as those where thinning may be inappropriate instead of providing a more or less random skeletonization. Not all valleys correspond to thin lines. The one shown with dots in Figure 2 is caused by touching (or nearly touching) blobs. We expect to reject such spurious lines from the topology of the "valley graph" and from the sharpness of the minimum.

Gray scale processing whether by the method of [WP86] or the new techniques is too slow for practical applications if implemented using high level languages on general purpose computers. However many of the processes are highly parallel and if the methods are indeed able to handle documents that are rejected by the current machines, then implementations using special purpose hardarware might be worthwhile.

Recognition of Text (Step 5)

Recognition of text has been the subject of intensive research for over thirty years and the fruits of that effort have been appearing in the market place over the last ten years. These include the postal code reading machines, both for handprinted (NEC, Toshiba, etc) and typed or machine printed addresses (AEG Telefunken, etc), numerous limited font reading machines (Cognex, Totek, Toshiba, Dest, etc), multifont trainable machines (Kurtzweil), and most recently omnifont machines requiring no training (Palantir). While the actual performance of a machine may not always meet all the claims of the manufacturer (a situation not limited to the OCR field only) OCR machines do offer a testimonial to the success of an application of pattern recognition. Most of the remaining problems seem to be of engineering rather than basic research nature. If the character set is either stylized handprinted (the usual case in engineering drawings) or it contains only similar fonts (for example, all of them with serifs), then recognition rates well above 99 per cent are definitely possible with the current technology [BKP86].

Probably the most important challenge in this area is the separation of text from graphics. (See [SZ86] for a survey of such techniques.) While some of the literature has focused on the separation of blocks of text from graphics the recognition of drawings entails the recognition of isolated letters or numbers that occur in the midst of graphics. Various heuristics based on size and overall shape have been proven useful, but the author's experience during the work reported in [BKP86] suggests that a classification should be attempted more often than not and only when the figure of merit (usually a logarithm of a posterior probability) is too low, then the shape will be assumed to be graphics. This is necessary because drawings often contain large lettering whose size is comparable to some graphics as illustrated in Figure 3 (left).

Figure 3

Another problem is the overlapping of text and graphics as shown in Figure 3 (right). There separation must be performed only after the recognition of the graphics. An interesting example of this approach can be found in [AAMS86] where characters are identifies as the leftovers after all lines have been extracted.

Normalization of Graphics (Step 6)

The process of normalization can be described mathematically as follows. We expect that certain geometric relations are likely to be present in a drawing. For example, groups of parallel lines, segments of equal lengths, etc. Also we expect that vertical lines are far more likely than lines making an angle of, say, 80 degrees with the horizontal. Let $\mathbf{x}_i = (x_i, y_i)$ denote the set of points in a drawing. Then in a well formed drawing we expect to have a set of relations of the form

$$R^k(\mathbf{x}_i) = 0 \quad \mathbf{x}_i \text{ in } I_k$$

where I_k is a set of indices. For example we may express the parallelism of two line segments by

$$(x_{i_1} - x_{i_2})(y_{i_3} - y_{i_4}) - (x_{i_3} - x_{i_4})(y_{i_1} - y_{i_2}) = 0$$

In a real drawing we search for groups of points which satisfy an equation of the form

$$R^k(\mathbf{x}_i) = \epsilon \quad \mathbf{x}_i \text{ in } I_k$$

for some small number ϵ. Then we attempt to modify the drawing so that the new point locations $\mathbf{x}_i{}'$ satisfy

$$R^k(\mathbf{x}_i{}') = 0 \quad \mathbf{x}_i{}' \text{ in } I_k$$

This process has been called *beautification*, and it has been investigated in [PV85]. A closely related process has been called *fair copy reproducing* (FCR) [MAS86]. FCR relies more on specific rules than beautification does and it is less general but somewhat more powerful in the things it does. For example FCR assumes that all straight lines are vertical or horizontal and any extension to other directions would require the addition of new rules. In contrast beautification, includes parallelism as a desirable relation supplemented by relations specifying parallelism to particular directions. On the other hand FCR includes some rules for symbol recognition that beautification does not.

The precise form of the relations must be chosen so it allows efficient computation. Some relations can be expressed as the difference between two numbers, each computed on a single geometric figure. The numbers are sorted, a clustering algorithm is applied, and then elements belonging to the same cluster are modified so that they have the same number. For example, to check parallelism we compute the angle θ that each line segment makes with the horizontal

$$\theta_{ij} = \tan^{-1}\frac{y_i - y_j}{x_i - x_j}$$

Then the exact relationship is

$$\theta_{i_{12}} = \theta_{i_{34}}$$

and the approximate relationship

$$|\theta_{i_{12}} - \theta_{i_{34}}| \leq \epsilon$$

A more interesting example is the search for lines that are tangent to circles (or circular arcs). For each circle we compute a bounding box and the distance of its center from each line segment intersecting the bounding. (Such bounding boxes may be replaced by a presorted list). Let (u, v) be the center of the circle and the equation of the line be

$$x \cos\phi + y \sin\phi - d = 0$$

(ϕ is the angle with the vertical and the distance of the line from the origin). Then the distance of the center will be

$$|u \cos\phi + v \sin\phi - d| .$$

To make a line tangent to a set of N circles we must solve N equations of the form

$$u_i \cos\phi + v_i \sin\phi - d = r_i \quad i = 1, 2, \cdots N$$

If the circles are fixed, then we can solve this system only for up to $N = 2$ (They will be 0, 1, or 2 solutions for ϕ and d in that case). For larger N usually we must modify the circles as well.

In addition to desirable relations, "beautification" also includes negative rules. For example parallelism is not enforced on intersecting line segments, even if they form a small angle. FCR seems to ignore that question. This is probably due to the use of FCR only in a laboratory environment while beautification has been in public use within AT&T as part of the graphics editor *ped* running under UNIX (TM). It was user feedback that motivated the inclusion of such relations.

The current implementation uses the following constraint relations among points, or lines, or points and lines. (With each positive constraint we list the corresponding negative constraint.) For brevity we use the term *side* for line segment.

1. A set of sides should lie at the same angle to the horizontal, unless they intersect. We modify this relation further so that clusters at angles near a preferred value (currently a multiple of $\pi/4$) are adjusted to have exactly that value.

2. A set of sides with similar slopes should be collinear, unless their projections perpendicular to the common slope intersect.

3. A set of sides should have the same length.

4. A set of points should be horizontally [vertically] aligned, unless they are also vertically [horizontally] aligned.

Examples are shown in [PV85]. Straighforward extensions of the implementation include additional relations such as.

1. A pair of adjacent sides of a polygon should meet at the same angle as another pair of adjacent sides of the same polygon.

2. A set of points or sides is horizontally or vertically symmetric.

3. A point lies midway (or in general, some fixed fraction of the way) between two other points.

4. A line is tangent to a circular arc

    etc.

A greater challenge is to include relations that cannot be expressed by the proximity of two numbers. For example,

1. Three lines pass through the same point.

2. Two curves are tangent.

3. A circle is inscribed in a polygon.

    etc.

While it is often possible to use a modified Hough transform for such problems, this may be too expensive computationally.

Recognition of Graphics (Step 7)

Recognition of graphics includes the recognition of geometric forms (for example, check whether a set of points or short line segments form a circular arc), the recognition of simple geometrical relations (whether a line is tangent on a circle, two lines are parallel, etc.), and the recognition of structures. Parts of the recognition process are closely related to normalization, for example whether a set of short line segments approximate a circular arc. The main difference is that in normalization an attempt is made to modify the drawing to make such relations exact while no such attempt is made in the recognition process.

(a) *Geometric forms:* Two classes of methods have been used in the past, one based on the Hough transform [BB82], the other on curve fitting [Pa77]. A third alternative is offered by regularization, and in particular its discrete version [LP87]. The method is related to curve fitting but it does not require the *a priori* selection of a family of curves. The data are approximated by splines and then the recognition of geometric form proceeds from the spline specification. Formally the problem can be stated as following. *Given a spline approximation of some digitized curves recognize the particular shapes that produced them.* For example, it can be shown that if a set of points lie on a circular arc, then the control polygon of the approximating spline is going to be nearly regular. The major challenge is the nonuniqueness of spline approximations to particular shapes. Therefore one must look for invariant properties of the approximations.

(b) *Geometric Relations:* Many such relations can be found during the first half of the normalization procedure, either through Hough transforms or through the techniques described in the previous section. For example, concentric circles, rectangles, etc. The critical question is what to search for. Human observers can extract information from widely dispersed data. For example, concentric circles are identified even though none of the drawn parts of the curves are near each other. (See Figure 4.) The problem is even more difficult if the circles are drawn by dashed rather than solid lines.

(c) *Structures:* The results of the previous steps will be a list of elements and relations. One difficult problem is that the structures may not be known in advance. This is true for drawings where constructions are taken from a large list, of which only a small fraction is present in each drawing. Many of the previously published techniques rely on a combination of clustering, graph matching, and ruled based recognition of details. Applications include geographic maps [HKLP82], electrical circuits [Bu82, Bl84, GSS85, GM86, KU86], flowcharts [AAMS86], mechanical (pipeline) drawings [FKE84], etc. Some of the work described in the literature is part of the normalization process and relatively is said about the recognition of graphics. [HKLP82] focuses on the recognition of polygons while [Bu82] does recognition of diode and resistor symbols by a decision tree. [AAMS86] seperates boxes from flow lines on the basis of their gross shape. Knowledge-based rules are emphasized in [Bl84] and [KU86]. Geometric similarity was used in [FKE84]. Graph matching in [GSS85, GM86].

Clearly, recognition of graphics must use graph matching techniques using knowledge based rules. The critical question is to find algorithms to do that efficiently while keeping the number of necessary rules within reasonable bounds. We distinguish structures in the following categories.

(c.1) *Inherently Geometrical:* An example of a simple inherently geometrical structure is offered by shading. A set of closely spaced parallel lines can be recognized as such but it is uncertain if similarly simple procedures could be used for other textures. Schematics can also be recognized by an extension of structural pattern recognition techniques of the kind used for the recognition of alphanumeric symbols. Examples of this approach can be found already in the work of [Bu82, Bl84, FKE84, KU86], etc.

(c.2) *Topological:* Inclusion and adjacency relations provide the basis for topological structures. The lack of local structure makes their recognition particularly difficult. Different parts of the structure may be not be geometrically connected and may even be quite far apart from each other geometrically. Figure 4 shows one of the more challenging examples. There a set of concentric circles is overlayed on a map to denote locations of equal distance from a center.

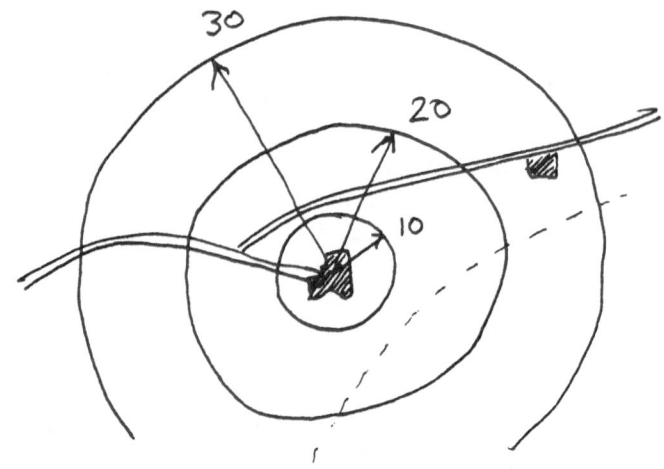

Figure 4

REFERENCES

[AAMS86] Abe, K. *et al* "Discrimination of Symbols, Lines, and Characters in Flowchart Recognition," *Proceeding of the Eight International Conference on Pattern Recognition*, Paris, 1986, pp. 1071-1074.

[BKP86] Baird,H. S., S. Kahan, and T. Pavlidis "Components of an Omnifont Page Reader," *Proc. 8th Intern. Conference on Pattern Recognition*, Paris, Oct. 28-31, pp. 344-348.

[BB82] Ballard, D. H. and C. M. Brown, *Computer Vision*, Prentice Hall, 1982.

[BJ86] Besl, P. J. and R. C. Jain, "Invariant Surface Characteristics for 3D Object Recognition in Range Images," *Computer Vision, Graphics and Image Processing*, **33** (1986), pp. 16-32.

[Bl84] Bley, H. "Segmentation of Electrical Schematics Using Picture Graphs," *Computer Vision, Graphics and Image Processing*, **28** (1984), pp. 271-288.

[BPYA85] Brady, M., J. Ponce, A. Yuille, and H. Asada, "Describing Surfaces," *Computer Vision, Graphics and Image Processing*, **32** (1985), pp. 1-28.

[Bu82] Bunke, H. "Automatic Interpretation of Lines and Text in Circuit Diagrams" *Pattern Recognition Theory and Applications*, Proc. of NATO Adv. Study, Inst. Oxford, March-April, 1981, J. Kittler, K. S. Fu, and L. F. Pau, eds. D. Reidel Publishing Co., 1982, pp. 297-310.

[Do84] Doster, W. "Different states of a document's content on the way from the Gutenbergian world to the electronic world," *Proceeding of the Seventh International Conference on Pattern Recognition*, Montreal, 1984, pp. 872-874.

[FKE84] Furuta, M., N. Kase, and S. Emori, "Segmentation and Recognition of Symbols for handwritten Piping & Instrument Diagrams," *Proceeding of the Seventh International Conference on Pattern Recognition*, Montreal, 1984, pp. 626-629.

[GSS85] Groen, F. C. A., A. C. Sanderson, and J. F. Schlag "Symbol recognition in electrical diagrams using probabilistic graph matching," **3** (1985) pp. 343-350.

[GM86]   Groen, F. C. A. and R. J. van Munster "Computer Aided Analysis of Schematic Diagrams," *Pattern Recognition in Practice II*, E. S. Gelsema and L. Kanal, editors, North Holland, 1986, pp. 363-372.

[HKLP82] Harris, J. F., J. Kittler, B. Llewwllyn, and G. Preston, "A Modular System for Interpreting Binary Pixel Representations of Line-Structured Data," *Pattern Recognition Theory and Applications*, Proc. of NATO Adv. Study, Inst. Oxford, March-April, 1981, J. Kittler, K. S. Fu, and L. F. Pau, eds. D. Reidel Publishing Co., 1982, pp. 311-351.

[HSK86] Hoshino, T., S. Suzuki, and M. Kosugi "Automatic Input Method for Large-scale Maps," *Proceeding of the Eight International Conference on Pattern Recognition*, Paris, 1986, pp. 449-453.

[JK84]   Jansen, H. and F-L. Krause "Interpretation of Freehand Drawings for Mechanical Design Processes," *Computers and Graphics*, **8** (1984), pp. 351-369.

[KU86]   Kuner, P. and B. Ueberreiter "Knowledge-Based Pattern Recognition in Disturbed Line Images using Graph Theory, Optimization, and Predicate Calculus," *Proceeding of the Eight International Conference on Pattern Recognition*, Paris, 1986, pp. 240-243.

[LC85]   Landy, M. S. and Y. Cohen "Vectorgraph Coding: Efficient Coding of Line Drawings," *Computer Vision, Graphics and Image Processing*, **30** (1985), pp. 331-344.

[LP87]   Lee, D. and T. Pavlidis "Smoothing Splines with Discontinuities for Image Analysis," submitted for publication.

[MN84]   Medioni, G. and R. Nevatia, "Description of 3-D Surfaces using Curvature Properties," *Proc. DARPA Image Understanding Workshop*, October 1984, New Orleans, LA, pp. 291-299.

[MAS86] Mino, M., Araki, M. and T. Sakai "Fair Copy Reproducing Algorithm from Roughly Sketched Diagrams," *Proceeding of the Eight International Conference on Pattern Recognition*, Paris, 1986, pp. 437-441.

[Na84]   Nackman, L. R. "Two-Dimensional Critical Point Configuration Graphs" *IEEE Trans. on Pattern Analysis and Machine Intelligence*, **PAMI-6** (1984), pp. 442-450.

[Pa77]   Pavlidis, T., *Structural Pattern Recognition*. Springer-Verlag, New York/Berlin, 1977.

[Pa82]   Pavlidis, T. "An Asynchronous Thinning Algorithm," *Computer Graphics and Image Processing*, **30** (1982), pp. 133-157.

[PV85]   Pavlidis, T. and C. J. Van Wyk "An Automatic Beautifier for Drawings and Illustrations," *Proceedings of SIGGRAPH'85*, San Francisco, July 22-26, 1985, pp. 225-234.

[Pa86]   Pavlidis, T. "A Vectorizer and Feature Extractor for Document Recognition" *Computer Vision, Graphics and Image Processing*, **35** (1986), pp. 111-127.

[PW86]   Pavlidis, T. and G. Wolberg "An Algorithm for the Segmentation of Bilevel Images," *Proc. IEEE Computer Vision and Pattern Recognition Conference*, Miami Beach, June 22-26, 1986, pp. 570-575.

[Pe81]   Peuquet, D. J. "An Examination of Techniques for Reformatting Digital Cartographic Data / Part 1: The Raster-to-Vector Process," *Cartographica*, **18** (1981), pp. 34-48.

[SZ86]   Srihari, S. N. and G. W. Zack, "Document Image Analysis," *Proceeding of the Eight International Conference on Pattern Recognition*, Paris, 1986, pp. 434-436.

APPLICATION OF STRUCTURAL PATTERN RECOGNITION IN HISTOPATHOLOGY

K. Kayser

Dept. of Pathology
Hospital for Thoracic Diseases
Amalienstr. 5
6900 Heidelberg
Germany

Abstract:

Structural pattern recognition in histopathology can be used for:

a) Assistance in difficult diagnoses
b) Measurements of interactions between different cell popula-
   tions
c) Estimation of proliferation activity of cancerous tissue in
   relation to survival of the patients.

The following system based upon interactive measurements was
developed:

Images of HE-stained and immuno-stained histopathological specimens were projected onto a graphic pad connected to a 4051
TEKTRONIX-computer. Coordinates of interesting structures
(glands, epithelial cancerous cells, positively immuno-stained
cells differentiating cancerous subpopulations) were marked
interactively. Centers of interesting structures were considered as vertices, neighboring structures as edges. A modified neighborhood condition based upon O'CALLAGHAN's definition was used. Measurements were performed at low and high
microscopic magnification. The following parameters were
measured:

Number of neighbors, cyclomatic number, n-simplices, n-stars,
distance between neighboring cells. Results were analyzed
by non-hierarchic discriminant analysis (test on cohesion and

NATO ASI Series, Vol. F45
Syntactic and Structural Pattern Recognition
Edited by G. Ferraté et al.
© Springer-Verlag Berlin Heidelberg 1988

centroids). The following results were obtained:

a) Measurements of healthy mucosa, tubulo-villous adenoma and
   highly to moderately differentiated adenocarcinoma of colon
   could be correctly separated and regrouped in 83% of the
   cases. 11/15 cases (73%) could be classified correctly in
   a prospective group. Similar percentage of correct separation and reclassification was obtained in a teaching set of
   20 cases and in a training set of 18 cases with metastatic
   adenocarcinoma of pleura of epithelial-biphasic mesothelioma.

b) Differences of tumour cell clones detectable by immunohistology were related to geometrical distance of positively
   stained tumour cells. Difference of nearest neighboring
   cells between negatively stained cells was undistinguishable
   for different antibodies opposite to significant differences in distance of positively stained nearest neighboring
   cells. The data indicate clonal origin of cells reacting
   positively to the applied antibodies.

c) Minimum distance of nearest neighboring cells was measured
   in order to determine proliferation activity of tumour
   cells. Survival of 60 patients with small cell anaplastic
   carcinoma of the lung showed close relation to distance
   of nearest neighboring cells ($p < 0.05$).

Introduction:

Histopathological images are difficult to describe and are
in general not standardized. Laboratory procedures as fixa-
tion and staining of tissue, and the morphological appearance
of a disease have a broad variability. Several groups have
applied morphometric measurements for supporting difficult
diagnoses (1, 2, 7, 18). BAAK et al. (1) found differences
in minimum diameter and area of glands of corpus endome-
trium related to certain disorders of the organ. YOUNG et al.
(18) found significant changes in rat urothelial cells in
development of urothelial cancer. KAYSER et al. (7) found

significant differences in minimum distance and minimum
diameter of neighboring glands in case of healthy, adenomatous
and carcinomatous colon mucosa. However, morphometric measurements do not relate to the underlying disease in every case.
KWEE et al. (10) were not able to demonstrate significant
differences of nuclear parameter between primary mesothelioma,
activated mesothelial cells and metastatic adenocarcinoma of
pleura.

In this paper a synopsis of measurements performed on different
organs using the approach of structural analysis of histopathological images is given. The measurements were performed
at different magnification. The results were related to different
clinical or scientific approaches. These include relation to
diagnosis (6, 9), geometrical description of biological properties of images (5) and analysis of clinical importance of
certain disesases, i.e. relation to prognosis.

Diagnostic algorithm:

The algorithms of diagnosis in histopathology is not a simple
recognition of a certain image or a simple description of
certain textures of an image context. It is more related to
an expert system including structure analysis of the image.
Additional information as sex and age of the patient, the
clinical history, biochemical findings and the performance
status are necessary prepositions in difficult diagnostic
cases. However, this additional information is not needed
for all diagnoses (4). The diagnosis "cancer" can be stated
by the histomorphological image in the majority of cases.
However, detailled clinical information is needed for correct
classification of the detected malignancy. The amount of
necessary additional information depends upon the size of the
analyzed tissue and upon the precision of the diagnosis
wanted by the clinician.

Basic assumptions:

Life is now understood as a dynamic process including death
and proliferation of cells in different organs. Regulation of
tissue growth in healthy human tissue can only be performed in
a steady state showing reproducible neighborhood conditions
or regularities. These regularities differ for various organs.
Disturbance of the steady state of living cells affect these
regularities either by disturbance of symmetry conditions
or by forming new geometrical structures. Benign or malignant
growth does not produce new basic structure elements (primitives) in every case but does basically form new geometrical
arrangement of the given structure elements of the affected
organ.

Analysis of tissue texture is the basic algorithm in histopathological diagnoses. The extraction of primitives of a
certain image in histopathology is very difficult due to its
broad variation in appearance. A useful diagnostic algorithm
cannot be based upon strictly defined primitives, and has to
be safe against variations of the primitives. Existing
regularities in the geometrical network of tissue structures
(histomorphological textures) can be used for construction
of new basic units at another magnification level (8). The
cells can be considered as primitive elements at high magnification level. In various organs a special kind of cells
tends to arrange themselves in rings forming oriented cylinders, in other organs honeycomb-textures of cells exist.
These regularities can be considered as primitive elements
at a low magnification level.

The geometrical network to be analyzed exists in the given
three-dimensional space. Analysis in histopathology is possible in any two-dimensional plane, i.e. the projection of the
geometrical structures onto this plane may not lead to actual
relations in the three-dimensional space but may only be related to different biological cohorts. Discrimination power
of the used descriptors at different magnifications is of

major interest for support in histopathological diagnosis.

Graph theory application:

Once the basic structure elements of an image are defined,
a geometrical network of these elements can be constructed by
application of graph theory (4, 8, 13, 14, 15). Histopathological images are embedded in a two-dimensional space and of
steady state. Depending upon the neighborhood condition the
obtained graph is an unoriented simple graph, and in the
majority of cases a connected graph. From the point of view
of biology it seems reasonable to analyze the network for its
regularities and for its connectivity. Simple parameters describing regularites and connectivity are the cyclomatic number,
the distribution of n-simplices and the number of neighboring
structures. Distance between nearest neighboring structure
elements should be included. A center of the geometrical network can be calculated defining the vertix at the crossing
point of paths from the right upper to the left lower corner
and from the left upper to the right lower corner due to the
fact that all vertices are unoriented. The obtained vertix
is not necessarily the geometrical center of the image. The
incidence function between neighboring structures was set to
be independent from additional parameters of the vertices,
i.e. to be constant. Morphometric measurements can be easily
included into this concept. In this case the measured properties of the vertices reflect to the incidence function, and
the concept of a cost function as described by SANFELIU (15)
is the appropriate approach for analyzing the image.

Neighborhood condition:

The obtained graph is strongly depending upon the introduced
neighborhood condition. VORONOI's neighborhood condition (16)
based upon the concept of the maximum area of non-overlapping

circles of arbitrarily distributed points in a two-dimensional
space is only appropriate in tissue with primitives being
directly connected and being of similar size. VORONOI's neighbor
hood condition (16) can be used in case of liver tissue, brain
or lymph nodes. It is not appropriate in case of glandular
tissues such as colon mucosa, lung, kidneys. Especially, neighboring higher order structures (glands, nerves, vessels, etc.)
have no common boundary. Their relation to each other depends
strongly upon their distance. O'CALLAGHAN's neighborhood condition (11) seems to be more appropriate describing the situation. It is based upon a "distant constraint" and a "direction
constraint" in order to exclude neighboring structures obviously hidden by other neighboring structures. The algorithm is
based upon Diriclet-cell and easy to implement on a computer.
It can be modified by introducing a lower limit of distance of
primitives to be located in or by satisfying additional conditions to be introduced into the incidence function. For example,
for defining the area supplied by oxigen of vessels the depth
of diffusion may be dependent upon the amount of fiberous
tissue between the vessels considered, i.e. dependent upon
direction or upon properties which may be measured by grey
value densities. In this case we are dealing with a neighborhood condition influenced by additional local conditions (17).

O'CALLAGHAN's neighborhood condition (11) can also be expanded
into an oriented neighborhood condition if orientations of the
structures considered are taken into account (for example,
direction of blood flow, tissue layers with non-isotropic
cell distribution, etc.).

The measurements were performed using O'CALLAGHAN's neighborhood condition (11) modified by an introduced lower limit in
order to exclude artifacts or overlapping of primitives. Lower
limit of distance constraint was set 1 μ and upper limit was
set 40 μ in case of low and high magnification.

Material and methods:

All measurements were performed interactively using a projection microscope and a graphic pad connected to a 4051 TEKTRONIX-computer. Measuring area was set 1 qmm in case of low magnification (glands or gland-like configuration) and 0.04 qmm in case of high magnification (cells). Programmes defining nearest, second nearest, etc. neighboring structures, cyclomatic number, distribution of n-simplices, distribution of n-stars and distribution of obtained disconnected graphs in case of nearest neighboring structures, second nearest neighboring structures, etc. were written in BASIC. Obtained results were analyzed by non-hierarchic discriminant analysis programmes (test on cohesion and centroids) obtained from SPSS (Statistical Package for Social Science). Four independent variables (cyclomatic number, number of n-simplices (n = 3, 4, 5); cyclomatic number, mean distance of nearest and second nearest neighboring structures; cyclomatic number at high and low magnification; mean of neighboring structures at high and low magnification) were entered into the analysis concurrently. The programmes created the discriminant functions from the entire set of variables. The discriminating power of each of the independent variables was disregarded. Each of the experiments consisted of a teaching set and a prospective set of additional cases.

Reproducibility of constructed graphs was measured $\geqslant 95\%$ for the following parameters: number of vertices, number of edges, cyclomatic number, n-simplices, average number of neighboring structures; mean distance between first, second, third, and fourth neighboring structures.

HE-stained routine specimens of each diagnostic group were chosen randomly from the files of the biopsy material of the Institute of Pathology, University Heidelberg. One characteristic area of morphological changes for each case was measured. Therefore, the measurements are not independent upon the prior knowledge of the observer.

In case of immunohistology specimens were stained with monoclonal antibodies using the indirect peroxidase method as pre-
described by the manufacturers. Fig. 1 and Fig. 2 examplarily
present obtained graphs in case of histopathological images
stained with antibodies against neuronspecific enolase (NSE,
Behring, Marburg).

In the experiment distinguishing mesothelioma from metastatic
adenocarcinoma only cases were taken into account with secon-
darily performed operation of the lung or finally performed
autopsy for validation of the diagnosis.

In case of small cell anaplastic carcinoma survival of the
patients was obtained from the files of the Hospital for Thoracic Diseases,Heidelberg-Rohrbach, or by writing to the house
physicians. Survival rates were estimated by the KAPLAN-MEIER
method (3) and were subjected to the log rank test (12).

Results:

Low magnification experiment:

Analysis of the geometrical network built by glands or gland-
like configuration of healthy colon mucosa, tubulo-villous
adenoma of the colon and high to moderately differentiated
adenocarcinoma of the colon was measured for 60 cases used
as a learning set and for additional 15 cases used as prospective set. Obtained cyclomatic number is given in Table 1.
It differs significantly within the three diagnostic groups.
The results of reclassification obtained by discriminant
analysis are given in Table 2. About 83% of the measured cases
were regrouped correctly in the teaching set and 11/15 cases
(73% in the prospective set (9).

High magnification experiment:

Geometrical arrangement of cells of different carcinoma of the
lung (epidermoid, adeno, small cell, large cell) including
healthy lung tissue was analyzed in 100 HE-stained specimens
used as a teaching set and in additional 25 cases used as
prospective set. The obtained results are given in Table 3.
Similarities of cyclomatic number and distance of neighboring
cells exist for epidermoid carcinoma and small cell carcinoma
as well as for adenocarcinoma and large cell carcinoma. Distinguishing these two groups against each other is difficult in
some cases in routinary histopathology. The result of reclassification is given in Table 4. Correct prediction of the teaching
set and the prospective set is similar and amounts to 90% -
95% (5).

High and low magnification experiment:

The method was tested for differentiating mesothelioma of the
pleura against metastatic adenocarcinoma in a teaching set of
20 cases and a prospective set of 18 cases. The period of the
prospective set covered more than 18 months. The probability
of classification for the prospective cases was included
into the clinical records. Results are given in Table 5. Error-
less classification of the prospective cases was performed in
18/18 cases although prediction probability was weak in some
cases (Table 6). Regrouping based upon measurements performed
at only one magnification level was unsuccessful (6).

Experiment on biological parameters:

Immunohistology with various monoclonal antibodies was performed on 60 cases of small cell anaplastic carcinoma of the
lung. Results are given in Table 7. Average distance of neigh-

boring cells and cyclomatic number is similar for negatively stained cells but revealed differences for positively stained cells with different monoclonal antibodies. Minimum distance between nearest neighboring cells is of prospective value. Patients suffering from tumours with a mean distance ≤ 8 μ between neighboring cells have a poorer survival than patients with a larger distance between nearest neighboring cells (Fig. 3; $p \leq 0.05$).

Discussion:

The measurements performed on different structure elements in histopathology revealed the following results:

a) The geometrical arrangement of the structure units contains information closely related to histopathological diagnoses. In case of low magnification the relation is weak to precise diagnostic groups and strong to generalize grouping of diseases as healthy - benign - malignant growth (9).

b) Structure analysis performed at high magnification contains information related to subgroups of different malignant tumours in case of lung carcinoma. Linear regression between cyclomatic number and distance of nearest neighboring cells in primary bronchus carcinoma revealed close similarities between the different diagnostic groups (5).

c) Structure analysis performed at two magnification levels was very useful for supporting the difficult diagnostic distinction between mesothelioma - metastatic adenocarcinoma in small biopsy specimens. A meaningful support was only found in the approach of combination of two different magnifications. This result reflects to the diagnostic procedure performed in routine histopathology analyzing specimens at various magnifications (6).

d) Analysis of simple descriptive parameters of the obtained
   graph seems to be sufficient for classification of histopathological images into certain diagnostic groups. Cyclomatic
   number and distribution of n-simplices were found to have
   the highest discrimination power (5, 6, 8, 9).

e) Structural analysis including measurements of distance of
   nearest neighboring cells seems to be related to biological
   behaviour of malignant growth in case of small cell anaplastic carcinoma of the lung. Patients with smaller distance
   between nearest neighboring cells had a significantly poorer
   survival than patients with larger distance. The findings
   may simply reflect to proliferation activity of tumours or
   to relations between textures of tumours and to the corresponding grade of malignancy.

Structural analysis of tissues affected by various diseases is
related to difficulties obtained in routine histopathology
indicating that similar algorithms may be performed by patholo-
gists. From the theoretical point of view biological behaviour
of diseases has to reflect to tissue structures. Various diseases are affected at various structure levels (Table 8).
Local disturbances of organ textures should be more easily
detected if area dependent neighborhood conditions are introduced. Whether they can be used as predictors for development
of malignant growth (precancerous lesions) has still to remain
open.

References:

1 Baak JPA, Oort J (eds) (1983) A manual of morphometry in diagnostic pathology. Springer Berlin Heidelberg New York Tokyo

2 Cornelisse CJ. (1983) Muscle: Morphometric analysis of biopsies. In: A manual of morphometry in diagnostic pathology. Baak, Oort (eds) Springer Berlin Heidelberg New York Tokyo, 142-148

3 Kaplan EL, Meier P (1958) Nonparametric estimation from incomplete observations. J Am Stat Assoc 53: 447-454

4 Kayser K, Höffgen H (1984) Pattern recognition in histopathology by orders of textures. Med Inform 9: 55-59

5 Kayser K, Kiefer B, Burkhardt HU, Shaver M (1985) Syntactic structure analysis of bronchus carcinoma - first results. Acta Stereol 4/2: 249-253

6 Kayser K, Kiefer B, Merkle NM, Vollhaber HH (1986) Strukturelle Bildanalyse als diagnostisches Hilfsmittel in der Histopathologie. TumDiagn Ther 7: 21-27

7 Kayser K, Modlinger F, Postl K (1985) Quantitative low-resolution analysis of colon mucosa. Anal Quant Cyt Hist 7: 205-213

8 Kayser K, Schlegel W (1982) Pattern recognition in histopathology: Basic considerations. Meth Inform Med 21: 15-22

9 Kayser K, Shaver M, Modlinger F, Postl K, Moyers JJ (1986) Neighborhood analysis of low magnification structures (glands) in healthy, adenomatous and carcinomatous colon mucosa. Path Res Pract 181: 153-158

10 Kwee WS, Veldhuizen RW, Golding RP, Mullink H, Stam J, Donner R, Boon ME (1982) Histologic distinction between malignant mesothelioma, benign pleural lesion and carcinoma metastasis. Virch Arch Pathol Anat 397: 287-299

11 O'Callaghan JF (1975) An alternative definition for neighborhood of a point. IEEE Trans Comput 24: 1121-1125

12 Peto R, Peto J (1972) Asymptomatically efficient rank invariant test procedures. J Roy Statist Soc A 135: 185-206

13 Prewitt JMS (1979) Graphs and grammars for histology. An introduction. Prov Third Ann Symp Comp Appl Med Care, Washington

14 Prewitt JMS, Barber A, Wu SC (1978) An application of pattern recognition to histology. In: Proc PRIP Chicago, 499-506

15 Sanfeliu A, Fu KS, Prewitt JMS (1981) An application of distance measure between graphs to the analysis of muscle tissue pattern recognition. Saragota Springs, New York, 86-89

[16]Voronoi G (1902) Nouvelles applications des paramètres continus a la théorie des formes quadratiques. Deuxième mémoire: Recherches sur les paralleloedres primitifs. J Reine angew Math 134: 198-287

[17]Voss K, Klette R (1986) Theoretische Grundlagen der digitalen Bildverarbeitung. IV. Orientierte Nachbarschaftsstrukturen. Bild und Ton 39: 213-219

[18]Young IT, Vanderlaan M, Kromhout L, Jensen R, Grover A, King E. Morphologic changes in rat urothelial cells during carcinogenesis: II. image cytometry. Cytometry 5: 454-462

Captions:

Fig. 1 : Graph obtained from small cell anaplastic carcinoma of of the lung. Vertices correspond to cells positively stained with NSE.

Fig. 2 : Graph obtained from small cell anaplastic carcinoma of the lung. Vertices correspond to cells negatively stained with NSE.

Fig. 3 : Survival rates of 60 patients suffering from small cell anaplastic carcinoma of the lung. Cohorts grouped according to minimum distance between nearest neighboring cells.

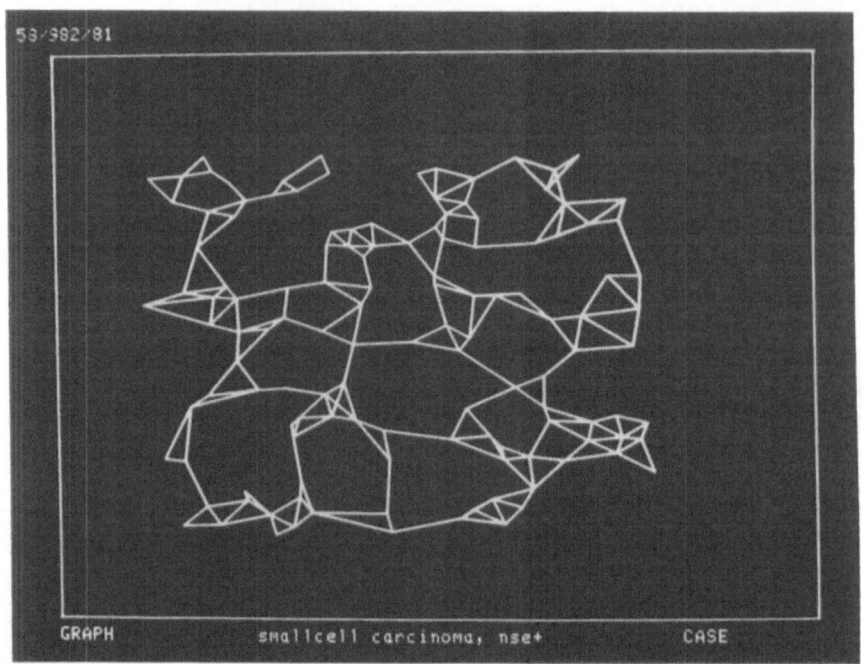

Fig. 1

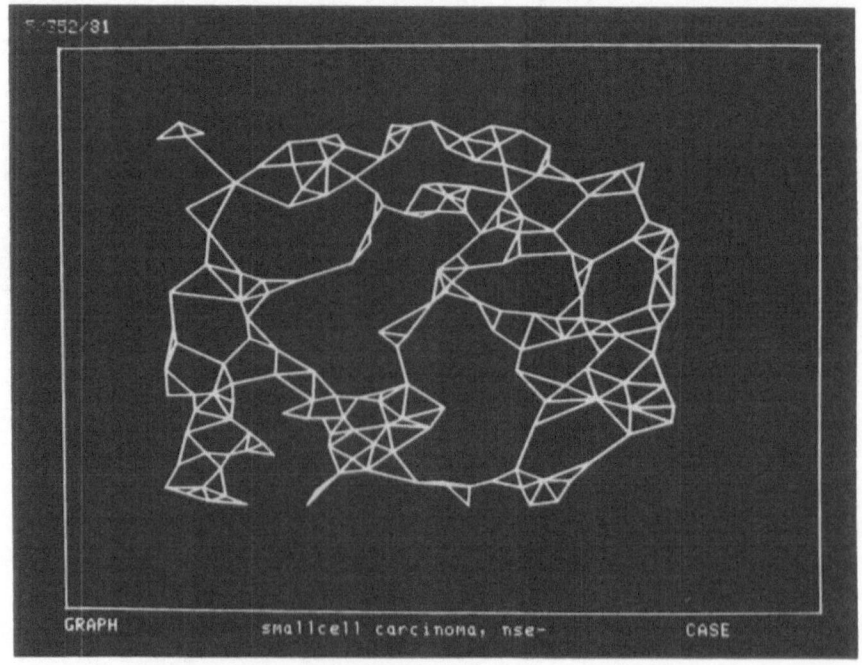

Fig. 2

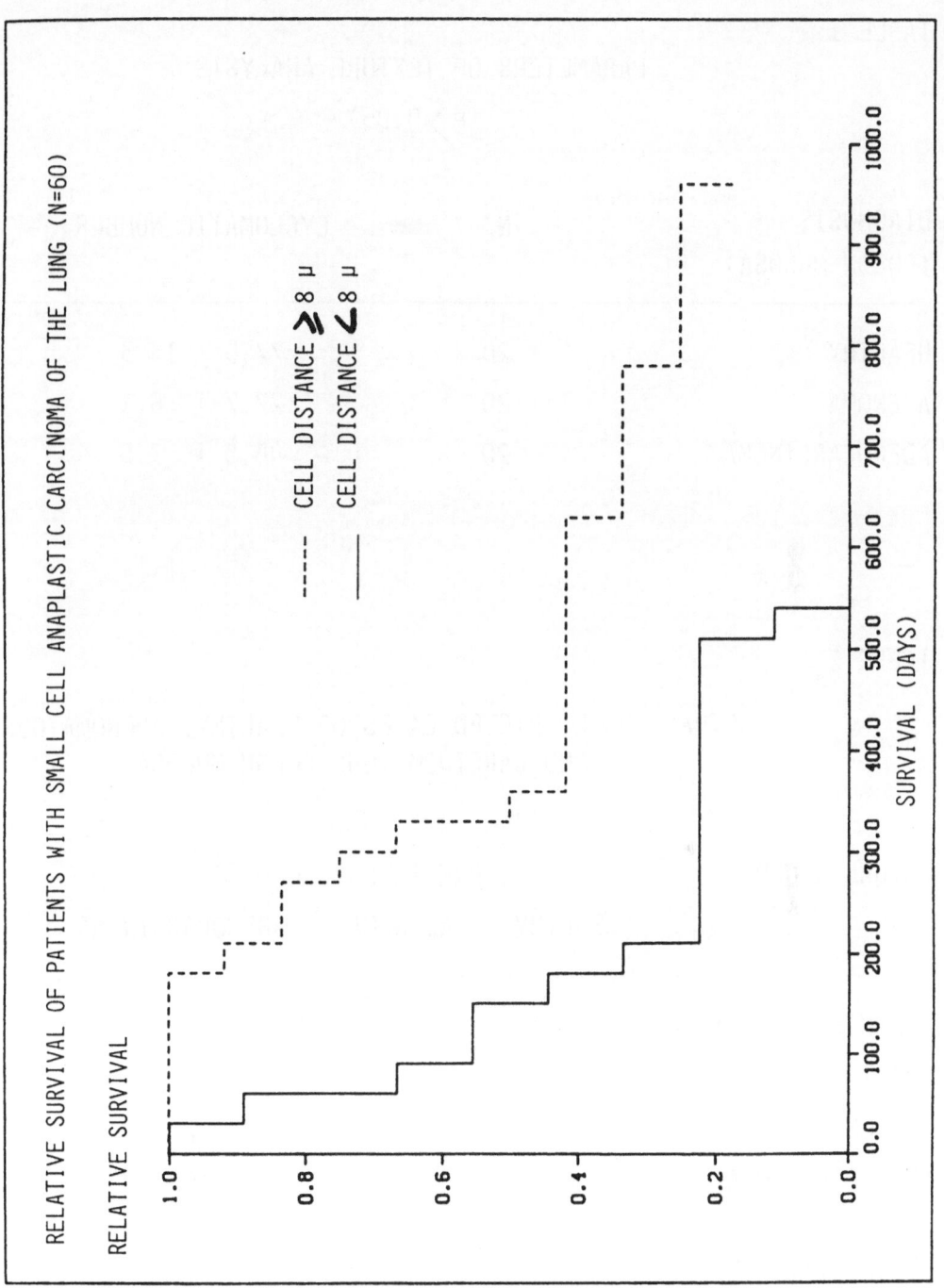

RELATIVE SURVIVAL OF PATIENTS WITH SMALL CELL ANAPLASTIC CARCINOMA OF THE LUNG (N=60)

RELATIVE SURVIVAL

CELL DISTANCE ≥ 8 μ
CELL DISTANCE < 8 μ

SURVIVAL (DAYS)

Fig. 3

TABLE 1

PARAMETERS OF TEXTURE ANALYSIS
$(P > 0.95)$

DIAGNOSIS (COLON MUCOSA)	N	CYCLOMATIC NUMBER
HEALTHY	20	77.5 ± 13.3
ADENOMA	20	27.7 ± 6.1
ADENOCARCINOMA	20	46.5 ± 7.5

TABLE 2

ACTUAL AND PREDICTED CASES OF HEALTHY, ADENOMATOUS AND CARCINOMATOUS COLON MUCOSA

ACTUAL GROUP	PREDICTED GROUP		
	HEALTHY	ADENOMA	ADENOCARCINOMA
LEARNING SET (N=60)			
HEALTHY	19	1	0
ADENOMA	1	17	2
ADENOCARCINOMA	0	1	19
PROSPECTIVE (N=15)			
HEALTHY	4	1	0
ADENOMA	0	4	1
ADENOCARCINOMA	0	2	3

TABLE 3

PARAMETERS OF TEXTURE ANALYSIS ACCORDING TO CELL TYPE

MEAN AND CONFIDENCE LIMITS, $p \gg 0.95$)

CELL TYPE	CYCLOMATIC NUMBER	DISTANCE OF NEAREST NEIGHBOR (μ)	DISTANCE OF 2ND NEAREST NEIGHBOR (μ)	DISTANCE OF 3RD NEAREST NEIGHBOR (μ)
HEALTHY	25±2.5	13.0±0.8	19.7±1.2	26.4±1.8
EPIDERMOID	265±25	8.4±0.4	10.7±0.6	12.7±0.7
ADENO	114±15	9.7±0.6	13.1±0.7	16.5±0.8
SMALL CELL	281±13	7.1±0.4	9.2±0.6	11.0±0.6
LARGE CELL	137±16	11.3±0.5	15.0±0.7	18.1±0.9

TABLE 4 SYNTACTIC STRUCTURE ANALYSIS

LEARNING SET (N=75)
PREDICTED CASES

ACTUAL CASES	HEALTHY	EPIDERMOID	ADENO	SMALL CELL	LARGE CELL
HEALTHY	15	0	0	0	0
EPIDERMOID	0	12	0	1	2
ADENO	0	0	13	0	2
SMALL CELL	0	2	0	13	0
LARGE CELL	0	0	0	0	15

PROSPECTIVE (N=25)
PREDICTED CASES

ACTUAL CASES	HEALTHY	EPIDERMOID	ADENO	SMALL CELL	LARGE CELL
HEALTHY	5	0	0	0	0
EPIDERMOID	0	4	0	1	0
ADENO	0	0	4	0	1
SMALL CELL	0	0	0	5	0
LARGE CELL	0	0	0	0	5

TABLE 5

SYNTACTIC STRUCTURE ANALYSIS FOR DISTINCTION OF
EPITHELIAL/BIPHASICAL MESOTHELIOMA - PLEURITIS CARCINOMATOSA

DIAGNOSIS DUE TO DISCRIMINANT ANALYSIS

DEFINITIVE DIAGNOSIS RETROSPECTIVE (LEARNING SET; N=20)	MESOTHELIOMA	PLEURITIS CARCINOMATOSA
MESOTHELIOMA	9	1
PLEURITIS CARCINOMATOSA	0	10
PROSPECTIVE (N=18)		
MESOTHELIOMA	7	0
PLEURITIS CARCINOMATOSA	0	11

TABLE 6

PREDICTION PROBILITY FOR METASTATIC ADENO-
CARCINOMA AND MESOTHELIOMA OF PLEURA

	NUMBER OF CASES	
PREDICTION	ADENOCARCINOMA	MESOTHELIOMA
> 0.95	11	10
> 0.90	4	2
> 0.80	2	2
> 0.70	2 + 1*	2
> 0.60	1	0
> 0.55	1	0
TOTAL	22 (21+)	16 (17+)

* MISCLASSIFIED CASE (ACTUAL MESOTHELIOMA)

+ NUMBER OF ACTUAL CASES

TABLE 7

DISTANCE BETWEEN NEIGHBORING TUMOUR CELLS AND NUMBER OF SUBGRAPHS IN SMALL CELLS CARCINOMA OF THE LUNG

ANTIBODY-STAINING

	CEA+	CEA-	NSE+	NSE-	406/14+	406/14-
1. NEIGHBOR						
DISTANCE (μ)	9.2 ± 0.7	8.5 ± 0.2	10.0 ± 0.7	8.5 ± 0.5	9.0 ± 0.5	8.5 ± 0.2
SUBGRAPHS	32 ± 3	55 ± 5	34 ± 3	53 ± 5	41 ± 4	54 ± 5
2. NEIGHBOR						
DISTANCE (μ)	13.5 ± 1	11.2 ± 0.5	13.7 ± 1	11.5 ± 0.5	12.1 ± 0.7	11.6 ± 0.5
SUBGRAPHS	17 ± 2	30 ± 3	18 ± 2	27 ± 3	22 ± 2	28 ± 3
3. NEIGHBOR						
DISTANCE (μ)	17 ± 3	14.3 ± 0.5	17.3 ± 1.3	14.5 ± 0.8	15.1 ± 0.8	14.3 ± 0.8
SUBGRAPHS	12 ± 1	22 ± 2	14 ± 1	22 ± 2	17 ± 2	22 ± 2
4. NEIGHBOR						
DISTANCE (μ)	20.2 ± 1.5	17.1 ± 0.5	20.2 ± 1.3	16.8 ± 0.8	18.0 ± 1.0	17.1 ± 0.8
SUBGRAPHS	10 ± 1	20 ± 2	11 ± 1	17 ± 2	12.5 ± 1	19 ± 2
CYCLOMATIC NUMBER	79 ± 8	152 ± 15	87 ± 9	144 ± 14	111 ± 12	149 ± 15

TABLE 8

RELATION BETWEEN STRUCTURE LEVELS IN GENERAL
DISORDERS IN HISTO-PATHOLOGY

DISEASES	LEVEL*			
	1	2	3	4
FUNCTIONAL DISORDER	-	+	+	?
INFECTION, UNSPECIFIC	+	+	?	?
INFECTION, SPECIFIC	+	+	?	?
BENIGN TUMOURS	-	+	+	+
CARCINOMA IN SITU	-	+	-	-
MALIGNANT TUMOURS	+	+	+	+

* 1 = EXTRACELLULAR MEMBRANES

  2 = CELLULAR LEVEL

  3 = GLANDS, VESSELS, ETC.

  4 = FUNCTIONAL COMPARTMENTS (I.E. PORTAL AREAS, ETC.)

APPLICATIONS OF MULTIDIMENSIONAL SEARCH TO STRUCTURAL FEATURE IDENTIFICATION

H.S. Baird

AT&T Bell Laboratories
Computing Science Research Cent.
600 Mountain Ave. 2c-557
Murray Hill, NJ 07974
U.S.A.

Abstract

Shape recognition by fast syntactic methods is possible when there exists a natural linear (one-dimensional) order on component shapes. This may not be available for "structural" shape descriptions taking the form of unordered, variable-length sets of simpler shapes. In this case, it is tempting to fall back on slower exhaustive correlation, graph matching, and relaxation methods. However, if the structural shapes are themselves "simple", it is possible to apply multi-dimensional search techniques for asymptotically fast feature identification. I exploit the fact that many simple shape types may be parameterized as points in low-dimensional spaces where distance models dissimilarity. During training, shapes are clustered heuristically within each class, then among all classes, giving a small set of characteristic shape distributions. Each of these is then associated with a binary feature variable taking the value one when any input shape falls within the distribution. This mapping from a structural description into a bit-vector is an example of a "feature identification" method. Selecting such a mapping is slow and heuristic, but fully automated, applicable uniformly to many shape types, and controlled by only a few natural statistical parameters. A mapping, once selected, can be applied quickly using kD-trees. Large-scale, statistically-significant trials have shown the technique to be superior to simpler fixed mappings, in an OCR context.

1. Introduction

In many pattern recognition applications a linear (*one-dimensional*) order on shape primitives is not available. As an example, consider "interior" features of shape such as voids or strokes making up the skeleton. Fast syntactic matching techniques require such an order. Many workers have therefore fallen back on (often slower) matching algorithms such as exhaustive correlation, graph-matching, and semantic relaxation.

This work focuses on an important special case of structural representations, where the component shapes are sufficiently simple that we can exploit *multi-dimensional* search algorithms for asymptotically fast feature identification and classification. The approach we describe is not a variation on syntactic methods, but rather gives a method for integrating structural descriptions with statistical classifiers.

In a structural approach, a pattern is represented as a set of simpler shapes: an unordered,

---

This paper is to be presented, by invitation, at the *Nato Advanced Workshop on Syntactic and Structural Pattern Recognition*, Barcelona, Spain, 23-25 October, 1986. An earlier version was presented at the *IEEE Computer Society Conf. on Computer Vision and Pattern Recognition*, Miami Beach, Florida, 22-26 June, 1986.

NATO ASI Series, Vol. F45
Syntactic and Structural Pattern Recognition
Edited by G. Ferraté et al.
Springer-Verlag Berlin Heidelberg 1988

variable-length list of geometric features of mixed type (see [Pa77]). Such representations are often intuitively information-preserving, in that a pattern reconstructed from its component shapes is readily recognizable by eye. Unfortunately, there is little theory supporting classification of unordered sets of mixed shapes, and few algorithms which are fast, trainable, and robust under noise.

In a statistical (or, "decision-theoretic") approach, a pattern is described by a fixed-length, ordered list of numeric features (see [DH73]). Statistical classifiers can be fast, trainable, and robust, but their performance depends strongly on the choice of a few "good" features.

One way to combine the strengths of the two approaches is to provide a translation between the two representations. I call this the *feature identification* problem: choose a mapping from variable-length, unordered sets of shapes, to fixed-length binary vectors, to meet some standard of performance. This paper describes such a method, so that:

- the choice is fully automated;
- computing a chosen mapping is usably fast; and
- classifier performance is improved, in comparison with simpler fixed mappings.

The method I present for *selecting* a mapping is heuristic and slow. However, it is guided by only a few "natural" statistical parameters (such as trim factors) whose exact values are not critical. It operates in a uniform way on all shape-types, and so it can be readily adapted to other schemes for structural analysis. It expects patterns to be isolated — but aside from this, it is not strongly specialized. Performance improvements, measured in large-scale trials, have been shown to be statistically significant, as well as practically important, in an OCR context.

2. The Engineering Context

This feature identification technique has been applied in an experimental mixed-font, variable-size character recognition (OCR) system [SPB86]. A bilevel image of a page is run-length-encoded and connected to form a "line-adjacency graph" [Pa82a]. Connected sub-graphs ("blobs") are rapidly traversed to find a set of simple component shapes, including strokes, holes, concave arcs, *etc.* [Pa85].

Training and classification is performed by a Bayesian technique using binary features [DH73], under the pragmatic assumption of class-conditional statistical independence among the features. This gives a fast single-stage classifier, widely-used in OCR [Na82].

I do not discuss here other methods to split touching characters, recombine fragmented characters, infer line- and page-structure, and exploit textual context [SPB86]. Therefore I assume that the characters are upright, unbroken, and not touching others, and that the classifier has no knowledge of font, point size, baseline location, or adjacent characters.

It is often possible to improve a classifier's performance by judiciously *splitting* a class into two or more variants. Most multifont OCR classifiers use extensive splitting (see, e.g., [YM78]) to help distinguish font-dependent shape differences. For the purposes of this paper, I allow practically none.

3. Prior Work

Recently, a Soviet group [LSVD85] has reported experiments with clustering of structural shapes for automatic selection of features, applied to a single font/size OCR system. Although many details of the clustering technique are not described, the approach appears to be roughly similar to mine, and also shows an improvement over manual feature selection.

Multifont recognition by an "integrated" classifier (with few split classes), rather than by multiple classifiers (one per font or font/size combination), is rarely emphasized in the OCR literature (see survey [Na82]). Most similar are some Japanese systems for handprinted Roman & katakana OCR (*e.g.* [YM78]). In these, feature identification is usually performed manually, *i.e.* the user specifies the mapping.

The use of structural analysis has, of course, been a recurrent theme in OCR [H72,YM78,SBM80,CG82]. Along with most Japanese researchers in handprinted Chinese OCR

[MYY84], I do not attempt combinatorial matching (such as subgraph-isomorphism or relaxation), since in the absence of a fixed ordering on the shapes, the number of combinations occurring is excessive. Instead, I match shapes geometrically in a manner (using kD-trees) that, to my knowledge, has not been applied in OCR before.

In many structural OCR systems, matching is performed by "correlating" the input pattern with a series of manually-chosen prototypes [SBM80]. The details of such "correlation" techniques, applied to structural descriptions, are often hard to justify theoretically. Most published methods are frankly heuristic and rely for good performance on hand-tuning. I use a Bayesian framework, in which all "features" are global, each contributing a weight to every class.

4. Structural Analysis

The structural analysis method [Pa85] I have used has five shape-types (see Figure 1):

stroke   straight line-segments approximating the skeleton;

hole   represented by center and radius;

arc   concave segment of the contour(s) (center, orientation, and depth);

crossing   where two strokes cross; and

endpoint   butting at the top (or bottom) of the bounding box (location, orientation, and width).

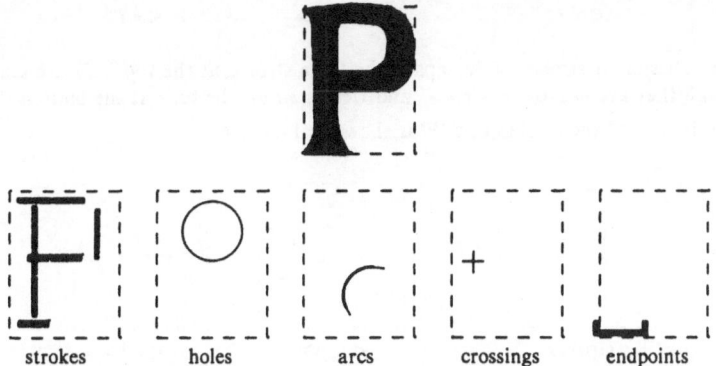

strokes   holes   arcs   crossings   endpoints

Figure 1: A structural analysis of a P, shown in its bounding box.

This choice of shape-types is somewhat arbitrary. Most have histories of effective use in OCR [H72,CG82,YM78] and other applications. To maintain size and location invariance all shape parameters are normalized with respect to the bounding box.

5. Parameterization of Shapes

Choosing a "parameterization" for each shape-type, is crucial. I wish to transform the given parameters of shape into new ones so that "perceived similarity" of shapes is modeled by a metric in the new parameter space. I require a representation with a "continuity" property: a small change in perceived shape should always result in a small displacement in parameter space, under some metric.

The details of parameterization are generally different for each shape-type: the principle can be illustrated using strokes. Each normalized stroke, given as two endpoints, can be represented as a 4-vector $<x,y,r,i>$ where $<x, y>$ gives the location of the stroke's center, and the complex number $<r, i>$ represents its length and orientation in a special way. Stroke-length is forced into the range $[0, 0.5]$, with 0.5 representing the maximum possible (the length of a diagonal of the bounding box). This scaled length is then represented by $|| <r, i> ||$. The angle of incidence of the stroke with the x-axis is doubled to give the angle of orientation $\tan^{-1}(i/r)$ of the complex number. This doubling is necessary since strokes are not directed, and so their range of orientation, $[0, \pi]$ radians, is discontinuous; doubling the angle gives a continuous range $[0, 2\pi]$. This has the desired continuity properties:

1)   similar strokes lie close together in parameter space (Figure 2); and

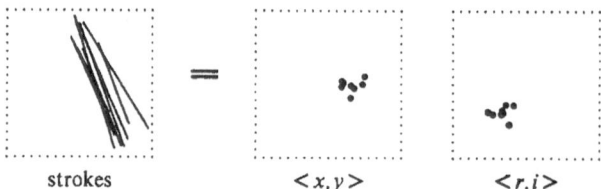

Figure 2:   Similar strokes lie close together (strokes are shown inside a bounding box, and the parameter space is projected onto its $x-y$ and $r-i$ planes).

2)   dissimilar strokes lie far apart (Figure 3).

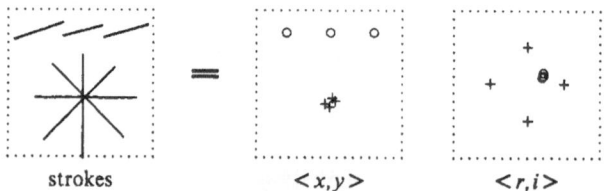

Figure 3:   Dissimilar strokes lie far apart: the three strokes at the top ("o") are scattered in $<x,y>$ even though they are nearby in $<r,i>$, and *vice versa* for the four at the bottom ("+").

The parameterization of the strokes of a 'P' is shown in Figure 4.

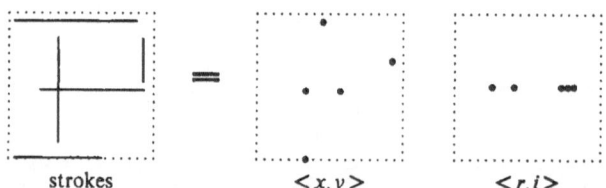

Figure 4:   Parameterization of the strokes of a 'P'.

There is a special problem with *arcs*, resulting from the fact that they lie along boundaries. Suppose two arcs are located at the same place, but face in different directions. If arc-depth and arc-orientation are represented by $<r,i>$ in the same way as for strokes, then, as their depths approach zero (as they "flatten out"), their $<r,i>$ both approach $<0,0>$ and become indistinguishable, even though they still face in different directions. I circumvent this problem by restricting $\| <r,i> \|$ for arcs to the range $[0.25, 0.5]$, where a value of 0.25 represents zero arc-depth.

Parameterization routines with the necessary properties must be hand-crafted for each shape-type. In doing so, I have made sure that each component of the new representation is constrained to the range $[-0.5, 0.5]$. Thus each shape becomes a point in a unit cube of known dimension ≤ 4. Such standardization makes it easy to ensure that all later processing can be applied uniformly to all shape-types.

6. Feature Mappings

The feature identification problem is to map a list of shapes into an binary feature vector. Let us somehow select a set of regions in the parameter spaces (within the unit cubes described above). Then associate with each region a binary feature variable, and let the variable take the value 1 if and only if some input shape falls in its region.

Such regions can be either specified by a hand-crafted partition, or selected automatically. An automatic method is suggested by the following experiment. When normalized shapes are superimposed for a single class, clusters often appear in parameter space (Figure 5).

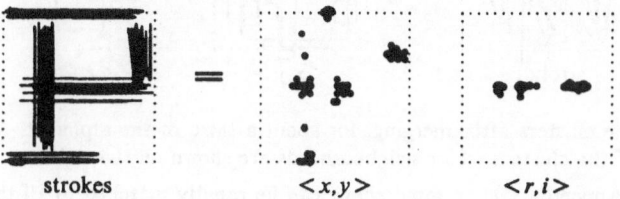

strokes $<x,y>$ $<r,i>$

Figure 5: Superimposed 'P' strokes, illustrating clustering.

For most classes I have observed a small number of tight, pairwise-disjoint clusters. For this reason they can be found robustly (during training) using a heuristic clustering algorithm (Appendix A). (There are parallels with Hough transform techniques [DH72,Ba81]; a difference is that clustering (an expensive step) needs to be performed only during training, not during matching.) I approximate clusters as rectangular neighborhoods centered at their means and with sides that are small multiples (2.25) of the component-wise standard errors (Figure 6).

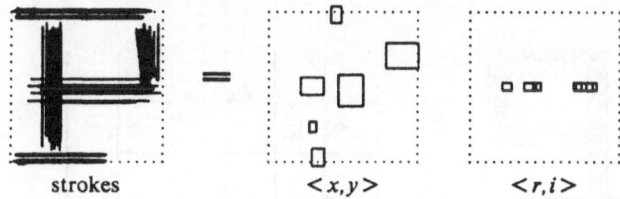

strokes $<x,y>$ $<r,i>$

Figure 6: Clusters approximated by rectangular neighborhoods.

This tends to preserve information in the sense that the set of cluster-means for a class is almost always immediately recognizable by eye (Figure 7).

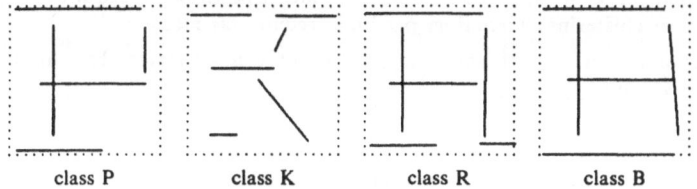

class P   class K   class R   class B

Figure 7: Stroke cluster means.

Almost all characters in the alphabet are as recognizable as 'P' is here.

Over the whole alphabet, many clusters are sufficiently similar that they can be combined (for example, the left vertical strokes in P, R, and B above). When this is done (see Appendix B), about 300 distinct ones remain (of which about 100 are stroke-clusters — 30 are shown in Figure 8).

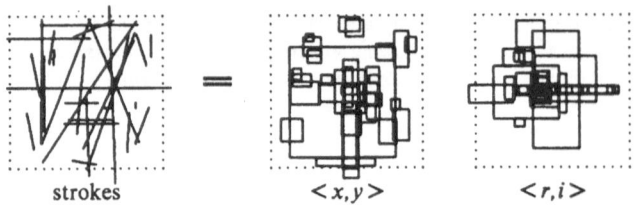

strokes $<x,y>$ $<r,i>$

Figure 8: Stroke clusters after merging, for about a third of the alphabet. The strokes on the left are the means of the clusters whose neighborhoods are shown on the right.

By using *kD*-trees (Appendix C), an input shape can be rapidly matched to all the regions containing it.

7. Simpler Mappings

The clustering technique may seem unnecessarily complex. A simpler method, using fixed partitioning of the parameter spaces, has been systematically compared to clustering.

Let "*k*-partitioning" divide up a parameter space by partitioning the range of each parameter into *k* equal-sized intervals, then forming the cross-product to partition the whole space. For example, strokes are 4-dimensional, so 3-partitioning divides their parameter space into $3^4 = 81$ regions (Figure 9).

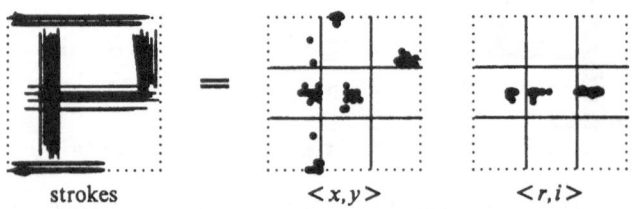

strokes $<x,y>$ $<r,i>$

Figure 9: A 3-partition, giving 81 disjoint regions.

The total number of features required, summed over all shape types, grows rapidly with increasing *k*: 288 for k=3, 864 for k=4, and 2050 for k=5. Thus 3-partitioning requires roughly the same number of features as clustering, while finer partitions require far more.

Actually computing a *k*-partitioned mapping is, of course, faster than for clustering, both asymptotically and absolutely.

8. Design of Experiments

All the algorithms were written in C and run under the UNIX† operating system on a DEC VAX 11/750 with floating-point accelerator. The input device is a 200 lines/inch scanner built by AT&T Teletype Corporation. This scans a page and does automatic thresholding in about five seconds to obtain a 1728 by 2048 binary matrix. A page is run-length-encoded into files of about 100 kilobytes.

A detailed description of the trials is given in [LB85]. Ten fonts were used in the experiments: Times Roman, **Times Bold**, *Times Italic*, Helvetica Light, **Helvetica Black**, *Helvetica Italic*, Palatino, Malibu Roman, Constant Width, and Printout. Six point sizes were tested: 8, 10, 12, 14, 16, 18. An alphabet of 70 characters was used:

ABCDEFGHIJKLMNOPQRSTUVWXYZ
abcdefghklmnopqrstuvwxyz

---

† UNIX is a Trademark of AT&T Bell Laboratories.

0123456789 & ()-$/[]*§

Characters omitted from the trials include: small punctuation, disconnected characters, ligatures, and characters not found in all the fonts. Since the classifiers were designed to be *insensitive* to font, size, and height above baseline, a few confusions were inevitable, and have been forgiven — they are:

C/c O/o S/s U/u V/v W/w X/x Z/z
0/O (zero/"oh") 1/l/I (one/"el"/"eye") 9/g

Classes were split in only these cases:

a/a g/g f/f Q/Q/Q

Print quality was excellent (original phototypesetter output); poor quality images, however, were provided by small point sizes (6 and 8 point) near the limit of scanner resolution.

Training and test sets were disjoint, selected in the ratio 1:3 in a randomized manner from the same population, so that the statistics would be unbiased estimates. I measured average top choice correct (and, separately, correct in top three choices), with each class given equal weight. (*Incorrect* choices are a mixture of substitution errors and rejects; I do not discuss here the question of the ratio between these types of error.)

To support my interpretations of the results, I show standard 95% statistical confidence intervals bracketing each measurement. For Bernoulli trials [F68] such as this, large numbers of samples are required to show statistically significant differences above 99%. For a statistic $\hat{p}$, I construct a 95% confidence interval for the transformed statistic $\log\left[\hat{p}/(1-\hat{p})\right]$ (assuming normality), and then back-transform to the original scale. This transformation is advisable [Pr85] since the normality assumption is suspect for $\hat{p}$ in the extremes of the distribution (near 0 or 1).

The smallest test set, for a single font/size mixture (see Figure 11), held 2100 samples, permitting distinctions above 99% (for example, between 100% and 99.3%). Larger multifont tests permitted finer discriminations (Figures 10, 12, 13).

9. Results of Experiments

The results of the large-scale trials were as follows.

a) Clustering is often markedly superior to fixed partitioning for mixtures of similar fonts (Fig. 10).

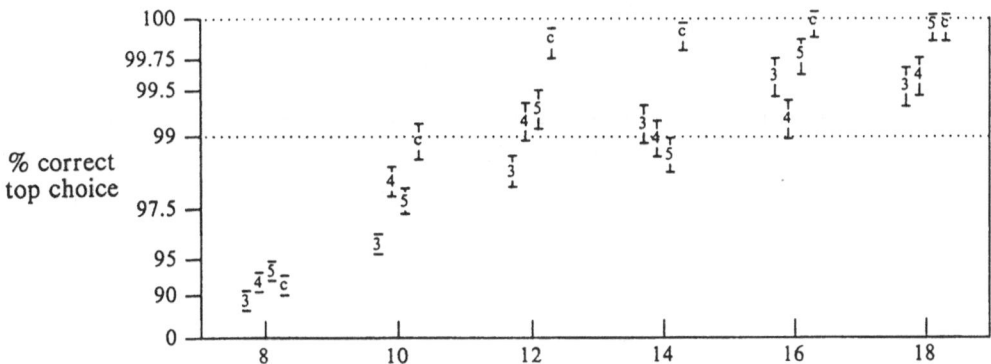

Figure 10: Percent top choice correct performance of classifiers trained on a mixture of the three similar but not identical fonts Times Roman, Palatino, and Malibu Roman, in mixed sizes 8-18 point and tested on mixtures of the same fonts at the sizes shown (for 3-, 4-, and 5-partitioning and clustering feature-identification methods; 6300 samples per datum).

b) Better than 99.5% top choice correct, averaged over point sizes 12 and up, is seen for mixtures of similar fonts (Fig. 10) and most single fonts (Fig. 11 is typical). Clustering classifiers trained and tested on *the same font* gave the following results (top choice correct, averaged over all classes and point sizes, 12 point and above, for a total of 8400 samples each):

Font	% Correct
Palatino	100.00
Constant Width	99.98
Times Italic	99.97
Printout	99.92
Malibu Roman	99.87
Times Roman	99.80
Helvetica Light	99.78
Helvetica Italic	99.78
Helvetica Black	99.51
Times Bold	97.56

The relatively poor showing of **Times Bold** is due entirely to confusions among I / [/] / l ("eye" / "left-square-bracket" / "right-square-bracket" / "el"). Errors on **Helvetica Black** are due largely to confusions among B / 8 / a / s (the a and s close up at small point sizes). This level of performance is close to commercially-quoted rates for single-font, variable-size classifiers.

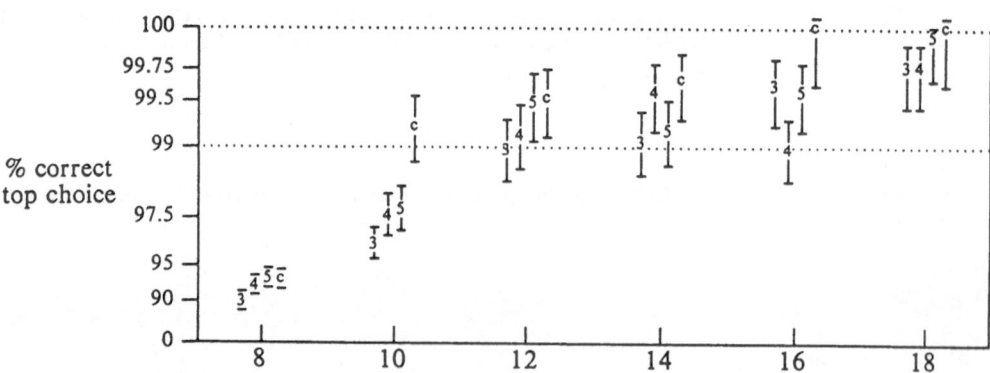

Figure 11: Percent top choice correct performance of classifiers trained on the Times Roman font for mixed sizes 8-18 point, and tested on the same font at the sizes shown (2100 samples per datum).

c)   Top choice performance declines for mixtures of dissimilar fonts, to 97.5% for six fonts (Fig. 12).

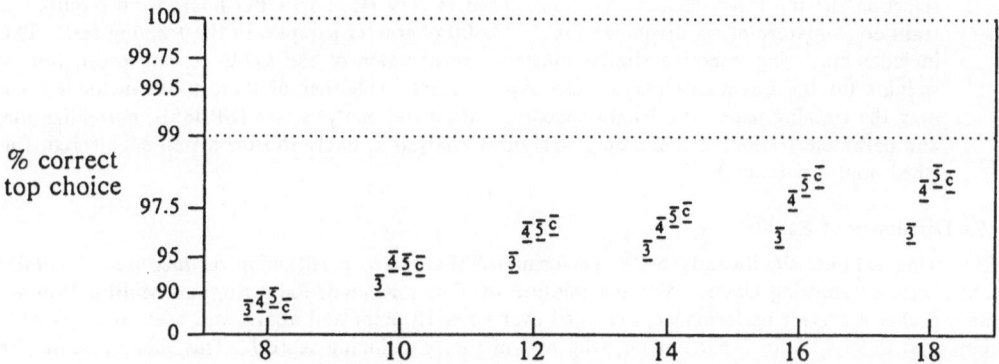

Figure 12: Percent top choice correct performance of classifiers trained on Times Roman, **Times Bold**, *Times Italic*, Helvetica Light, *Helvetica Black*, and *Helvetica Italic* fonts for mixed sizes 8-18 point, and tested on mixtures of the same fonts at the sizes shown (12,600 samples per datum).

d)   Under the laxer standard *some correct in top 3*, which is appropriate where linguistic context can help correct a top choice [SPB86], good performance is still possible on mixtures of up to six fonts (Fig. 13).

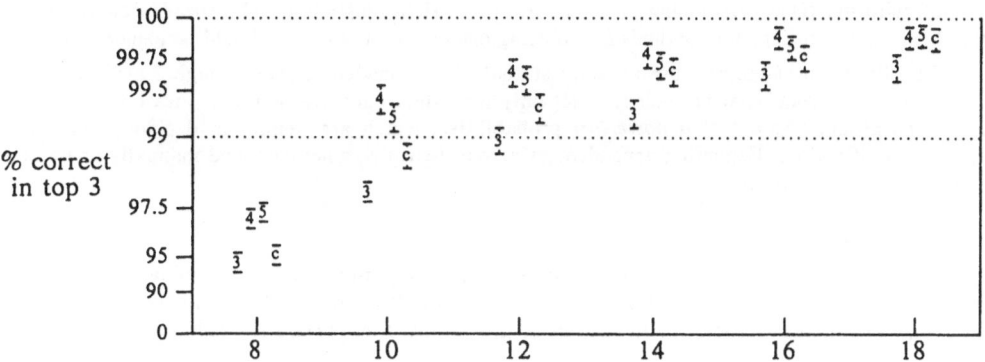

Figure 13: Performance (percent correct in top 3 choices) of classifiers trained on Times Roman, **Times Bold**, *Times Italic*, Helvetica Light, **Helvetica Black**, and *Helvetica Italic* fonts for mixed sizes 8-18 point, and tested on mixtures of the same fonts at the sizes shown (12,600 samples per datum).

e)   Classification speed of 6 characters/second is adequate for large-scale experimentation, but at least an order of magnitude too slow for routine use. This includes mapping normalized shapes to binary feature vectors, computing *a posteriori* probabilities of class membership, and selecting the top three choices. Training speed is very slow: 16 CPU hours were required to train on a mixture of six dissimilar fonts (75,600 character samples in the training set). This includes clustering, merging similar clusters, construction of the *kD*-trees, and estimation of weights for the Bayesian classifier (see Appendices). (Neither of these times includes scanning the training pages, run-length-encoding, structural analysis (see [SPB85]), normalization, and parameterization; of these, only structural analysis is likely to slow a well-engineered, finished implementation.)

10. Discussion of Results

Clustering exhibits significantly better performance than fixed-partitioning on mixtures of similar fonts, without splitting classes. For the mixture of Times Roman, Palatino, and Malibu Roman, the top choice correct performance, averaged over sizes 12 point and above, was seen to be 99.93%, compared to 99.43% for 5-partitioning (the best of the partitioning scores). This shows a *reduction of the error rate by a factor of 4.0, at 95% confidence.* (Similar results were measured for other mixtures, including (i) Times Roman and Helvetica Light, and (ii) *Times Italic* and *Helvetica Italic*.) I feel this is evidence of a "generalizing" power which will be important in holding down the number of split-classes as more fonts are added.

The normalization method performs consistently well across a wide range of point sizes (12-18 point; below 12 point, scanner quantization noise intrudes). The generally good performance on single fonts (12-up) suggests that the set of shape-types is adequate, and that the structural analysis algorithm [Pa85] produces consistent analyses. Where this fails, it appears [SPB86] that a few details of boundary shape can complete discrimination.

I have refused to split classes during these experiments, even to accommodate drastic font style changes, with few exceptions. This has held the speed of the classifiers roughly constant, but has contributed to an increased error rate on large mixtures of dissimilar fonts. For these mixtures, without splitting, fixed-partitioning performs no worse than clustering. The speed/accuracy trade-offs of selective splitting, and methods for splitting classes automatically, should be explored.

The Bayesian classifier I use assumes statistical independence, even though I am certain the features used are somewhat dependent. Not only may this reduce recognition rates but it seems to distort the absolute values of *a posteriori* probabilities, which are helpful in deciding whether to reject a classification. Exploiting dependencies among features, whether found manually or by clustering, may improve results.

11. Conclusion

It can be difficult to make strong claims for pattern recognition systems as complex as those for OCR. There is no accepted model of shape-formation for Roman alphabets, and there is little consensus on performance standards. In support of my claims, I emphasize the following aspects of the work.

- *Simple, fully-automated methods.* Many reported OCR systems depend for good performance on an accumulation of hand-crafted rules. In my opinion, this will always be required to some degree. However, the search for simple, fully-automatic methods is important theoretically (as it may reveal models of shape formation) and practically (by reducing the labor of improving classifiers). Although the clustering method is heuristic and its implementation complex, I can claim that it is *simple* in the sense that it is controlled by a small number of natural statistical parameters whose exact values are not critical (see Appendices). Also, aside from the requirement that the patterns be isolated, and that shape-types can be normalized appropriately, the technique is not strongly specialized for OCR.

- *Carefully-designed, large-scale trials.* I have restricted my claims to those I can demonstrate statistically at 95% confidence. A usable OCR system must meet high standards, well above

99% correct. In these circumstances, very-large-scale trials are required to show improvements. In addition, I have tried to simplify the experimental design and disclose all restrictions, so that useful comparisons with other work (always difficult) may be at least possible.

12. Acknowledgements

Larry Wolfrum of AT&T Teletype designed and built an interface for the scanner to the VAX UNIBUS, and Lorinda Cherry wrote the UNIX operating system driver. The structural analysis program was developed by Theo Pavlidis. J. B. Kruskal provided background on clustering algorithms. Jon Bentley helped me decide how best to use kD-trees. Daryl Pregibon advised on the computation of confidence intervals. Susanna Lam collected the test database, automated the test procedures, and ran many tests. Theo Pavlidis, Simon Kahan, and Doug McIlroy made valuable comments on an earlier draft.

Appendix A. **Clustering Shapes**

The clustering method is an adaptation of the sequential, agglomerative k-means algorithm given in [H75]. The goal is to find a small number of clusters, approximated by component-wise mean & standard error, satisfying the conditions:

a) *Maximum cluster size*: the width of a cluster neighborhood must be less than half (.54) the width of the bounding box (otherwise it might match too many points); recall that the standard errors are multiplied by 2.25 to define neighborhoods, and so the upper bound on component-wise standard errors is 0.12.

b) *Minimum cluster size:* no cluster owning fewer than 10 points will be reported.

c) *No clustering within patterns*: no two points from the same pattern may be in the same cluster; this restriction is uncommon in published clustering applications;

d) *Satisfactory coverage*: quit when 90% of points have been assigned to some cluster.

This is computed in two stages to reduce processing time. The first stage projects all points onto each component axis in turn, and clusters these one-dimensional distributions separately; then the cross-product of these sets of clusters are used as seeds for the next stage, which operates on the full space.

The first stage is designed to overestimate the number of component-wise clusters. First, the range of values seen is partitioned into intervals no wider than 1/3 the maximum standard error allowed. Then pairs of intervals are merged, in a manner used in the second stage (described below).

The second stage is iterative. At the outset, all points are unassigned. For each cluster (inherited from the previous iteration), rectangular neighborhoods are computed, as follows: at the first iteration, use 1.5 times the standard errors of the seed clusters; at each iteration thereafter, volumes grow by a factor of 1.25 over the prior volume. Then each hitherto unassigned point is assigned to the cluster whose neighborhood encloses it, if any. If there are two or more, the point is assigned to the one closest in L_∞ distance (scaled component-wise by standard errors). If no neighborhood encloses the point, it remains unassigned. A point, once assigned, is never reassigned. At each iteration, assignment of points to clusters is followed by merging of similar clusters. Pairs of clusters are merged if (a) they own no two points from the same pattern, and (b) their merged size (number of points) does not exceed the maximum size. Merging occurs in decreasing order of the number of merged points (so that large clusters tend to swallow small ones). Once a cluster owns at least 5 points, its statistics are recomputed at each iteration. This proceeds until satisfactory coverage is achieved, or more than three iterations have occurred.

Each shape type is of course clustered separately, but the same parameters are used for all the shape types. Coverage is usually good: in each class on average, 87% of the points have been assigned to some cluster (for the font mixture of Figures 12 & 13). Clusters tend to be small: the average standard error (over all components of all clusters of all shape types) is 0.03, about a quarter of the maximum 0.125. Within each class, clusters tend to be disjoint: for an average class, 81% of the patterns have no shape lying in more than one cluster's neighborhood. My experience is that small changes in the guiding parameters seldom change these results significantly.

The program is 1200 lines of C and run time grows roughly quadratically with the number of points. In a typical computation, 2000 points can be clustered in 2-4 CPU minutes. Total clustering time makes up the overwhelming majority of total training time.

Appendix B. **Merging Clusters**

Since many shape distributions are very similar to those in other classes, the total number can be reduced by a merging phase, with little effect on classification accuracy. Recall that each cluster has been approximated by a rectangular neighborhood whose center is the mean of the cluster and whose half-widths are fixed multiples (x2.25) of the component-wise standard errors. To reduce computation time, merging two or more neighborhoods is performed by finding the smallest neighborhood enclosing all of them (rather than recovering the constituent points and recomputing merged statistics).

Merging two or more clusters usually produces an "increase", defined as the largest change in "size" undergone by *any* of the clusters involved in the merge. "Size" is defined in a dimension-free manner as the dth root of the neighborhood's volume, where d is the dimension of the parameter space. The algorithm guarantees that no increase will occur greater than a factor of 2.25. Pairs are merged in ascending order of resulting increase. As many as six iterations over all pairs are usually required. Typically between 2 and 5 neighborhoods are merged into one, depending on shape-type. The program is 1000 lines of C, computation time grows quadratically with the number of clusters, and a typical computation (such as the one above) requires 22 CPU minutes.

Appendix C. kD-trees

Locating all rectangular neighborhoods, among a large number, that enclose a given point can be accomplished quickly using a variation of kD-trees. These are generalizations of binary search trees to multi-dimensional data [BF79]. In this adaptation, the "cut" dimension cycles with increasing depth in the tree, and splitting values are chosen so that an equal number of intervals are isolated on each side of the cut. The resulting trees are roughly balanced, so search time increases asymptotically logarithmically with the number of neighborhoods.

In practice, the large size of the trees forces a time/space engineering tradeoff. A complete kD-tree grows exponentially with the dimension. I limit this growth by imposing a coarse resolution on the data (4 bits). This gives a typical total tree size of 400K bytes, summed over all shape-types. Such a set of trees, on average, can locate 7 candidate neighborhoods at the cost of 15 "$\leqslant$" comparisons. On average 5 of these 7 are correct when checked at full resolution. The kD-tree-search implementation runs about a factor of three faster than optimized sequential search, for a typical large mixture (as for Figure 12). The program to build the trees is 1400 lines of C, and a typical computation requires 1 CPU minute.

13. References

[Ba81]   Ballard, D. H., "Generalizing the Hough Transform to Detect Arbitrary Shapes", *Pattern Recognition* 13, 1981, pp. 111-122.

[BF79]   Bentley, J. L., Friedman, J. H., "Data Structures for Range Searching," *Computing Surveys*, Vol. 11, no. 4, December 1979, pp. 397-409.

[CG82]   Cox, C. H., P. Coueignoux, B. Blesser, and M. Eden "Skeletons: A Link between Theoretical and Physical Letter Descriptions," *Pattern Recognition*, 15 (1982), pp. 11-22.

[DH72]   Duda, R. O., and Hart, P. E., "Use of the Hough Transform to Detect Lines and Curves in Pictures", *Graphics and Image Processing* 15, 1972, pp. 11-15.

[DH73]   Duda, R. O., and Hart, P. E., *Pattern Classification and Scene Analysis*, Wiley (New York, 1973), Sects. 2.9-2.10.

[F68]   Feller, W., *An Introduction to Probability Theory and Its Applications*, Wiley (New York, 1968), Chapter VI.

[H72]   Harmon, L.D., "Automatic Recognition of Print and Script," *Proceedings of the IEEE*, vol 60, October 1972, pp. 1165-1176.

[H75]   Hartigan, J. A., *Clustering Algorithms*, Wiley (New York, 1975), Chapter 4.

[HTW78]   Hattich, W., Tropf, H., and Winkler, G., "Combination of Statistical and Syntactical Pattern Recognition - Applied to Classification of Unconstrained Handwritten Numerals", *Proceedings of the Fourth International Conference on Pattern Recognition*, Kyoto, Japan, November 7-10, 1978, pp. 786-788.

[Fu74]   Fu, K. S., *Syntactic Methods in Pattern Recognition*, Academic Press (New York, 1974).

[Fuk72]   Fukunaga, K., *Introduction to Statistical Pattern Recognition*, Academic Press (New York, 1972).

[LSVD85]   Lashas, A., Shurna, R., Verikas, A., & Dosinas, A., "Optical Character Recognition Based on Analog Preprocessing and Automatic Feature Extraction", *Computer Vision, Graphics, and Image Processing*, vol. 32, pp. 191-207, 1985.

[Ka85]   Kahan, S. "Problems in Recognizing Handprinted Characters," *M. S. thesis*, Dept. of EECS, Univ. of California, Berkeley, Febr. 1985.

[LB85]   Lam, S., and Baird, H. S., *Performance Testing of Mixed-font, Variable-size Character Recognizers*, AT&T Bell Labs, in preparation.

[MYY84]   Mori, S., Yamamoto, K., Yasuda, M., "Research on Machine Recognition of Handprinted Characters," *IEEE Transactions on Pattern Analysis and Machine Intelligence*,, vol PAMI-6, no 4, July, 1984, pp. 386-405.

[Na82]   Nagy, G. "Optical Character Recognition- Theory and Practice" *Handbook of Statistics* (P. R. Krishnaiah and L. N. Kanal, eds) vol.2, North-Holland, 1982, pp. 621-649.

[Pa77]   Pavlidis, T., *Structural Pattern Recognition*, Springer-Verlag (New York, 1977).

[Pa82a]   Pavlidis, T., *Algorithms for Graphics and Image Processing*, Computer Science Press, Rockville, Maryland, 1982.

[Pa83]   Pavlidis, T., "Effects of Distortions on the Recognition Rate of a Structural OCR System," *Proceedings, Computer Vision and Pattern Recognition '83*, Washington, D.C., June, 1983, pp. 303-309.

[Pa85]   Pavlidis, T. "A Vectorizer and Feature Extractor for Document Recognition" (submitted for publication).

[Pr85]   Pregibon, D., AT&T Bell Laboratories, personal communication, June, 1985.

[Sc82]   Schurmann, J. "Reading Machines," *Proc. 1982 Intern. Conference on Pattern Recognition*, Munich, W. Germany, October 1982, pp. 1031-1044.

[SBM80]   Suen, C. Y., M. Berthod, and S. Mori "Automatic Recognition of Handprinted Characters - The State of the Art," *IEEE Proceedings*, 68 (April 1980), pp. 469-487.

[SPB86]   Kahan, S., Pavlidis, T., Baird, H. S., "On the Recognition of Printed Characters of Any Font and Size", AT&T Bell Labs, Murray Hill, NJ (submitted for publication).

[Ul76]   Ullmann, J. R. "Picture Analysis in Pattern Recognition," *Digital Picture Analysis*, A. Rosenfeld, editor, Springer, 1976, pp. 295-343.

[YM78]   Yamamoto, K. and S. Mori, "Recognition of Handprinted Characters by Outermost Point Method", *Proc. Fourth Intern. Joint Conf. Pattern Recognition*, Kyoto, Japan, Nov. 7-10, 1978, pp. 794-796.

IV. GRAMMATICAL INFERENCE AND CLUSTERING

LEARNING FROM EXAMPLES IN SEQUENCES AND GRAMMATICAL INFERENCE

L. Miclet(*), J. Quinqueton(**)

(*) E.N.S.T. (CNRS UA 820)
46 Rue Barrault
75634 Paris Cédex 13, France
(**) CRIM
Monte de St. Priest
34100 Montpellier, France

Keywords

Grammatical inference / Logical learning / Non monotonous logic / Learning systems / Structural pattern recognition / Artificial intelligence / Markov model.

Abstract

The purpose of this paper is to compare two methodologies devoted to an *intelligent* analysis of sequences: Learning from Examples and Grammatical Inference.

For each of them, various techniques and algorithms are presented and we try to point out the similarities and differences. We present an example in Biology in order to compare the two approaches.

NATO ASI Series, Vol. F45
Syntactic and Structural Pattern Recognition
Edited by G. Ferraté et al.
© Springer-Verlag Berlin Heidelberg 1988

1.INTRODUCTION

The purpose of this paper is to compare two methodologies devoted to an "intelligent" analysis of sequences: Learning from Examples and Grammatical Inference.

For each of them, various techniques and algorithms are presented and we try to point out the similarities and differences. We present an example in Biology in order to compare the two approaches.

Several Learning techniques will be reviewed, not for the purpose of exhaustivity, but for exploring the various "typical" features of these techniques: static/dynamic, data/model driven.

In the field of Grammatical Inference, we quote algorithms for induction of regular languages from examples, and the Markov model approach as an hybrid (statistical and structural) technique.

2.STATEMENT OF THE PROBLEM.

2.1.Some Principles of Learning

A **Learning Problem** starts from a set of examples (and, possibly, counter-examples), called the **Learning Set**, described in a given **Language** (in our case, these examples are sequences). The aim of a Learning Mechanism is to state the learning set as a logical statement (or a grammar) understandable without the examples.

Most Learning Problems end when an **Action**, result of what has been learnt, can be executed by a machine. In our case, such an action is the result of applying the learnt rule(s) to new sequences.

Then, a learning system appears as made of two parts: an **Inductive** mechanism and a **Deductive** one. We can also define a third one, the **Argumentation** mechanism.

If we try, using a learning system, to find a hidden rule, like for instance in the **Eleusis** game [Dietterich et al 1985], the system cannot find the rule, unless we give it a space of admissible rules. This can be done with semantic constraints (i.e. general knowledge about the learning set), or syntactic constraints (i.e. a logical structure without semantics).

But in the Eleusis game, there are several players. Then, to find a rule has no sense since we are unable to **explain** it to the others. This remark also holds for Learning Problems, and the explanation has to be clear for a human being (the user), or for a machine (to make an action).

2.2. What is a sequence?

A sequence is, according usual dictionaries, a (possibly infinite) chain of elements, each one being taken in a finite set of symbols, called the **alphabet**.

Then, the most "natural" way of describing a sequence is to state it as a word (or sentence) in a language: such a description, called **grammatical**, takes fully in account the structure of the sequence.

But other descriptions are possible, based upon **logical** formalisms, by giving properties of each element (of the alphabet) and relations between elements of the sequence.

For instance, a sequence like "cauu", using the alphabet {a, u, c, g} of nucleotides in Biology, can be viewed as a word in this alphabet (of course!) or as a structure made of 4 elements {e1, e2, e3, e4}, related by the "before" relationship:

"is-c(e1) and is-a(e2) and is-u(e3) and is-u(e4) and before(e1,e2) and before(e2,e3) and before(e3,e4)".

If we want to use non redundant binary properties, they have to be given by the user, according his knowledge of the domain. In Biology, for instance, the usual properties of the nucleotides are: to be a purin (a, g) or a pyrimidin (c, u), and to have 2 (a,u) or 3 (c,g) H links. In the above example, it leads to the following description:

"pyrimidin(e1) and 3-h-links(e1) and purin(e2) and 2-h-links(e2) and pyrimidin(e3) and 2-h-links(e3) and pyrimidin(e4) and 2-h-links(e4) and before(e1,e2) and ..."

2.3.How to learn?

A learning problem can be stated according the definitions of set theory. There are two ways of defining a set: the **extensive** one (enumeration of the elements) or the **intensive** one (a characteristic property).

But the intensive definition of a set is equivalent to the extensive one: it is a necessary and sufficient condition for an object to belong to the set.

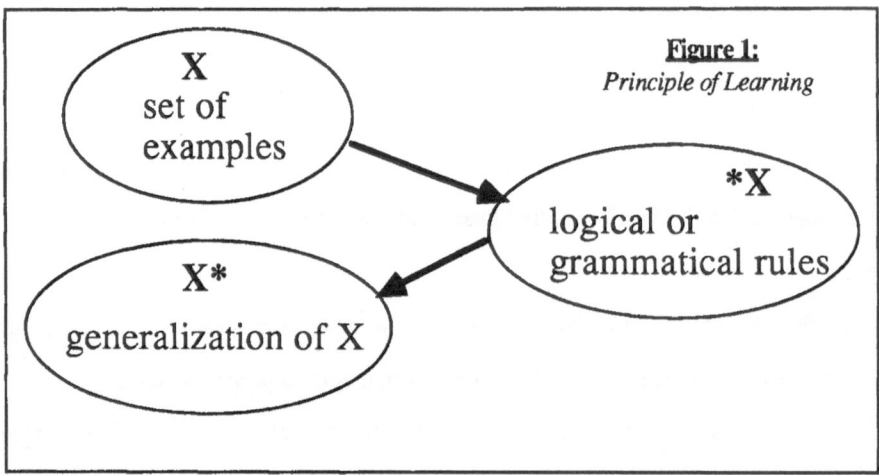

Figure 1:
Principle of Learning

Then, we must define a **generalization** procedure (figure 1), in order to reach new objects with the learnt rules. There are two kinds of methods regarding the way of performing this task: **Model Driven** techniques and **Data Driven** techniques.

3.LEARNING METHODS.

Learning methods can be classified according the kind of knowledge used to build the generalization, and the way the examples are given to the system.

The result of Learning may be (or not) sensitive to the order in which the examples are given to the system (as in program construction from examples [Jouannaud et al 1980]), or to their multiplicity (several examples with the same description, as in statistical methods).

The methods can be classified as **static** or **dynamic** regarding this feature. In the dynamic methods, the result will depend on the order in which the examples are given: the user has to be a good "teacher".

The examples are supposed to be given in a logical language. Then, each example is a conjunctive statement of properties and relations, and the learning set is a disjunctive statement of its elements. This leads to a statement identifying exactly the learning set.

Then, two ways of generalization are possible. The **data-driven** way (figure 2) consist on using additionnal knowledge (taxonomy of predicates, counter examples,...) to perform the generalization [Kodratoff 1985]. The **model driven** approach tries to use non classical disjunction and conjunction to obtain a generalizing intensive statement of the learning set [Quinqueton et al 1986].

The description of the examples gives the space in which the rules are looked for. This space is structured by the possible relationships between the attributes. Such a topological space can be very important for the interpretation of the rules. It is the basis of the research on associative memory [Kohonen 1983], neuronal networks [Fogelman et al. 1986], and more generally the research on memorization processes [Schank 1980].

But, more generally, this is important for the **explanation** of the learnt concept, as defined above in the example of Eleusis game: the rules can be "self explanatory", or the system can provide examples to illustrate them.

Another important feature in Learning systems is the way the learnt knowledge is used. Various solutions have been proposed.

In the SEEK system [Ginsberg et al 1985], the learning process do not create new rules, but suggests modifications of existing rules.

In EURISKO [Lenat 1983] and PLAGE [Gascuel 1986], the system has to discover new concepts, and the researchers have to estimate their interest. This is only possible in a given experimental context.

In the WMP project [Carbonell et al 1985], the learning system is interpreted as a robot which learns to act in a simplified newtonian world, which control the learning process, which is **dynamical** in this case. The MACHIN system [De Sainte Marie 1986] is also of this kind. Its aim is to learn heuristics for problem solving.

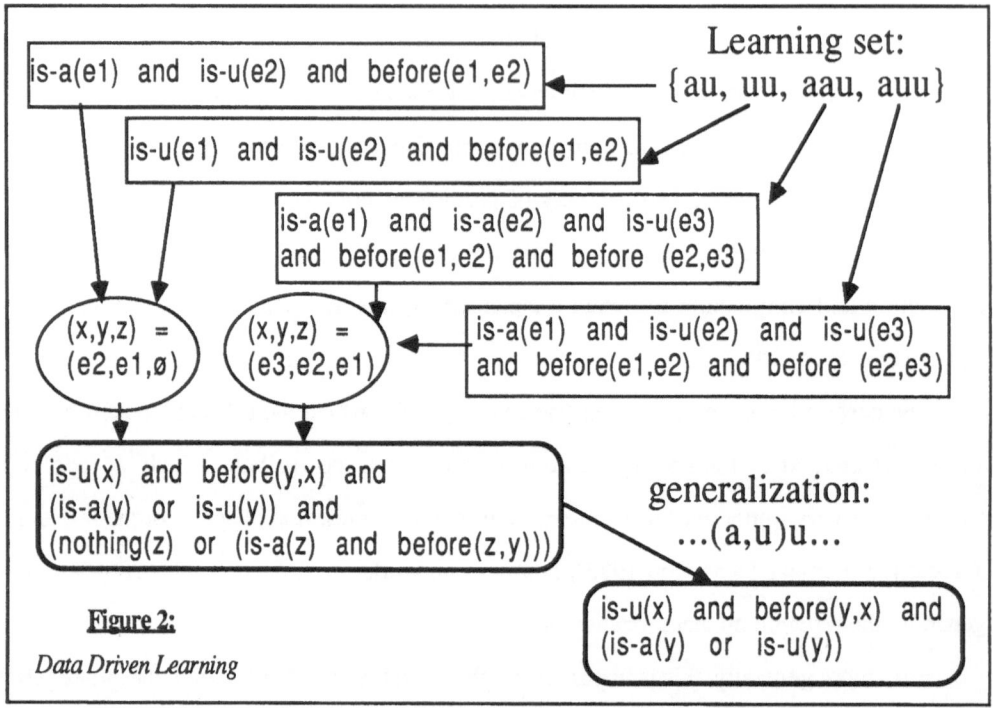

Figure 2:

Data Driven Learning

Kodratoff [Kodratoff 1985] proposes a general data driven technique, AGAPE, to learn first order logic statements, in a Knowledge base made of trees of predicates and predefined links, which represent the basic knowledge of the domain. In this case, the learnt rules can be used by any Prolog (for instance) interpreter.

INDUCE2 [Michalski 1983] is a learning mechanism oriented towards the search of formulas with a given syntax: he asked to psychologists a suitable syntax for 1st order logic statements to be easily understood by human experts.

Mitchell proposes a **Learning Apprentice** [Carbonell et al 1983], based upon the "Version Space" technique [Mitchell 1982], which tries to learn explanation mechanisms. His system is then oriented towards pointing out what has been typical of a given reasoning, in a given knowledge base and experimental context.

We have proposed a learning technique, CALM [Quinqueton et al 1986] based upon the "majority logic" [Von Neumann 1952], which is typically a model driven technique.

The principle is the following. Each object of the learning set E is supposed to be described by a set of properties. The learnt concept *E will be the list of the properties which hold for more than a given number of examples.

Then, the generalisation E* is the set of the objects on which more than a given threshold of the statements in *E are true. It is easy to see that E* ≠ E in this case.

4.GRAMMATICAL INFERENCE.

Its principle is to extract from a set of sentences, the sample or learning set, a grammar which generates a set of sentences containing the sample. A lot of algorithms have been proposed, mainly in the seventies [Fu 1982, Miclet 1984].

They have been applied in various domains of Pattern Recognition: speech recognition, biomedical signal processing, texture in images, boundaries of objects,...

In the case of regular grammars [Muggleton 1984], the problem can be stated as the building of a finite "prefix tree acceptor", which represent a grammar recognizing no more and no less than the sequences belonging to the learning set (figure 3).

This acceptor, called the **canonical acceptor**, is then transformed, by merging some of its states, into acceptors which accept more and more sequences.

The generalization process is then performed according a heuristic rule, chosen by the designer: it is a knowledge given to the system. Then, such a generalization technique is basically data-driven. A lot of heuristics have been proposed. The main proposed heuristics have been [Pao et al 1978, Biermann et al 1972, Miclet 1980, Angluin 1982].

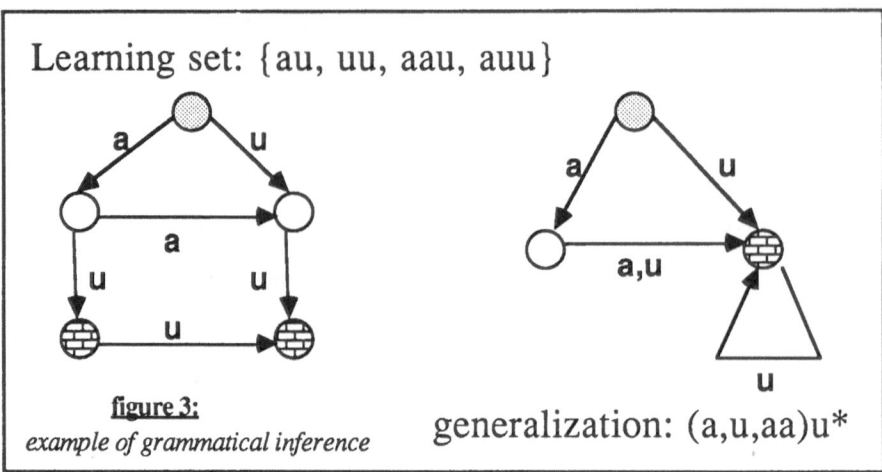

Learning set: {au, uu, aau, auu}

figure 3:
example of grammatical inference

generalization: (a,u,aa)u*

Another regular grammatical inference scheme is that of Hidden Markov Model inference. It is a Model Driven technique, in the sense that the number of states and the basic structure of the acceptor is fixed, and that the learning which has to be done is that of the probabilities of transition between the states, and the probabilities of emitting a letter of the alphabet from a given state. The sample set is made of sentences over this alphabet, assuming they have been produced by the model to discover.

The learning of these probabilities may be realized by at least two different methods, but basically the technique is to use the sample set to lead the probabilities to convergence, starting from some random initialization. Good operational results of such methods have been recently obtained in Speech Recognition.

For a more accurate description of these models and their applications, references [Rabiner et al 1986, Bahl et al 1983, Bourlard et al 1985] can be consulted.

5.EXAMPLE IN BIOLOGY.

We present in this section a small example, taken from Molecular Biology, on which we try both the CALM technique as presented at the end of section 3, and a Markov model as presented at the end of the previous section.

The concept to learn is the intron-exon junction [Biochimie 1985] in nucleic acids, which are represented as sequences in the alphabet {a,u,g,c}. The learning set is made of 38 examples, obtained by a 12-length window around the junction (4 before and 8 after) in several sequences taken from a data bank. It is shown on figure 4.

mouse beta globin
(1) uucc auucuuaa
(2) guuu cacucaga
(3) gucc aaccauag
(4) gucc cacucaga

mouse alpha globin
(5) uucc acucuugu
(6) guuc cauacgcg

chicken epsilon globin
(7) gguc cauccaga
(8) gucc cacucuac

chicken rho globin
(9) gguc cacccaga
(10) gucc cacucuac

rabbit beta globin
(11) cguc caaccaua
(12) gucc cacucaaa
(13) gucc aaccauag
(14) gucc cacucaaa

human beta globin
(15) gucc aaccauag
(16) gucc cacucaga

mouse metallo thionein
(17) uggc cauucuga
(18) uucu cacucaac

human pepro glucagon
(19) gguc cauaauuu
(20) uguc cauucuca
(21) cucu cauucaga

pro alpha 2(1) collagen
(22) cuuc cauucguc
(23) uucc cauuccuu

chicken ovalbumin
(24) ucca cucggauc
(25) agug caugucuu
(26) uacc auuccauc
(27) acuc cauauacc
(28) cguc cauaccgg

mouse oncogen c-fos
(29) uguc cacucaaa
(30) cguc cacucguc
(31) ucgc cauccaac

human oncogen c-myc
(32) agac cauucgcu

chicken oncogen c-myb
(33) uucc cauucguc
(34) cugc cauuacug
(35) uuua cauucgag
(36) aguc caucuauu
(37) uggu cauucuaa
(38) gagg cauucaug

Figure 4:

Learning set

We use a 100-length sequence (50 before and 50 after the known junction) of rat actin gene as test sequence. It is presented on figure 5.

figure 5: *test set (rat actin)*

acuucggagugaaggaugggagccguggucccggucucagucucgucguc
caucccaccuccaccccucccacuggaccucugggucgucucuuuggauua

5.1.Learning using CALM.

We use the learning technique on the examples described using the biological attributes language presented in section 2.2: 2 binary attributes for each letter of the alphabet. Then, "a" is described as "purin and not 3-h-links", "c" as "not purin and 3-h-links", "g" as "purin and 3-h-links" and "u" as "not purin and not 3-h-links".

With this language, each example will be described with 24 binary attributes, each one corresponding to a property in a given position (from -4 to +8). For instance, the first example of figure 4 is described as:

> *"not purin-4 and not 3-h-links-4 and not purin-3 and not 3-h-links-3 not purin-2 and 3-h-links-2 and not purin-1 and 3-h-links-1 and purin+1 and not 3-h-links+1 and not purin+2 and not 3-h-links+2 and not purin+3 and not 3-h-links+3 and not purin+4 and 3-h-links+4 and not purin+5 and not 3-h-links+5 and not purin+6 and not 3-h-links+6 and purin+7 and not 3-h-links+7 and purin+8 and not 3-h-links+8"*

We learned 3 concepts, each one corresponding to a threshold: 90% (more than 34/38), 80% (more than 30/38) and 70% (more than 26/38). They are given below:

90%: *"not 3-h-links+2 and not purin+3"*

80%: *"not purin-2 and not purin-1 and 3-h-links-1 and not purin+1 and 3-h-links+1 and purin+2 and not 3-h-links+2 and not purin+3 and not purin+4 and not purin+5"*

70%: *"not purin-2 and not purin-1 and 3-h-links-1 and not purin+1 and 3-h-links+1 and purin+2 and not 3-h-links+2 and not purin+3 and not purin+4 and not purin+5 and 3-h-links+5 and not 3-h-links+6"*

There is (intuitively) a clear jump between 90% and 80%: the length of the learnt rule increases quickly (2 to 10), and then only a little (10 to 12) between 80% and 70%. This is, to our opinion, a kind of "percolation" (critical phenomenon, in the sense of Physics), which suggests that the 80% threshold is a good one.

This remark can be studied more formally using an information measurement on the learnt rules [Quinqueton et al 1985].

The learnt concept (80%) is then applied to the learning set, by measuring the "score" of each example. We use a threshold of 80%, i.e. at least 8/10 attributes must be true. 7 examples are "rejected" by this control. For each of them, we performed a shifting of 1 or 2 positions on the left and on the right. The results are given on figure 6. They were interpreted by a biologist as the fact that the glogin genes are known to have a slightly "fuzzy" intron-exon junction.

The application of the concept to the test set is shown on figure 7, and we can see that the first recognized position is the good one which is considered as a good result from biological point of view.

```
mouse beta globin                  human beta globin
(1)    uucc auucuuaa = 7           (15) gucc aaccauag = 7
       -uuc cauucuua » 10               -guc caaccaua » 9
       --uu ccauucuu » 5                --gu ccaaccau » 3
       ucca uucuuaa- » 4                ucca accauag- » 3
(3)    gucc aaccauag = 7
       -guc caaccaua » 9
       --gu ccaaccau » 2          chicken ovalbumin
       ucca accauag- » 3          (24) ucca cucggauc = 5
                                       -ucc acucggau » 5
                                       --uc cacucgga » 10
mouse alpha globin                     ccac ucggauc- » 2
(5)    uucc acucuugu = 6          (26) uacc auuccauc = 7
       -uuc cacucuug » 10              -uac cauuccau » 9
       --uu ccacucuu » 5               --ua ccauucca » 5
       ucca cucuugu- » 6               acca uuccauc- » 3


rabbit beta globin
(13) gucc aaccauag = 7                      Figure 6:
     -guc caaccaua » 9
     --gu ccaaccau » 3            Control of the learning set
     ucca accauag- » 3
```

```
acuucggagugaaggaugggagccgugguccggucucagucucgucguc
.caucc.caccuc.caccc.cucc.cacuggaccucugggucgucucuuuggaua
```

figure 7:   *result on test set*

5.2.Learning using the Markov Model.

To learn a Markov model on our learnig set of figure 4, we start from the frequency table of each letter in each position:

position	g	c	u	a
-4	15	6	13	4
-3	13	3	20	2
-2	5	17	15	1
-1	2	30	4	2
1	0	32	0	6
2	0	1	3	34
3	0	15	22	1
4	2	11	21	4
5	1	27	5	5
6	6	4	11	17
7	9	4	12	13
8	7	10	6	14

Then, we cluster the 12 positions into 8 states, according this frequency table. The emission probabilities in each state are given by the following table:

state	positions	g	c	u	a
s1	{-4,-3}	.37	.12	.43	.08
s2	-2	.13	.45	.39	.03
s3	-1	.05	.79	.11	.05
s4	1	0	.84	0	.16
s5	2	0	.03	.08	.89
s6	{3,4}	.03	.34	.56	.07
s7	5	.03	.71	.13	.13
s8	{6,7,8}	.19	.16	.26	.39

The recognition process consist on computing the probability for an object to have been generated by the list of transitions:

$$->s1->s1->s2->s3->s4->s5->s6->s6->s7->s8->s8->s8->$$

If we perform the same kind of control as in section 5.1 on the learning set (figure 6), we have to estimate the probabilities of the shifted lists of transitions for each example: the results are given in the following table, for the examples in which the maximum probability was not obtained by the unshifted sequence, and for the examples with very low probability.

example	shift −2	shift −1	unshift	shift +1
(1)	9.06×10^{-9}	$\mathbf{2.27 \times 10^{-4}}$	8.23×10^{-7}	0.0
(3)	1.50×10^{-9}	$\mathbf{1.08 \times 10^{-5}}$	2.33×10^{-6}	4.72×10^{-10}
(5)	1.79×10^{-9}	$\mathbf{4.47 \times 10^{-5}}$	1.00×10^{-7}	4.10×10^{-8}
(13)	1.51×10^{-9}	$\mathbf{1.08 \times 10^{-5}}$	2.33×10^{-6}	4.73×10^{-10}
(15)	1.51×10^{-9}	$\mathbf{1.08 \times 10^{-5}}$	2.33×10^{-6}	4.73×10^{-10}
(24)	$\mathbf{1.82 \times 10^{-5}}$	1.29×10^{-8}	3.87×10^{-10}	0.0
(25)	0.0	0.0	1.02×10^{-8}	$\mathbf{1.35 \times 10^{-8}}$
(26)	1.04×10^{-9}	$\mathbf{4.40 \times 10^{-6}}$	3.43×10^{-7}	0.0
(27)	0.0	1.32×10^{-9}	$\mathbf{1.12 \times 10^{-7}}$	2.45×10^{-9}
(28)	0.0	2.04×10^{-10}	$\mathbf{1.64 \times 10^{-6}}$	5.30×10^{-8}
(32)	7.32×10^{-12}	8.45×10^{-11}	$\mathbf{9.23 \times 10^{-7}}$	2.56×10^{-9}
(34)	0.0	9.26×10^{-10}	$\mathbf{1.28 \times 10^{-6}}$	1.19×10^{-7}
(38)	0.0	0.0	$\mathbf{6.17 \times 10^{-7}}$	1.76×10^{-8}

If we compare these results to those of figure 6, we can see a very strong similarity, except that some examples (the end of the previous table) appear to have a very low probability.

If we apply our Markov model to the test set, it appears as very accurate in recognition. The results are shown on figure 8, where the positions corresponding to the highest probabilities are displayed.

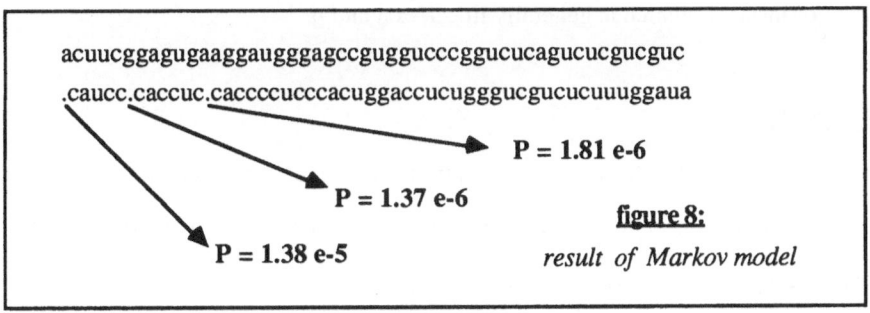

acuucggagugaaggaugggagccguggucccggucucagucucgucguc
.caucc.caccuc.caccccucccacuggaccucugggucgucucuuuggaua

P = 1.81 e-6

P = 1.37 e-6

P = 1.38 e-5

figure 8:
result of Markov model

6.COMPARISON ACCORDING VARIOUS CRITERIA.

We try in this section to point out similarities and differences between the Grammatical Inference and the A.I. Learning Techniques.

6.1.Complexity

The complexity of the learning algorithms are to be studied versus the number of examples given to the system and the length of the description language.

In the case of learning systems, if n is the number of examples and p is the length of the description language, the algorithm has a complexity bound by $\emptyset(n \times p^{2\ or\ 3})$ generally. Some techniques, which use structural languages, are more dependant on the number of examples, by $\emptyset(n^2 \times p^2)$.

In the case of grammatical inference, the complexity is generally bound by $\emptyset(q^{2\ or\ 3})$, where q is the number of prefixes, i.e. of different substrings which appear in the examples. It is clear that q is to be compared to n x p above.

Then, all these algorithms appear as low order polynomial ones. In the problems with a large number of examples, it is clearly preferable to preprocess the data using a statistical method (clustering, for instance) which is generally linear vs n and p.

6.2.Stability

The learnt concept must be stable if we learn it again on the objects that it has recognized as correct, but which are not in the learning set: *(E*) = *E. In terms of Topology, it means that the concept is a topological "closure" of the learning set.

In data driven techniques, this property is obvious, both for learning systems and grammatical inference.

In model driven techniques, it has to be controlled regarding the parameters of the method. A proof in the case of the CALM method is given in a previous work [Quinqueton et al 1985].

6.3.Use of negative examples

In the case of grammatical inference, no practical algorithm has tried to use negative samples, though it could be a very good help for the basic heuristic on the positive sample. The problem of taking in account a negative sample in the inference algorithm itself has been prooved to be NP-complete [Gold 1978]. In the case of Markov Models, it is not clear how such a negative information could be used.

In A.I. Learning Techniques, the use of negative examples is related to two logical properties: contradiction and completeness.

The completeness of the set of examples regarding the concept to be learnt means that it is a good illustration of the concept. Then, the whole information is supposed to be given by the examples.

This condition is requested by Soldano [Belaid et al 1984], Quinqueton and Sallantin [Quinqueton et al 1983, Quinqueton et al 1984] and, more generally, in model driven methods, where the control of the generalization is not based upon negative examples.

The completeness may also be defined as the statistical one: this is requested by the Qualitative Regression method [Daudin 1979].

In fact, the needed completeness decrease with the strength of the constraints on the space of rules.

In the case we use negative examples, if a positive example has the same description as a negative one, there is a **contradiction**.

This can be forbidden, as in Mitchell [Mitchell 1982], Kodratoff [Kodratoff 1985], Michalski [Michalski 1983] and, generally, in data driven methods, where the negative examples are used to bind the generalization of the positive examples.

The contradiction can also be managed by the system, as in model driven methods. This point is, in fact, closely related to the problem of learning with noisy data.

6.4.Argumentation

The problem here corresponds to the explanation problem in deductive expert systems. A learning system has to defend the rules it has learnt, when they are contested by the user. This point is related to the use of counter examples.

In grammatical inference, the only argumentation is a message pointing out where the proposed sequence is wrong according the learnt grammar. This has been widely used in compilers, which are a very old kind of decision rule based upon grammars.

In data driven methods, the basis of argumentation is the "near misses", which are negative examples very closed to positive ones. They are taken in account, in the learning step, as "critical" elements of knowledge. They are the basis of argumentation in these methods.

In model driven methods, the argumentation is performed through a "contestation" process, during the decision step: the system proposes modifications of the submitted sequence in order it is well recognized.

6.5.Noise tolerance

The main difference here is between data driven methods and model driven ones, regardless they belong to the grammatical inference group or the logical learning group.

In the case of low level noise, the data driven methods are generally disturbed, in the sense they can miss a concept because of only one noisy example.

The model driven methods will learn however the concept, and point out the noisy examples as bad, like in the biological example given in section 5.

In the case of high level moise, the data driven methods will learn only noise, but the model driven ones will say "I can't learn anything".

The basic problem with noise is that it is not always negative: in the previous example of section 5, the elements of the learning set which are not recognized by the rules can be considered as noise (typing errors, for instance) or as singular examples (like the globin genes, for instance).

In this last case, noise can be considered as a suggestion of a discovery.

7.CONCLUSION.

It is more and more difficult to put a boundary between Pattern Recognition and Artificial Intelligence, because of the lot of existing research projects which involve both these domains.

Structural Pattern Recognition, because of its descriptive nature, finds in automatized perception a very good field of application. The complexity of these problems leads the structural methods to an evolution towards A.I. approach, where it is easier to take in account all the levels of a perception problem together.

Then, the conclusion of this work can be, once we pointed out very strong similarities between grammatical inference and logical learning, that a learning problem depends mainly of the existing knowledge in its domain.

8.REFERENCES.

[Angluin 1982] D.ANGLUIN, "Inference of Reversible Languages", Journal of the ACM, vol 29, pp 741-765, july 1982.

[Bahl et al 1983] L.R.BAHL, F.JELINEK, R.L.MERCER, "A Maximum Likelihood approach to Continuous Speech Recognition", IEEE Trans on PAMI, Vol PAMI 5, no 2, March 1983.

[Belaid et al 1984] S.BELAID, J.SALLANTIN, H.SOLDANO, "Use of Learning Techniques for Structure Elucidation in Organic Chemistry using 13 C NMR Spectra", ECAI 84 proceedings, pp 508-509, Elsevier Science Pub, 1984.

[Biermann et al 1972] A.W.BIERMANN, J.A.FELDMAN, "On the Synthesis of Finite-State Machines form Samples of Their Behaviour", IEEE Transactions on Computers, vol C 21, pp 592-597, 1972.

[Biochimie 1985] BIOCHIMIE, special issue on "The Application of Computer Methods to Molecular Biology" (in English), vol 67, no 5, may 1985.

[Bourlard et al 1985] H.BOURLARD, Y.KAMP, H.NEY, C.J.WELLEKENS, "Speaker Dependant Connected Speech Recognition via Dynamic Programming and Statistical methods", in "Speech and Speaker Recognition", M.Schroeder ed., Karger Pub., 1985.

[Carbonell et al 1985] J.G.CARBONELL, G.HOOD, "The World Modeller Project: Objectives and Simulator Architecture", proceedings of the 3rd Int. Machine Learning Workshop, june 1985.

[Carbonell et al 1983] J.G.CARBONELL, R.S.MICHALSKI, T.MITCHELL, "Machine Learning", Tioga Publication Co, Palo Alto, 1983.

[Daudin 1979] J.J.DAUDIN, "An Iterative Method for Analysis of Variance", Int. Journal of Biomedical Computing, Vol 10, 1979, pp507-518.

[De Sainte Marie 1986] C.DE SAINTE MARIE, "The Necessity of Learning while Doing", European Working Session in Learning, Orsay 1986.

[Dietterich et al 1985] T.G.DIETTERICH, R.S.MICHALSKI, "Discovering Patterns in Sequences Events", Artificial Intelligence 25, 1985, pp187-232.

[Fogelman et al 1986] F.FOGELMAN SOULIE, P.GALLINARI, Y.LE CUN, S.THIRIA, "Automata Networks and Artificial Intelligence", in Computation on Automata Networks, Manchester University Press.

[Fu 1982] K.S.FU, "Syntactic Pattern Recognition and Applications", Prentice Hall, 1982.

[Gascuel 1986] O.GASCUEL, "Plage: a way to Give and Use Knowledge in Learning", European Working Session in Learning, Orsay 1986.

[Ginsberg et al 1985] A.GINSBERG, P.POLITAKIS, S.WEISS, "An overview of the SEEK 2 Project", Technical Report, Dept of Computer Sci., Rutgers University, NJ, 1985.

[Gold 1978] E.M.GOLD, "Complexity of Automaton Identification from Given Data", Information and Control, 37, pp 302-320, 1978.

[Jouannaud et al 1980] J.P.JOUANNAUD, Y.KODRATOFF, "An Automatic Construction of LISP Programs by Transformation of Functions Synthesized from their Input Output behavior", Internat. Journ. of Policy Anal. and Information Systems, vol 4 (1980), pp331-358.

[Kodratoff 1985] Y.KODRATOFF, "Une Theorie et une Methodologie de l'Apprentissage Symbolique", Invited Paper, Cognitiva 85, pp639-651, CESTA, Paris, june 1985.

[Kohonen 1983] T.KOHONEN, "Associative Memory, A System Theoretical Approach", Springer-Verlag, 1983.

[Lenat 1983] D.B.LENAT, "The Role of Heuristics in Learning by Discovery: three case studies", in [Carbonell et al 1983].

[Michalski 1983] R.S.MICHALSKI, "A Theory and Methodology of Inductive Learning", Artificial Intelligence 20, pp 111-162, Feb 1983.

[Miclet 1980] L.MICLET, "Regular Inference with a Tail Clustering Method", IEEE Transactions on System, Man and Cybernetics, Vol SMC 10, pp 737-743, 1980.

[Miclet 1984] L.MICLET, "Méthodes Structurelles pour la Reconnaissance des Formes", Eyrolles, Coll. Scientifique et Technique des Telecoms, Paris 1984, English version "Structural Methods for Pattern Recognition", North Oxford Academic 1986.

[Mitchell 1982] T.M.MITCHELL, "Generalization as Search", Artificial Intelligence 18, pp203-226, 1982.

[Muggleton 1984] S.MUGGLETON, "Induction of Regular Languages from Positive Examples", Turing Institute Research Memo 84-009, December 1984.

[Pao et al 1978] T.W.PAO, J.W.CARR III, "A Solution of the Syntactical Induction-Inference problem for Regular Languages", Computer Languages, Vol 3, pp 53-64, January 1978.

[Quinqueton et al 1983] J.QUINQUETON, J.SALLANTIN, "Algorithms for Learning logical Formulas", IJCAI 1983, Karlsruhe (FRG), pp476-478.

[Quinqueton et al 1985] J.QUINQUETON, J.SALLANTIN, "CALM (Contester pour Apprendre en Logique Modale)", proceedings AFCET-INRIA Congress on PR and AI, Grenoble (France), November 1985.

[Quinqueton et al 1986] J.QUINQUETON, J.SALLANTIN, "CALM: Contestation for Argumentative Learning Machine", in Machine Learning - A Guide to Current Research, Kluwer Academic Publishers.

[Rabiner et al 1986] L.R.RABINER, B.H.JUANG, "An Introduction to Hidden Markov Models", IEEE ASSP Magazine, January 1986.

[Schank 1980] R.C.SCHANK, "Language and Memory", Cognitive Science 4, 1980.

[Von Neumann 1952] J.VON NEUMANN "Probabilistic Logics and the Synthesis of Reliable Organisms from Unreliable Components", Lectures at the California Institute of Technology, January 1952.

AN EFFICIENT ALGORITHM FOR THE INFERENCE
OF CIRCUIT-FREE AUTOMATA

H. Rulot (*), E. Vidal (**)

(*) Centro de Informatica
Dr. Moliner s/n
46100 Burjasot, Valencia, Spain
(**) Facultad de Informatica
Dept. Sistemas y Computacion
46071 Valencia, Spain

ABSTRACT

In this paper, a recently introduced grammatical inference method is reviewed. In this method, a non left(right)-recursive regular grammar is built in an incremental way: as each training sample is presented, it is parsed by the current (error-correcting extended) grammar, minimizing explicitly, by dynamic programming, the number of error-rules needed. These error-rules are then added to the grammar. This procedure has proved to be well suited for capturing the relevant information associated with the lengths of the substructures of the patterns to be analized, and with their concatenation. A stochastic extension of the method is presented, and some alternative approaches for estimating the probabilities of both the error and non-error rules are discussed. Finally, the results of some experiments with speech samples, which show the capabilities of the proposed method, are summarized.

1.- INTRODUCTION.

Recently, a new Grammatical Inference (GI) method has been introduced which aims at adequately accounting for all the relevant variability present in the local (sub)structures ((sub)strings) of the patterns being considered and in their concatenation (Rulot,86). In particular, it has proved to be especially well suited for syntactic modelling of constraints which are based on the lengths (extents) of these (sub)structures. Classically, within the Syntactic approach to Pattern Recognition, these (often important) constraints are introduced through what is often called "hibrid approaches", in which some kind of complementary numerical attributes are attatched to the structural models (grammars). The best known of these approaches is Stochastic Syntactic Pattern Recognition, as well as the closely related Hidden Markov Modelling. Within this framework, length modelling is accomplished through the use of probabilities to represent the likelihood of repetition of the different left(right)-recursive grammar rules (and/or the corresponding automata loops). However, it has been widely claimed that, in most cases, this length modelling is quite inappropriate, because it leads to geometric (exponential) length probability distributions (Juang,85) (Rusell,85) and, consequently, other ad-hoc refinements must be introduced. Also, other ad-hoc methods, such as the introduction of "counters" and "length thresholds" in the states of automata, have been extensively utilized. In fact, all these methods can be seen as particular instances of using Attributed Grammars. It has been shown (Fu,83) that, in this case, arbitrary tradeoffs between structural (grammar-rules) and semantic (rule-attributes) complexities can be achieved for any given PR problem. Therefore, the obvious alternative to using attributes to represent length constraints, is to represent them by syntax.

However, this kind of purely syntactic modelling involves the use of non left(right)-recursive grammars or loop-free automata and, consequently, most previous approaches to GI (Fu,82) (Hunt,75) (Itoga,81) (Miclet,80) (Richetin,84) (Vernadat,84) result inapplicable. In fact, if R_+ is the (positive) training sample, all these methods produce inferred grammars G for which R_+ is a subset of L(G), and L(G)-R_+ (the "extralanguage") is infinite; more specifically, it is composed of structures (strings) which derive from those of R_+ by arbitrarly repeating certain substructures (substrings) of its patterns. Although this characteristic is often assumed to represent a "natural" abstraction ability, it usually leads to "too permissive" grammars for many Pattern Recognition problems. In particular, these grammars are rather badly suited to represent any (sub)string-length-based discriminative features of the patterns to be recognized.

In this paper, the new GI method introduced for solving the proposed problem will be reviewed, and current improvements will be presented. Also, some results of the application of a real Automatic Speech Recognition task, showing the capabilities of the proposed method, will be summarized.

2.- THE PROPOSED GI ALGORITHM.

The new GI method, achieves the construction of a finite-state automaton (or its corresponding regular grammar) through an incremental procedure which considers the strings in the training-sample R_+ sequentially and with no repetitions. As each new string is presented, the "learning" procedure attempts its recognition through the current automaton. If it fails, the automaton is updated by adding the necessary states and arcs so as to enable the recognition of the new string. The number of added states and/or arcs is explicitly minimised by applying a Dynamic-Programming-based algorithm. The resulting ("minimum") automaton is (explicitly) free of circuits, and accepts all the strings in R_+ as well as other strings composed of adequate concatenations of certain subset of their substrings. The (normalized) minimum number of additional states required to accept a given input string can be considered as a measure of the dissimilarity between that string and the nearest string acceptable by the current automaton. Consequently, this dissimilarity can be used to classify the input string with respect to different classes represented by the corresponding automata. Furthermore, the minimization algorithm can incorporate any other "classical" error-correcting parsing schemes with full specification of substitution, deletion and insertion error costs; this allows more appropriate dissimilarity measures to be defined for recognition in specific applications. Therefore, the method can perform inference and recognition in an integrated and simultaneous way, which is an essential feature of any proper learning methodology.

Given the equivalence of finite-state automata and regular grammars, only the inference of the latter will be presented formally. The corresponding automata can also be directly inferred by (essentially) the same algorithm.

The inferred grammar G=(N,V,P,S) is <u>regular</u>, <u>non-deterministic</u>, <u>circuit-free</u> (in that $\not\exists A \overset{+}{\Rightarrow} \alpha A$, $A \in N$, $\alpha \in V^*$), and has the property:
$\forall A,B,C \in N$, $\forall b,a \in V$, <u>if</u> $(B \to aA) \in P$ <u>and</u> $(C \to bA) \in P$ <u>then</u> b=a; i.e., the same terminal is associated with all the rules which have the same non-terminal in their right-hand side. This allows us to label the <u>states</u> (instead of the arcs) of the equivalent automaton with terminal symbols.

For the description of the algorithm we will make use of some error-correcting definitions. These definitions are similar to the usual ones (Fu,82), though adapted to our purposes:

<u>Definition 2.1</u>. <u>Error-rules</u>: Every rule in P has the following error rules associated (ε is the "nil" symbol):
Insertion (of a): $A \to aA$ $\forall(A \to bB) \in P$; $A \to ab$ $\forall(A \to b) \in P$ $\Big\}$
Subst.(of a for b): $A \to aB$ $\forall(A \to bB) \in P$; $A \to a$ $\forall(A \to b) \in P$ $\Big\}$ $\forall a \in V$
Deletion (of b): $A \to B$ $\forall(A \to bB) \in P$; $A \to \varepsilon$ $\forall(A \to b) \in P$

<u>Definition 2.2</u>. <u>Expanded grammar</u> of G: Is the (non regular) grammar $G'=(N',V,P',S)$ obtained by adding the corresponding error-rules to P.

<u>Definition 2.3</u>. <u>Optimal error-correcting derivation of</u> $\underline{x \in V^*}$: Is the derivation of x, $D_G(x)$, that use a minimum number of error-rules of P'.

In order to obtain $D_G(x)$ $\forall x \in V^*$, a Viterbi-like dynamic programming procedure (Forney,73) is utilized, which finds the "best alignment" of x with the nearest string in L(G). The principles upon which this procedure is based are not new, and they can be found in a number of matching algorithms, many of which being related with Speech Recognition (see e.g. (DeMori,81) - chapt.VIII). A related procedure based on the same principles has also been proposed recently (Thomason,86).

The Error-Correcting Grammatical Inference Algorithm (ECGIA) can then be outlined as follows:

ALGORITHM ECGIA (Error-Correcting G.I. Algorithm).

INPUT: Set of training strings ($R_+ \subset V^*$).

INITIALIZATION: if $a_1 \ldots a_i \ldots a_n \in R_+$ is the first training string, then the set of productions of the <u>initial</u> grammar G is:
$$P = \left\{ S \rightarrow \varepsilon A_0 ; \quad A_{i-1} \rightarrow a_i A_i , \quad i = 1..n; \quad A_n \rightarrow \varepsilon \right\}.$$

INFERENCE: for all (the remaining) training strings $y = b_1 \ldots b_i \ldots b_\ell \in R_+$ do:
1) Parse: With the current grammar, obtain the optimal error-correcting derivation $D_G(y) = r_1, \ldots, r_k$, $r_i \in P'$, $i = 1..k$.
2) Build: for every pair of <u>non-error</u> rules r_i, r_j of $D_G(y)$ such that $r_i = (C_{i-1} \rightarrow c_i C_i)$, $r_j = (C_{j-1} \rightarrow c_j C_j)$, $i < j$, between which there are (in $D_G(y)$) <u>only</u> error-rules, add to the current grammar the following m+1 rules (and the corresponding non-terminals):
$$C_i \rightarrow x_1 X_1 ; \quad \ldots \quad ; \quad X_{m-1} \rightarrow x_m X_m; \quad X_m \rightarrow c_j C_j .$$
where it is assumed that $y = b_1 \ldots c_i \alpha c_j \ldots b_\ell$, ($\alpha = x_1 \ldots x_m \in V^*$).
if $\alpha = \varepsilon$ then only the rule $C_i \rightarrow c_j C_j$ is added.

RESULT: The inferred grammar G is the current grammar.

END-of-ECGIA.

It follows from the inference procedure that the inferred grammar contains the <u>canonical</u> grammar (it generates all the positive samples in R_+). Also, since only those (sub)strings which are not already modelled by the current grammar lead to the addition of new rules (and non-terminals), the average number of rules added for each training sample is directly related to the <u>variability</u> of R$_+$. Consequently, although the size (number of rules and/or non-terminals) of the inferred grammar is not limited a priori, in practice, due to the fact that the variability of any reasonable positive sample is not unlimited, the grammar grows more and more slowly with each new training sample, with an asymptotical tendency to a constant maximum. This has been verified empirically and the corresponding results can be found in (Rulot,86). On the other hand, the size of the language L(G) generated by the inferred grammar is <u>finite</u>, but it tends to grow <u>exponentally</u> with the number of rules, allowing a large extralanguage, of samples similar (in the sense of the error-correcting dissimilarity utilized) to those in R$_+$, to be generated by G (or accepted by the equivalent automaton) (Rulot,86).

3.-<u>STOCHASTIC EXTENSION TO THE ECGIA</u>.

The stochastic extension of the proposed method of GI and string recognition involves: a) statement of the recognition process within the stochastic framework; and b) corresponding formulation of the probability estimation of the rules of the inferred grammars.

a) <u>Stochastic recognition</u>.

Let G$_c$ be a inferred characteristic grammar; i.e. a non-stochastic grammar obtained with the ECGIA. If we assume that the probabilities of using the rules of G$_c$, as well as those of their corresponding error-rules are known, then the maximum likelihood that a string x $\in$ V* be generated by the stochastic (error-correcting) extended grammar of G, G', is given by:

$$(3.1) \qquad p_{G'}(x) = \max_{\forall D'_G(x)} \prod_{\forall r_i \in D'_G(x)} p(r_i)$$

where D$'_G$(x) is any (not necessarily optimal) error-correcting derivation of x form G, and p(r$_i$) is the probability of the i-th rule of D$'_G$(x). In this case, the same Viterbi-like algorithm as that used for the non-stochastic (error-correcting) parsing, can be utilized for maximizing (3.1) and to obtain the maximum-likelihood derivation $D_G(x)$. However, the algorithm must now be based on the local maximization of the probabilities of <u>all</u> the rules required for the parse, instead of the local minimization of the number of error rules.

If a C-class PR problem is considered, with each class C_i represented by a grammar G_i, the generation probabilities (3.1) can be utilized for classification in the usual way (Fu,82):

$$(3.2) \quad \forall x \epsilon V^*, \; x \epsilon C_i \; \underline{iff} \; p_{G_i'}(x) \cdot p_i > p_{G_j'}(x) \cdot p_j \quad i \neq j; \; j = 1..C$$

where p_j , j=1..C is the a priori probability of class C_j.

b) <u>Estimating the probabilities of the rules</u>.

Following a fairly classical paradigm of stochastic GI (Fu,82), the characteristic grammar G_c can be inferred first, on the basis of a given positive sample R_+, and then the probabilities of the rules can be estimated from the frequency of use of each rule in the parse of the strings in the same sample R_+. It comes from well known results (see e.g. Fu,82) that under suitable conditions on the training data, and assuming that G_c is unambiguous, the probabilities estimated in this way approaches the true probabilities of the rules as the number of training samples approaches infinity. Alternatively, another approach (which is mandatory in the case that G_c is ambiguous) consist of using methods similar to the classical probability estimation techniques of Hidden Markov Models; namely, Viterbi estimation and/or the Backward-Forward algorithm (see e.g. Jelinek,76). These techniques involve an arbitrary (task-adequate) initialization of the probabilities of the rules, followed by a probability reestimation loop, in which all the strings in R_+ are repeatedly (maximum-likelihood) parsed with the current stochastic grammar G, until convergence is achieved.

Note that such procedure(s) can be effectively carried out since all the strings in R_+ can be <u>exactly</u> parsed by G_c without using any error-rule. However, when a string of V^* not in R_+ is submited for recognition, some error-rules of G'_c may be required to achieve a parse. Therefore, the probability of the error-rules of definition 2.1 must also be estimated.

The problem of estimating the probabilities of all error-rules, as defined in 2.1, requires some attention. First, since every $x \epsilon R_+$ can be parsed by G_c without the use of error-rules, there is no method for this estimation unless new training strings are utilized; and second, given that the number of error-rules of G' can be rather large ($\sim(1+2|V|)|P|$), a prohibitive amount of new training strings would be necessary to allow for a reasonable estimation reliability. Therefore, some approximation seems to be necessary.

A first step is to simplify the stochastic error-correcting scheme by assigning the same probability to every error-rule of each type (insertion, deletion, substitution), regardless of the non-error rules it is associated to. In this way, the number of probabilities to be estimated is reduced by a factor of $|P|/|V|$, resulting in a quite classical stochastic error-correcting scheme (Fu,82):

$$(3.3) \quad \begin{array}{ll} \text{Insertion} & p_I(a,b) \\ \text{Substitution} & p_S(a,b) \\ \text{Deletion} & p_D(b) \end{array} \left. \begin{array}{l} \\ \\ \end{array} \right\} \forall a \epsilon V \left. \begin{array}{l} \\ \\ \\ \end{array} \right\} \forall b \epsilon V$$

which corresponds to a total of $|V|+2|V|^2$ probabilities to be estimated.

Besides reducing the requirements in the number of training strings, this approximation allows the same strings of the original positive sample R_+ to be used for the estimation of the probabilities (3.3). To do so, a "rigid" (archetipical) characteristic grammar $\hat{G}_c$ can be inferred first from a small subset of "good" strings of R_+ (e.g. some centroid(s) with respect to the error-correcting (Levenshtein) distance), and the rest of the strings in R_+ can be utilized for estimating the probabilities (3.3) through the frequency of use of the corresponding type of error-rule in the parse of these strings.

This error-correcting scheme can be further simplified by assigning, for each type of error, exactly one probability to the symbol producing the error (like "b" in (3.3)), resulting in 3IVI probabilities to be estimated. Alternatively, one can go further and ignore absolutely what the symbols involved in the errors are, and assign the <u>same</u> probability to all the types of error-rules associated with each (non-error) rule of the grammar, thus giving importance to the "position" of the errors rather than to the symbols actually involved. In this case the number of probabilities to be learned would be IPI. An approach similar to the latter is followed in a method recently proposed for inference of Markov chains (Thomason,86).

The probability estimation procedure just outlined requires the training positive sample to be considered at least twice. Although this is very usual in stochastic GI (Fu,82), and can be considered convenient in many practical situations, such an approach would prevent us from achieving the goal required by a proper learning methodology; i.e., performing both inference and recognition in a simultaneous and integrated way. Nevertheless, if the (training) strings are assumed to be presented in (infinite) sequence following a <u>textual presentation model</u> in the sense discussed in (Feldmann,72), a probability estimation procedure based on the same principles stated above can be easily devised which can be carried out sequentially and in parallel with the inference of the characteristic grammar.

This procedure consists of recursively updating the frequency of use of each rule $r \in P$, and each non terminal $A \in N$, as each new (training) string x is presented:

$$(3.4) \quad p_t(r) = n_t(r) / m_t(A_r) \quad \forall r \in P \ (A_r \ N \text{ is the left-side non terminal of } r)$$

$$n_t(r) = n_{t-1}(r) + N_r(x); \quad n_{t_0}(r) = 1 \quad \forall r \in P$$

$$m_t(r) = m_{t-1}(r) + M_A(x); \quad m_{t_0}(A) = 1 \quad \forall A \in N$$

where t is the step at which x is being presented (x is the t-th. training sample), t_0 and t_0', respectively, are the steps at which the rule r and the non terminal A were created, and $N_r(x)$ and $M_A(x)$ are the number of times (0 or 1) that r and A, respectively, have been used in the parse of x.

On the other hand, some of the above indicated reduced sets of probabilities of error rules can also be estimated sequentially and simultaneously with the other inference and recognition activities. This can be done by extending the recurrence (3.4) with the corresponding updating of the frequency of use of the error rules in the error-correcting parsing of each new string. Interestingly eanough, these probabilities are expected to decrease progressively as the corresponding error-rules are being incorporated to the set of non-error rules of G, with a tendency towards zero as all the variability in the target language has been sufficiently represented through the sequence of strings presented to the stochastic ECGIA.

4.-EXPERIMENTAL RESULTS AND CONCLUSIONS.

The ability of the proposed GI method to model (sub)string-length-based constraints has been verified through some simulation experiments. The results have shown that it does effectively capture these constraints. Also, the ECGIA has been applied to the task of inferring the automata corresponding to different Automatic Speech Recognition applications. One of these applications (single-speaker isolated-word Spanish-digits recognition), was presented in (Rulot,86) in which very good results were reported. Using a very crude representation of speech, a 81% recognition rate was obtained if no durational information was made available to the ECGIA (deleting all successive repetitions of the same symbol) while, without deleting this information, a 98% rate was reached. These results show how the durational information is very relevant for the proposed task, and how the proposed

method does effectively take advantage of this fact. In
general, the results show the inferred automata to perform
better than those constructed by hand by experienced speech
researchers (Vidal,85).

These results are expected to be improved further by
applying the stochastic extension presented in this paper. Some
experiments on this issue, involving multispeaker tasks, are
now in progress in our laboratories, and formal results will be
reported when they be available. Another (hopefully) improving
extension would be to allow the recognition of "fuzzy-symbols"
as in (Vidal,85), as well as to achieve both inference and
recognition directly from a "vector-string" (as opposed to
symbol-string) representation of the samples. Finally, though
the inference procedure has been explicitly restricted to
generate circuit-free automata, it is quite easy to modify it,
if required, to allow single-state loops when (a large enough
number of) terminal repetitions are to be modelled.

5.- REFERENCES.

(DeMori,81) R. De Mori: "Computer Models of Speech using Fuzzy
    Algorithms", Plenum Press, New York, 1981.

(Feldman,72) J.Feldman: "Some decidability results on
    grammatical inference and complexity". Information and
    Control, Vol.20, pp. 244-262, 1972.

(Forney,73) G.D.Forney: "The Viterbi algorithm". IEEE Proc. 3,
    pp. 268-278. 1973.

(Fu,82) K.S.Fu: "Syntactic Pattern Recognition and
    Applications". Ed. Prentice-Hall. 1982.

(Fu,83) K.S.Fu: "A Step Towards Unification on Syntactic and
    Statistical Pattern Recognition". IEEE Trans. Pattern
    Analysis and Machine Intelligence. Vol.PAMI-5, N.2,
    pp.200-205. 1983.

(Hunt,75) E.B.Hunt: "Artificial Intelligence". Academic Press,
    New York. 1975.

(Itoga,81) S.Y.Itoga: "A New Heuristic for Inferring Regular
    Grammars". IEEE-Trans. on Pattern Analysis and Machine
    Intelligence, Vol. PAMI-3, N.2, 1981.

(Jelinek,76) F.Jelinek: "Continous Speech Recognition by
    Statistical Methods". Proc. IEEE, Vol. 64, N.4, pp.
    532-556, 1976.

(Juang,85) B.H.Juang, L.R.Rabiner, S.E.Levinson, M.M.Sondhi:
    "Recent developments in the application of Hidden Markov
    Models to Speaker-Independent Isolated Word Recognition".
    IEEE-ICASSP, pp. 9-12, 1985.

(Miclet,80) L.Miclet: "Regular Inference with a Tail-Clustering Method". IEEE Trans. on Syst. Man and Cybernetics, SMC-10, pp. 737-747, 1980.

(Richetin,84) M.Richetin, F.Vernadad: "Efficient Regular Grammatical Inference for Pattern Recognition". Patt. Recognition, Vol. 17, N.2, pp.245-250, 1984.

(Rulot,86) H.Rulot, E.Vidal: "Modelling (sub)string-length-based constraints through a Grammatical Inference method". NATO-ASI on Pattern Recognition and applications. Spa (Belgium). June, 1986.

(Rusell,85) M.J.Rusell, R.K.Moore: "Explicit modelling of state occupancy in Hidden Markov Models for automatic speech recognition". IEEE-ICASSP, pp. 5-8, 1985.

(Thomason,86) M.G.Thomason, E.Granum, R.E.Blake: "Experiments in dynamic programming inference of Markov networks with strings representing speech data". Pattern Recognition, Vol.19, N.5, pp. 343-352, 1986.

(Vernadat,84) F.Vernadat, M.Richetin: "Regular Inference for syntactic pattern recognition: a case study". IEEE-ICPR, pp. 1370-1372, 1984.

(Vidal,85) E.Vidal, F.Casacuberta, E.Sanchís, J.Benedí: "A General Fuzzy-Parsing Scheme for Speech Recognition". NATO-ASI New Systems and Architectures for Automatic Speech Recognition and Synthesis. R.DeMori, C.Y.Suen (eds.). Springer-Verlag. pp. 427-446, 1985.

VORONOI TREES AND CLUSTERING PROBLEMS

F. Dehne[1], H. Noltemeier[2]

Abstract:

This paper presents a new data structure called Voronoi tree to
support the solution of proximity problems in general pseudo
metric spaces with efficiently computable distance functions.
We analyse some structural properties and report experimental
results showing that Voronoi trees are a proper and very efficient tool for the representation of proximity properties and
generation of suitable clusterings.

1.Introduction

Cluster analysis is an important pattern recognition technique.
It may be characterized by the use of resemblance or dissemblance
measures between objects to be identified. The objective of a
cluster analysis is to uncover natural groupings, or types, of
objects. [DS], [DJ]

In this paper we are concerned with the problem of how to support
proximity and clustering problems in pseudo metric spaces.

A set E is called pseudo metric space if there is a distance
function $d:E^2->R_+$ with the following properties (for any
e,e',e"∈E):

   (1) d(e,e)=0

   (2) d(e,e')=d(e',e)   (symmetry)

   (3) d(e,e")≤d(e,e')+d(e',e")   (triangle inequality)

However, it may happen in pseudo metric spaces that there are two
different elements e,e'∈E with d(e,e')=0.

Most of the literature on nearest neighbor problems is dealing
with finite dimensional real spaces with some L_p norm - very few

---

[1] School of Computer Science, Carleton Univ., Ottawa, Canada K1S5B6
[2] Lehrstuhl für Informatik I, Univ. of Würzburg, Würzburg, W.-Germany

NATO ASI Series, Vol. F45
Syntactic and Structural Pattern Recognition
Edited by G. Ferraté et al.
© Springer-Verlag Berlin Heidelberg 1988

are considering more general spaces.

One of these few more general approaches is the recent work of I.Kalantari and G.McDonald [KMcD]. The data structure they propose is a straight forward generaliztion of binary search trees (which can support nearest neighbor search in E^1) and is applible to any normed space if the norm is computable effectively. The next section will briefly summarize their approach but will also point out some disadvantages.

In section 3 we propose a slightly different structure (which we call 'Voronoi tree') which, however, will show significant structural advantages.

Some experimental results will be stated in section 4. Due to their structural advantages it turned out that Voronoi trees are a proper and very efficient tool for the representation of proximity properties and solution of clustering problems.

2. Bisector Trees

Let E be an arbitrary space, $d:E^2 \to R_+$ a mapping which induces a metric on E, and $S=\{e_1,...,e_n\}$ a finite point set in E. [KMcD] represents S using a binary tree (called 'bisector tree' or 'bs-tree' for the remaining of this paper) as follows:

  (α) each node of the actual tree T_i (representing the actual set $S_i := \{e_1,...,e_i\}$, $0 \le i \le n$) contains at least one (p_L) and at most two elements (p_L and p_R) of S_i.

  (β) to insert a new element e_{i+1} into T_i (yielding T_{i+1}) start at the root node of T_i as follows:

  - if the actual node v contains only one element then insert e_{i+1} into v

  - if (otherwise) v contains two elements p_L and p_R then recursively
    insert e_{i+1} into the left subtree (which has root p_L) if $d(e_{i+1},p_L)<d(e_{i+1},p_R)$
    or into the right subtree (which has root p_R) if $d(e_{i+1},p_R)<d(e_{i+1},p_L)$
    and arbitratrely if the distances are equal.

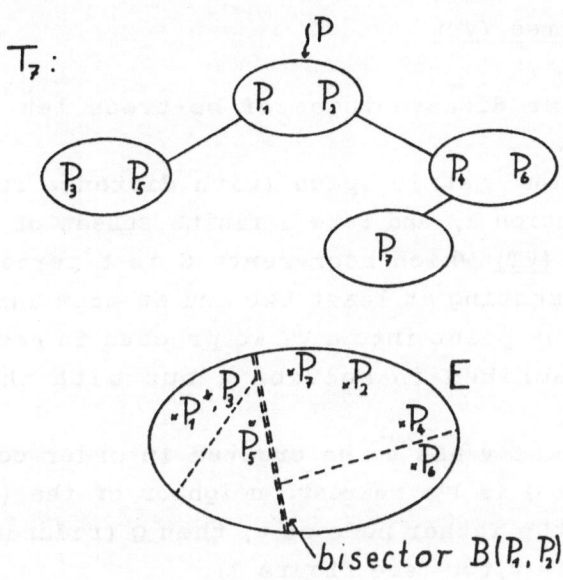

T_7 :

bisector $B(P_1, P_3)$

Figure 1

To support nearest neighbor queries each node v of the bs-tree T storing two elements p_L, p_R with left subtree T_L and right subtree T_R additionaly stores the following two values:

LRADIUS(v) := max{d(p, p_L)/p is stored in T_L}   (left radius of v)

RRADIUS(v) := max{d(p, p_R)/p is stored in T_R} (right radius of v).

Thus, given a query point P∈E and a bs-tree representing a set S⊆E in order to search for the nearest neighbor of p in S, the search process can prune the left [right] subtree T_L [T_R] of an actual node v if

   (*)   d(p, p_L) - LRADIUS(v) $\geq$ DACTUAL

      [d(p, p_R) - RRADIUS(v) $\geq$ DACTUAL]

holds with DACTUAL denoting the distance of p to its actual nearest neighbor in S (for more datails see [KMcD]).

However, we have to state the following remarks:

(R1) A son may have larger radia than his father ('excentric son').

(R2) There are cases in which a bs-tree has to be searched exhaustively even if we consider only balanced bs-trees (having logarithmic height).

(It is easy to construct examples for such cases even in E^2 and is left to the reader.)

3. The Voronoi tree (VT)

To overcome some disadvantages of bs-trees let us introduce
'Voronoi trees'.

Let E be any pseudo metric space (with distance function d) as
described in section 1, and S be a finite subset of E.

A Voronoi tree (VT) which represents S is a ternary tree with
each node representing at least two and at most three points of
S. To insert a new point into a VT we proceed in essentially the
same way as decribed in section 2 but with the following
difference :

(**) If a new leaf v has to be created in order to store a new
point P and Q is P's nearest neighbor of the (three) points
stored in the father node of v, then Q (redundantly) has to
be stored in v,too (see figure 2).

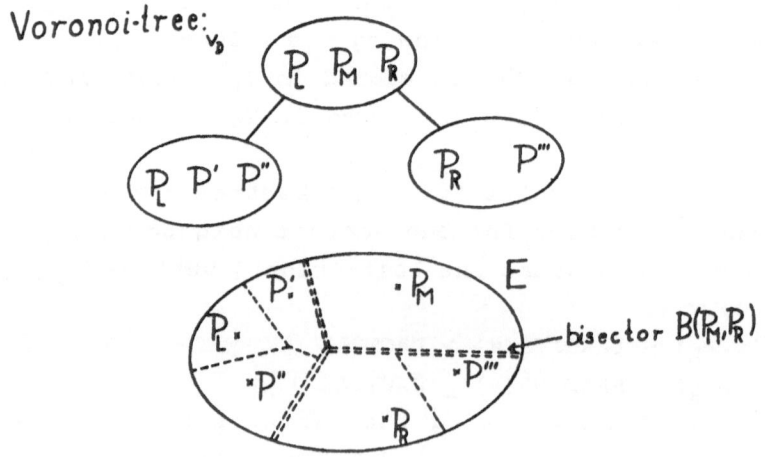

Figure 2

With this we can prove the following

Theorem:

1. Voronoi trees do not have excentric sons.
2. A Voronoi tree is transitive in the following sense:
   $d(P, father(P)) = min\{ d(P,P')/ P' \in ancestors(P) \}$ (with
   ancestors(P) denoting the set of all elements stored in those
   nodes which are on the path from the root to P).

3. If we search for nearest neighbors of points which are stored in the root node, then this search is possible in time O(h) with h denoting the height of the tree.

Proof: Part 1 follows immediatly from the definition of Voronoi trees where we have to distinguish LRADIUS(v), MRADIUS(v), and RRADIUS(v), respectively. Part 2 and 3 are a consequence of (**).

Remark: The term 'Voronoi tree' is chosen due to the fact that all triples of bisectors exactly determine the vertices of all order k Voronoi diagrams (cf. [D1], [SH]).

4. Experimental Results

In this section we will present experimental results to demonstrate that VT-trees can be efficiently used to represent proximity properties in pseudo metric spaces (in general no linear spaces) and specially that VT-trees are an efficient tool for the solution of clustering problems.

Our test data were taken from two very different areas:

1) Randomly generated points in the unit cube of the d-dimensional real space R^d where the metric can be chosen from arbitrary L_p norm (f.e. d=2,3,4,5 and p=1,2,∞)

2) A set of 15,000 elements and three different pseudo metrices which were induced by a given Gozinto graph (see [DN3], [N]) alternatively gave the basis for our field experiments.

In each case the Voronoi tree was constructed (with a wide range of different parameters and number of points in the first environment) and the computation time, storage requirement, height of the VT-tree, scattering of the heights of the leafs, etc. was analysed and compared with experimental results from bs-trees and ternary bs-trees (bs-trees with at most 3 points stored in each node but without condition (**), k=3 in [KMcD]).

Additionally, the VT could be modified by choosing some rule how to place a copy of a father P' in his son node with respect to condition (**), i.e. fixed place, cyclic placement during the construction, randomly chosen, or "at the same place as in the father node" (hereditary property).

We furthermore studied how the order in which the points are inserted into the VT-tree influences the final result, especially some (dynamic) selforganizing principles.

The following figures illustrate a small but typical sample from our test series.

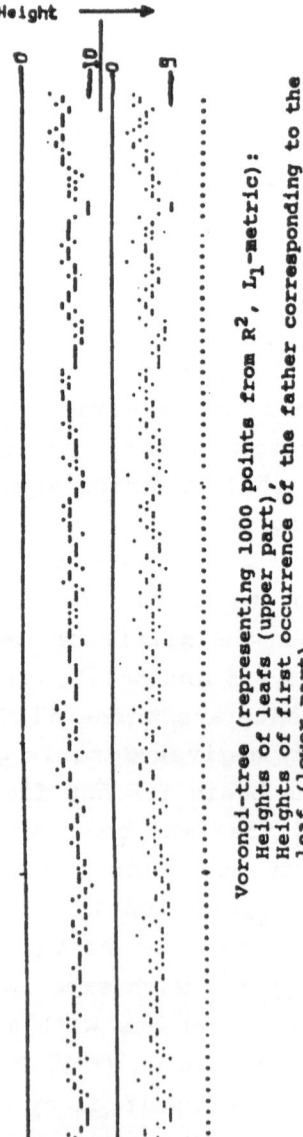

Voronoi-tree (representing 1000 points from R^2, L_1-metric):
Heights of leafs (upper part),
Heights of first occurrence of the father corresponding to the leaf (lower part)

Figure 3

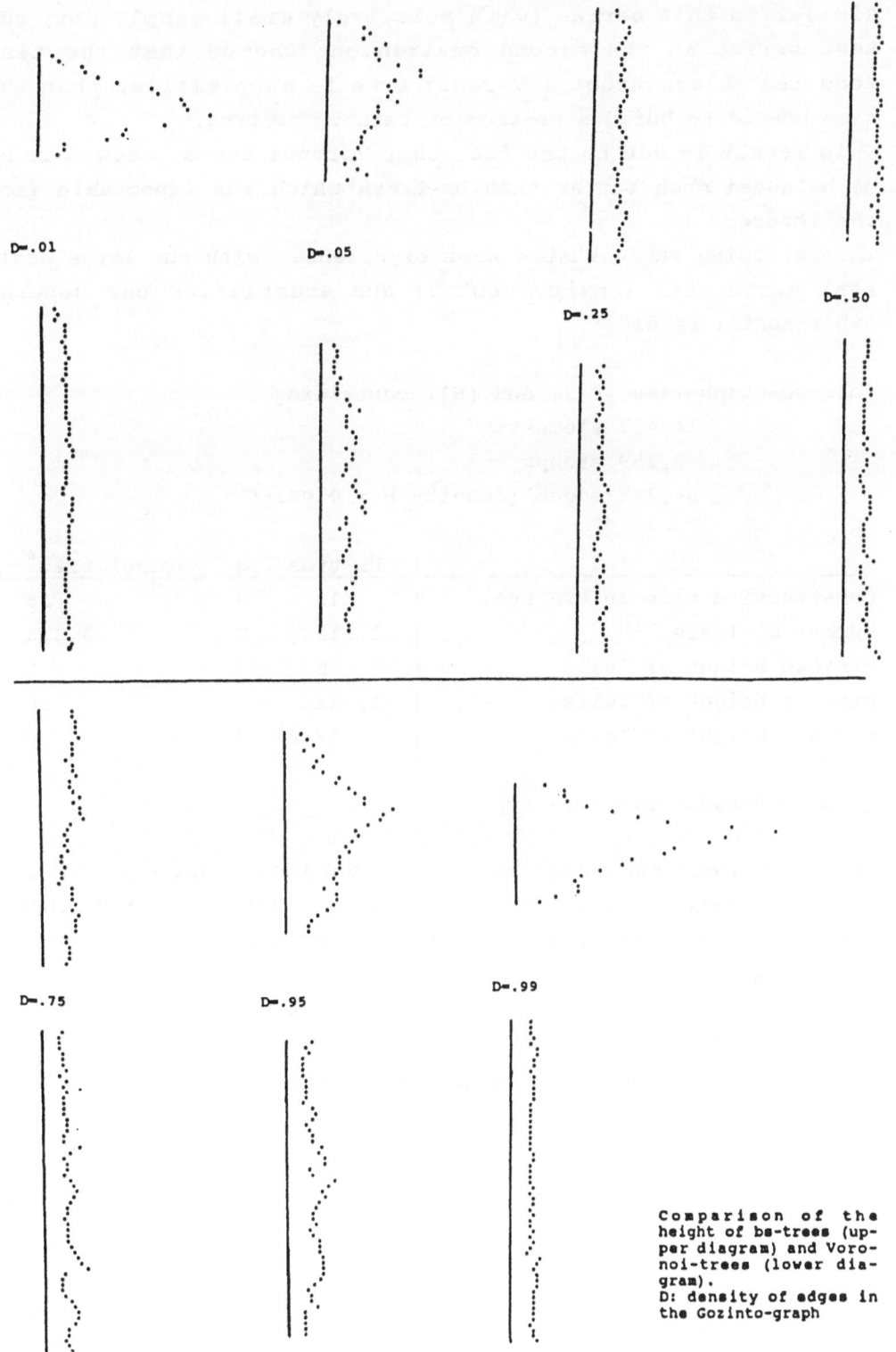

Comparison of the height of bs-trees (upper diagram) and Voronoi-trees (lower diagram).
D: density of edges in the Gozinto-graph

Figure 4

Similar to this series (with relatively small sample set) our
test series in the second environment showed that the time
required to construct a Voronoi tree is much smaller than the
time needed to build a bs-tree or ternary bs-tree.

This result is due to the fact that Voronoi trees showed to be
be balanced much better than bs-trees which was expectable from
the theorem.

The following table states some experiences with our large scale
real world data (environment 2) and examplifies our general
experimental results:

Gozinto-Graph (see [DN3] and [N]) containing

          14,457 elements
           2,259 groups
          54,735 edges (density D = 0.00168)

	bs-tree	Voronoi tree[*]
Construction time in CPU sec.	513	35
Number of leafs	1,512	4,273
Minimum height of leafs	5	5
Maximum height of leafs	1,211	16
Average height of leafs	427.27	10.24

([*] with hereditary property)

Applied to our test data the use of Voronoi trees did not only
reduce computation time dramatically, but did also induce
feasible clustrings of large point sets in pseudo metric spaces.

It showed that subtrees of a Voronoi tree represent subsets of
points with certain proximity properties (with resp. to the
chosen quasi metric of course).

How good these proximity properties are represented by Voronoi
trees is exemplified in figure 5.

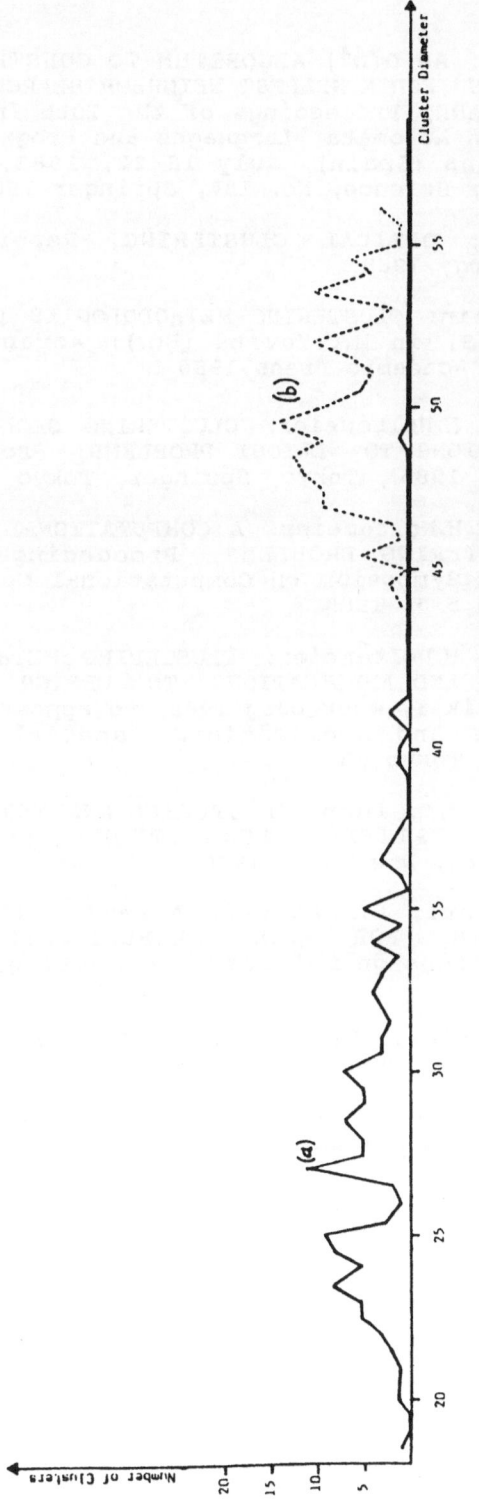

Distribution of the diameters of clusters induced by Voronol-
trees (a) and randomly generated (b).
Sample set: 5000 points in E^5, Cluster size between 30 and 50

Figure 5

References

[D1]   F.Dehne: AN $O(n^4)$ ALGORITHM TO CONSTRUCT ALL VORONOI
       DIAGRAMS FOR K NEAREST NEIGHBOR SEARCHING IN THE EUCLI-
       DEAN PLANE, Proceedings of the 10th International Collo-
       quium on Automata, Languages and Programming (ICALP '83),
       Barcelona (Spain), July 18-22, 1983, Lecture Notes in
       Computer Science, No. 154, Springer 1983

[D2]   F.Dehne: OPTICAL CLUSTERING, Report, Informatik I,
       Wuerzburg, 1985

[DJ]   Dubes,Jain: CLUSTERING METHODOLOGIES IN EXPLORATORY DATA
       ANALYSIS, in M.C.Yovits (Ed.): Advances in Computers,
       Vol.19, Academic Press,1980

[DN1] F.Dehne, H.Noltemeier: CLUSTERING GEOMETRIC OBJECTS AND
       APPLICATIONS TO LAYOUT PROBLEMS, Proceedings 'Computer
       Graphics 1985', Tokyo, Springer, Tokyo 1985

[DN2] F.Dehne, H.Noltemeier: A COMPUTATIONAL GEOMETRY APPROACH
       TO CLUSTERING PROBLEMS, Proceedings of the 1st ACM
       SIGGRAPH Symposium on Computational Geometry, Baltimore,
       MD, June 5-7, 1985

[DN3] F.Dehne, H.Noltemeier: CLUSTERING METHODS FOR GEOMETRIC
       OBJECTS AND APPLICATIONS TO DESIGN PROBLEMS, Report,
       Informatik I, Wuerzburg 1985, to appear in: IEEE Computer
       Graphics and Applications, special issue: Computer
       Graphics Tokyo 85

[DS]   E.Diday, J.C.Simon, CLUSTERING ANALYSIS, in K.S.Fu (ed.),
       DIGITAL PATTERN RECOGNITION, Springer, Berlin,
       Heidelberg, New York, 1980

[KMcD] I.Kalantari, G.McDonald: A DATA STRUCTURE AND AN
       ALGORITHM FOR THE NEAREST POINT PROBLEM, IEEE
       Transactions on Software Engineering, Vol. SE-9, No.5,
       Sept.83

[N]   H.Noltemeier: DISTANCES IN HYPERGRAPHS, Report,
       Informatik I, Wuerzburg 1985

[SH]   Shamos, Hoey: CLOSEST POINT PROBLEMS, Proc. 16th Ann.
       IEEE Symp. on Found. of Comp. Sci., 1975

V.   IMAGE UNDERSTANDING

HOUGH-SPACE DECOMPOSITION FOR POLYHEDRAL SCENE ANALYSIS

H.-P. Biland, F.M. Wahl

Communications and Computer Science Dept.
IBM Zurich Research Laboratory
8803 Rüschlikon
Switzerland

Abstract

This paper describes a new method to recognize 3d polyhedral objects on the basis of single 2d images. The recognition is performed by transforming the captured image into the Hough space and a subsequent interpretation in this domain. The new technique is capable of splitting up more complex scenes into their components and of decomposing these into object primitives stored in an extendable object library. The method is exceptionally robust with respect to noise and the data structures of the objects in Hough space are very compact and easy to handle.

Introduction

The discipline of machine vision is concerned with the automated understanding of complex real-world scenes by computers. An intermediate step towards this end is the recognition of objects based on some kind of a priori knowledge in constraint environments (e.g., industrial applications). Figure 1 illustrates the machine vision principle that we are following. A 3d physical scene containing an object is captured by a sensor. The captured image is processed by a computer to extract image features of the 2d representation of the 3d scene such as object edges or vertices. On the other hand, abstract object models can be represented as CAD data structures. A polyhedron, for example, can be modeled in 3d coordinate space by coordinate triples of its vertices and their associated spatial interrelations. By applying

---

Based on "Decomposition of Polyhedral Scenes in Hough Space" by F.M. Wahl and H.P. Biland appearing in the Proceedings of the 8th International Conference on Pattern Recognition, Paris, France, October 1986, Vol. 1, pp. 78-84.

NATO ASI Series, Vol. F45
Syntactic and Structural Pattern Recognition
Edited by G. Ferraté et al.
Springer-Verlag Berlin Heidelberg 1988

3d coordinate transformations such as translation, rotation, skewing, scaling and a subsequent projection onto a plane, a 2d computer image can be generated. This 2d representation can be used to fit the object model with the image features extracted from the real scene image. The matching process involves a topological and a geometrical match. The topological match decides what type of object is present, while the geometrical match delivers the transformation parameters which have to be applied to the object model to exactly fit the 3d scene.

With this in mind we can state the problem of identifying objects as follows: Given a real image and given some object models, find the model which under

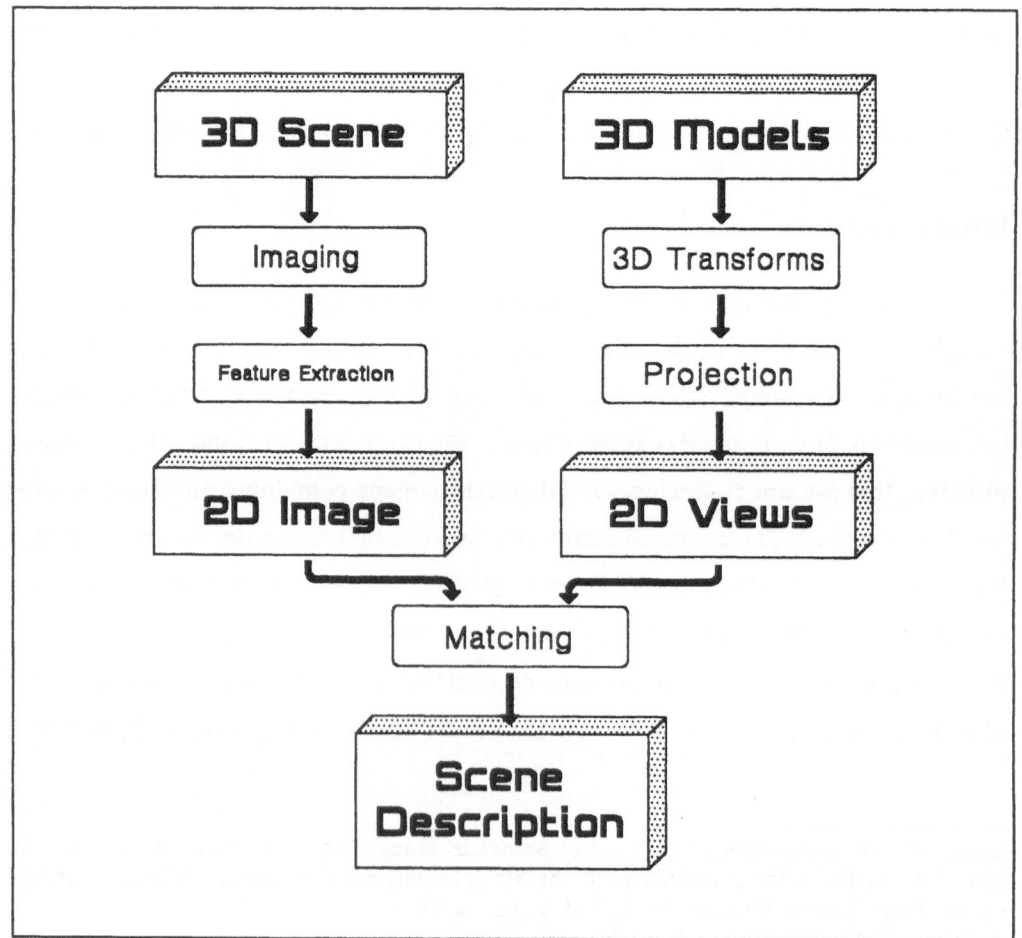

Figure 1. Machine Vision Principle. © 1986 IEEE.

some coordinate transforms (with parameters to be determined) best fits the real image. For polyhedral objects, it is advantageous, with respect to computational cost, to use vertex points extracted from the scene image and to relate these to object model vertices. There are numerous published articles on how to extract edges and vertices in digital images to produce line drawings. Most of them incorporate interactively or heuristically chosen parameters which may vary from scene to scene. Often the proposed methods fail owing to random grey-level variations (noise) in digitized images. On the other hand, feature space transforms like the Hough transform are robust with respect to noise, and do not need calibrated parameter settings.

The Hough Transform and Some of its Properties

As is well-known, the Hough transform is a noise-insensitive method for detecting colinear points in images, e.g., straight lines ([HOUG62], [DUDA72]). Using one of several possible parametrizations, a line can be denoted by the equation $y = ax + b$ with the two parameters a,b for slope and intercept, respectively. Resolving this equation with respect to b yields $b = -xa + y$. This means, that an image point with coordinate values (x,y) is mapped into a line in a,b -space, denoted as Hough space. The Hough transform accumulates lines in an accumulator array with slopes and intercepts corresponding to the x,y-coordinate values of individual pixels in image space. The very useful property is that colinear image points correspond to lines in Hough space which intersect at exactly one distinct location, called cluster. In return, the position and orientation of a line in image space is determined by its corresponding cluster location in Hough space. To avoid problems with nearly vertical lines and hence nearly infinite slopes, we compute the Hough transform twice as proposed in [ROSE69]: first it is applied directly to the image to be transformed; the second time it is applied to the version of the image rotated by 90°. This yields the so-called *twin Hough-space* representation of the image. The a axis is scaled from -1 to $+1$, or from $-45°$ to $+45°$ respectively, while the a' axis is scaled from $+1$ to -1, or from $+45°$ to $+135°$, respectively. Figure 2 shows a processing example. A simple edge detector (Sobel) and a subsequent thresholding operation has been applied to a digital image containing a cube. The resulting binary gradient image is the input to the Hough transform. The resulting two accumulator arrays of the Hough transform are shown in the second row of Figure 2. Although the input of the Hough transform is noisy and some lines include gaps, three distinct clusters can be seen in the first accumulator array (left) and six of them in the second (right). They can be easily detected as shown in in the third row of Figure 2. The inverse Hough transform, which is identical to the forward Hough transform, applied to the extracted clusters results in a representation of the original image shown at the bottom of Figure 2. Evidently, the Hough-space representation on its own does not deliver object vertices required for vertex matching in object recognition. We are now ready to understand and appreciate the central new idea of this paper: The interpretation of cluster patterns in Hough space as vertices of 3d objects. Before explaining the new interpretation technique, some important properties of clusters in Hough space are described.

Properties of Clusters in Hough Space

The clusters in Hough space are not at all a set of unrelated points:

- Clusters corresponding to lines with same slope in image space (parallel lines) are aligned vertically in Hough space.
- Clusters corresponding to lines with the same intercept in image space are aligned horizontally in Hough space.
- n lines intersecting each other at one distinct point in image space (vertex) correspond to a colinear arrangement of n clusters in Hough space.
- Two or more cluster colinearities intersecting each other at one distinct cluster location in Hough space correspond to vertices sharing one common edge in image space.

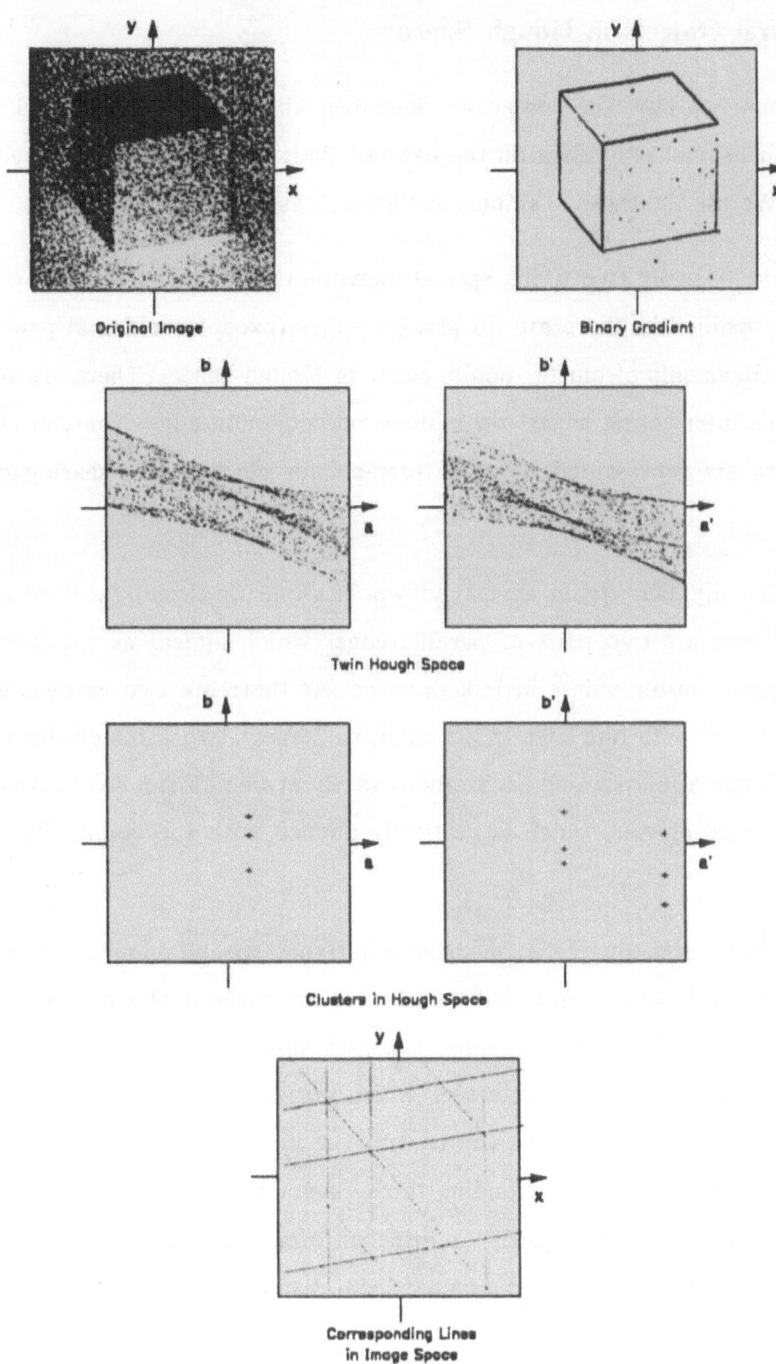

Figure 2. Example of a Hough Transform and its Inverse.

Polyhedral Objects in Hough Space

Let us now see how the properties described above can be employed to describe three-dimensional polyhedra on the basis of their two-dimensional image representations. We use the three examples in Figure 3 as illustration.

Tetrahedron: Usually two (from special viewpoints one or three) faces of a tetrahedron are visible. As there are no parallel edges (except accidental parallelism), no vertical alignments of cluster points occur in Hough space. There are two vertices with three intersecting edges; these share one common edge. Thus, in Hough space tetrahedra are represented by two three-cluster colinearities sharing exactly one cluster.

Prism: Usually two (from special viewpoints one or three) faces of a prism are visible. There are two pairs of parallel edges which appear as two pairs of vertically aligned cluster points in Hough space. As there are two vertices where three edges intersect with one edge in common, we expect two three-cluster colinearities in Hough space intersecting each other exactly at one cluster location and an additional isolated cluster, which is vertically aligned with the colinearity intersection point.

Parallelepiped: Usually three quadrilateral (from special viewpoints one or two) faces of a parallelepiped are visible. As there are three triples of parallel edges, we expect three vertical cluster alignments with three clusters each. There are four vertices where three edges intersect which correspond to four three-cluster colinearities in Hough space. As the central (or inner) vertex point shares its edges with three others, the corresponding three-cluster colinearities intersect the central three-cluster colinearity exactly at three different cluster locations. The resulting structure recalls a z-shape with an additional diagonal crossing it.

The examples demonstrate, how distinct vertex points of polyhedral objects can be identified easily in Hough space (see [WAHL85]) and thus can be related to model vertices which is the prerequisite for object matching by means of geometric vertex fitting as described, e.g., in [ROBE65]. We shall show next, how the polyhedral objects of Figure 3 can be used to decompose complex polyhedral scenes and objects into object primitives.

Prerequisites

There are usually several ways to decompose a polyhedron into its constituting parts. The results obtained depend heavily on the *decomposition strategy*. Before we give a formulation for such a decomposition strategy, some prerequisites have to be discussed.

We distinguish between the *principal* views $P_1^k \ldots P_{i_k}^k$ of an object k and the *degenerate* views $D_1^k \ldots D_{j_k}^k$ with $k = 1 \ldots n$ (n is the number of different object primi-

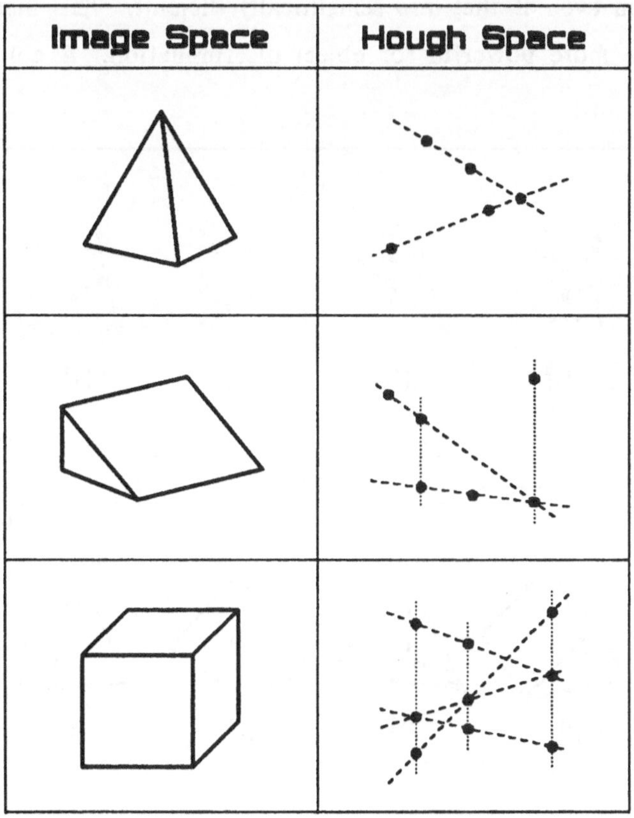

Image Space	Hough Space

Figure 3. Polyhedral Objects in Image Space and Hough Space. © 1986 IEEE.

tives). A view is a principal one when there is an $\varepsilon > 0$ such that for all parameter changes (rotation angles, translation etc.) within $\pm \varepsilon$ the view remains qualitatively unchanged, i.e., no vertex nor edge changes its visibility. Otherwise, we call it a degenerate view. In other words, the colinearities of a principal view change continuously in an ε-hypercube in transformation parameter space (n-dimensional cube with sidelength 2ε). The clusters in Hough space together with their colinearities are called a *Hough net*. The Hough nets of the principal views of object primitives are called *Hough Net Primitives (HNPs)*, denoting the HNP of view P_j^i by H_j^i.

When analyzing the Hough net of a scene we distinguish between a *v-colinearity* (vertical alignment in Hough Space) and a *g-colinearity* (general nonvertical alignment). As several objects are usually located on the same supporting planar surface, they are likely to have their vertical edges in parallel, i.e., they share the vertical direction even if they are semantically disjoint. This means that g-colinearities are the more powerful for object discrimination. A colinearity with *m*

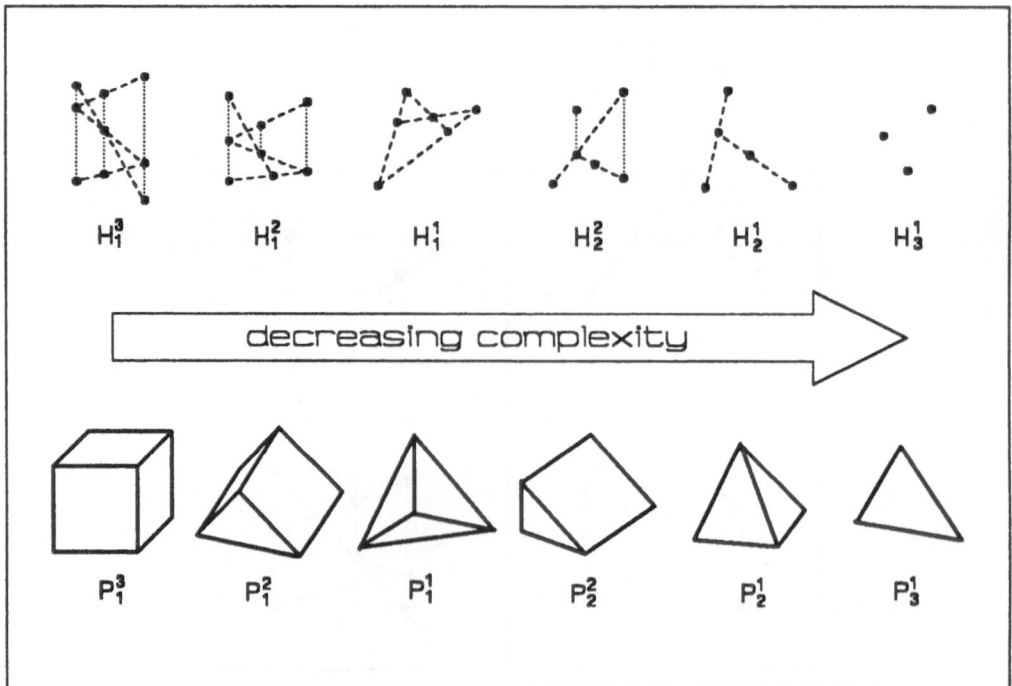

Figure 4. Complexity Ordering of Principal Views. © 1986 IEEE.

clusters is called *g-m-colinearity* or *v-m-colinearity* respectively. The number of clusters constituting a colinearity is called the *order* of the colinearity; g-colinearities must have at least order three whereas v-colinearities of order two are already meaningful.

For decomposition it is necessary that the HNPs of all object primitives have to be ordered by decreasing complexity. The primary sorting key for this is the highest order of colinearities $o_{max}(H_j^i)$ given by the HNP. The second criterion is the number of colinearities of this order $c(H_j^i, o_{max})$. If these numbers are the same for two HNPs then the next lower order will be examined. The third key is the total number of colinearities in the HNP $c_{tot}(H_j^i)$. Applying these ranking criteria to the principal views of the tetrahedron, prism, and parallelepiped we get the complexity ordering illustrated in Figure 4. The HNPs shown actually will be used as object-model primitives for decomposition.

Each cluster point $\pi_1 \dots \pi_i$ in a Hough net has several attributes stored with it:

- its center coordinates *a,b* in Hough space
- a list of g- and v-colinearities π_k is member of
- a flag indicating whether π_k is a genuine (i.e., found in the original image) or cluster point augmented later
- a real number γ ("coverage" number, $0 \leq \gamma \leq 1,$) giving a weight with which π_k has to be covered by some HNP in the ongoing matching process
- a list of object labels (typically one or two) of which π_k is a part

The New Technique

Figure 5 gives an overview of the decomposition system. The list of clusters found in the Hough accumulator array is fed into a *feature extractor* which computes cluster colinearities, and builds up intersection tables. This results in the

Hough-net representation of the scene. The *decomposition control program* analyzes and modifies the Hough net by invoking several *expert modules*, one for each principal view of the object primitives in question. The output of the decomposition control program is a list of image vertices related to object primitive (model) vertices.

For the decomposition strategy the following design criteria have been taken into account:

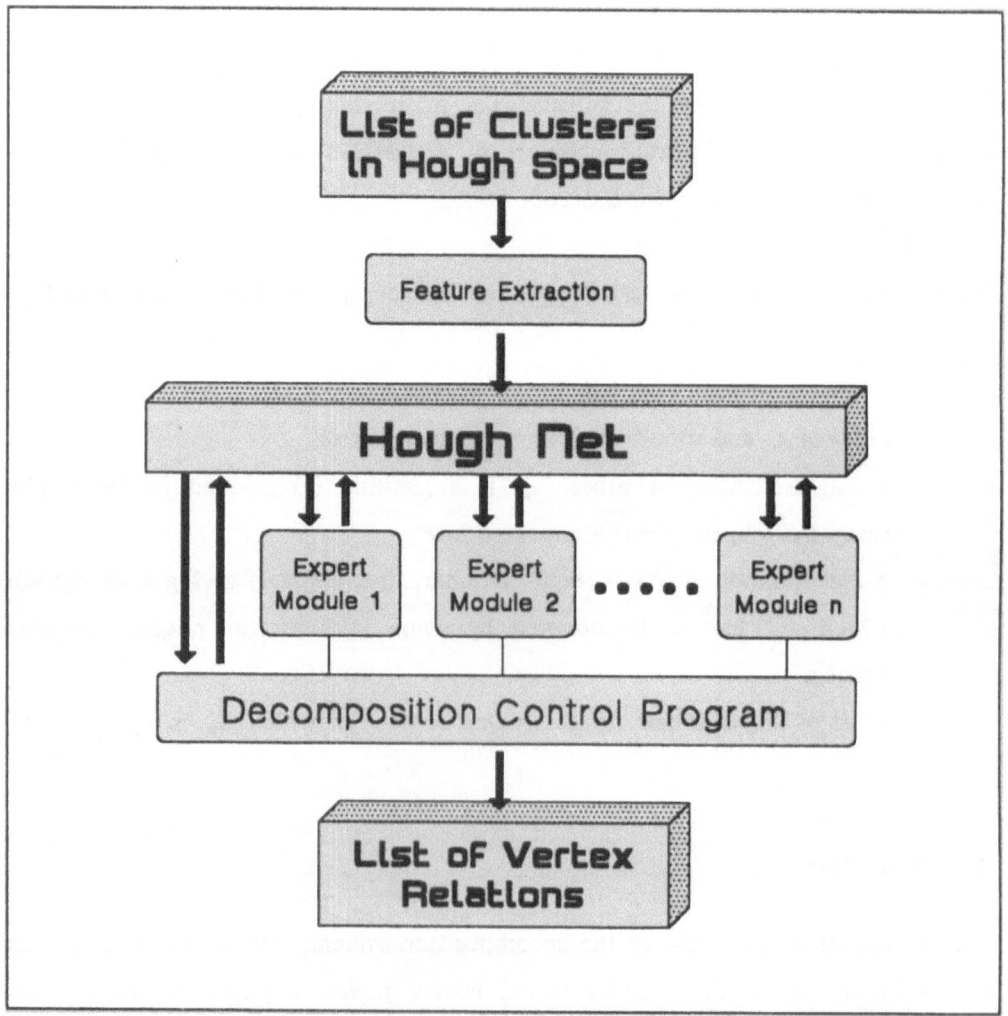

Figure 5. Decomposition Process.   © 1986 IEEE.

- robustness with respect to missing clusters (e.g., very short edges)
- robustness with respect to superfluous clusters (e.g., caused by shadows)
- decomposition with as few and as complex as possible object primitives
- handling of parallel and perspective projection
- modest computational time consumption

The decomposition procedure comprises the following steps:

1. Split up the Hough net into smaller subnets
2. For each of these subnets:
   a. Call the expert modules $E_1 \ldots E_m$
   b. Decide, basing on the experts' results, which match to choose
   c. Modify and label the Hough net accordingly
   d. Repeat steps a. to c. until all genuine cluster points are labeled
   e. Verify the solution in image space by comparing vertices defined by the HNPs with image vertices; if unsuccessful repeat steps a. to d.

Once the decomposition of the scene into object primitives has been completed, the following *post*-processing steps are executed to yield a scene description (cf. [ROBE65]):

1. Relate the vertices of each object primitive to the vertices found in the image
2. Calculate the transformation matrix for each matched object
3. Calculate the transformation parameters for each matched object

It is beyond the scope of this paper to describe the decomposition steps in detail (an extended description can be found in [BILA86]). Some more qualitative explanations of individual, processing steps mentioned above are given below, and will next be illustrated with a decomposition example.

Splitting Up the Hough Net: The Hough net has to be separated into subnets corresponding to semantically (and physically) disjoint objects. A labeling process groups clusters together which are interconnected by g-colinearities. If clusters remain unlabeled (isolated clusters), v-colinearities will be considered for assigning these to related groups. If the isolated cluster aligns vertically only with clusters of one single group, it will be assigned to this group. If it vertically aligns with clusters from several groups, we shall assign it temporarily to any of these. Finally, if the isolated cluster does not vertically align with clusters at all, it will get an own group label.

Decide how to proceed: We want to avoid fitting too simple HNPs instead of augmenting the Hough net with additional clusters and then fitting a more complex HNP. For example H_3^1 always fits perfectly well any Hough net with three cluster points or more. But in a larger Hough net, we certainly do not want to match some three cluster points to a tetrahedral face. On the other hand, if *only three* unmatched cluster points remain, we really might want them to be matched this way. So we have to minimize the number of augmented clusters and the "uncoveredness," i.e., the sum of weights $\Sigma \gamma_i$ of clusters remaining uncovered after modification of the Hough net. The weigthed sum of both criteria is called target function.

Modifying and Labeling the Hough Net: Whenever a match occurs, the Hough net has to be modified. First, clusters corresponding to hidden lines of the matching object primitive have to be *added* in Hough space. Clusters corresponding to edges totally disjoint with hidden edges have to be *removed* from Hough space. When these modifications have been completed, the Hough net has to be re-evaluated. Cluster points that have been augmented (visible lines of matching object primitives) get the highest coverage number γ, clusters that have been added (corresponding to formerly hidden lines) are assigned a lower γ and the rest gets the lowest γ.

Search Tree Depth: The depth of the search tree has to be limited to an integer number. Its value depends on the complexity of the scene and its objects and should be at least equal to the highest *expected number of object primitives* of which an individual scene is composed. Without such a limiting factor, solutions would be likely with one object decomposed into one complex plus numerous simple object primitives instead of a decomposition into, maybe, two medium complex primitives.

Verify the solution: When a match has been established, it is still possible that the matched object primitive cannot be found in image space. Such a false solution stems from the fact that in the Hough-transform representation, each line segment is extended to virtual infinite length (cf. backtransformed picture in Figure 2). If several such lines accidentally intersect at one location not corresponding to an image vertex, false interpretations of the Hough space might be possible. To eliminate incorrect solutions, the *image space* has to be analyzed within local windows at positions given by the vertex locations of object primitives found by the decomposition process.

The Expert Modules: For each principal view of an object primitive, a procedure for the corresponding Hough-subnet matching has to be provided. The procedural descriptions for finding the HNPs in the Hough net of the scene are called *Expert Modules* $E_1 \ldots E_m$. When an expert module E_k is initiated by the decomposition control program, it tries to fit its own HNP with the Hough net. If no match can be established, the expert module will allow one cluster point to be augmented in the Hough net of the scene; if this is unsuccessful, a second cluster point may be augmented to fit the HNP, and so on. Eventually, this *must* lead to a match. The expert then report to the decomposition control program the augmentation number and the number of cluster points remaining uncovered together with their coverage numbers γ .

A Decomposition Example

Figure 6 shows an example scene of modest complexity. It consists of two solid objects, one partially occluded by the other. One of the objects is already a primitive, the other is not, and hence will have to be decomposed. For didactical reasons, a rotational offset and scaling have been applied to the original image to make all clusters appear in the Hough-space window. Furthermore, the clusters are now numbered allowing them to be related to corresponding edges in image space. The short line segments below the Hough space indicate the slope of the corresponding edges in image space. The object primitive library is assumed to contain a tetrahedron, a prism, and a cube. Thus the six HNPs shown in Figure 4 are necessary for decomposition.

The initial scene and its corresponding Hough net is shown in Figure 7 (a) and (b). The splitting process applied to (b) yields two disjoint subnets, shown in (d) and (f). Applying the experts to Hough net (d) a tetrahedron is yielded (compare (d) with H_2^1 in Figure 4). No further decomposition is necessary nor possible for subnet (d). The composite object (e) is up to decomposition next.

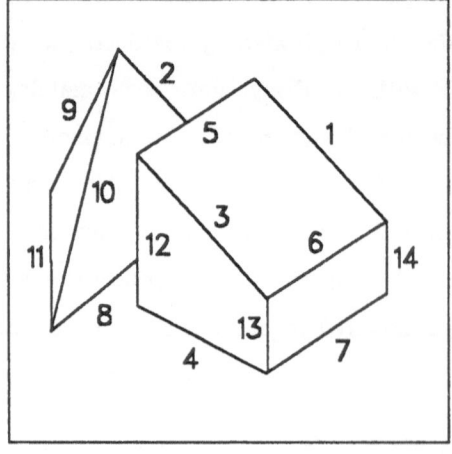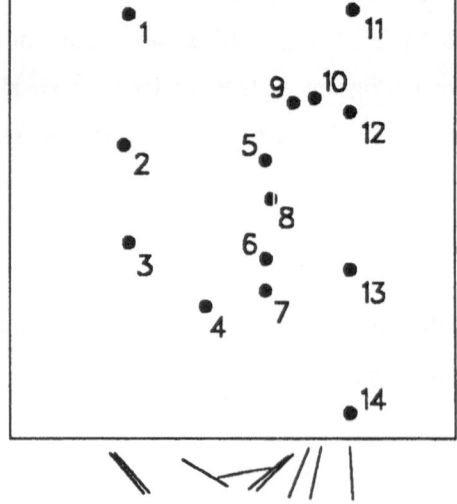

Figure 6. Example Scene. © 1986 IEEE.

Consulting the experts for (f) yields a minimum target function for the parallelepiped shown in (i) of Figure 8. H_1^2 covers all but one cluster (j), requiring only one cluster augmentation marked with a star in (h). The augmented cluster corresponds to the bold line in image space (g). The Hough net has now to be modified. In (l), the stars mark the clusters newly introduced, corresponding to the hidden lines of the parallelepiped. These lines are drawn with dashed lines in (k). The clusters in (l) indicated by arrows have to be removed. They correspond to the bold lines in (k).

The Hough net in (n) of Figure 9 illustrates the new situation. The size of the bullets indicates the coverage number values γ. (m) shows the corresponding situation in image space. Bold, dotted, and dashed lines indicate coverage numbers with high, medium, and low values, respectively. Applying the experts to the intermediate decomposition result (n), two alternative solutions H_2^2 with identical target function values are yielded, shown in (o) and (q). However, only (p) corresponding to (o) passes the verification stage; (q) has to be rejected, as the lower right vertex has no corresponding vertex in image space. The decomposition process stops here.

The final result of the decomposition are three identified object primitives with vertex coordinates given by the intercepts and slopes of corresponding cluster colinearities in Hough space. The vertex coordinates can then be used to calculate the transformation matrix, as well as the transformation parameters according to [ROBE65]. A complete scene description is the result of the decomposition process.

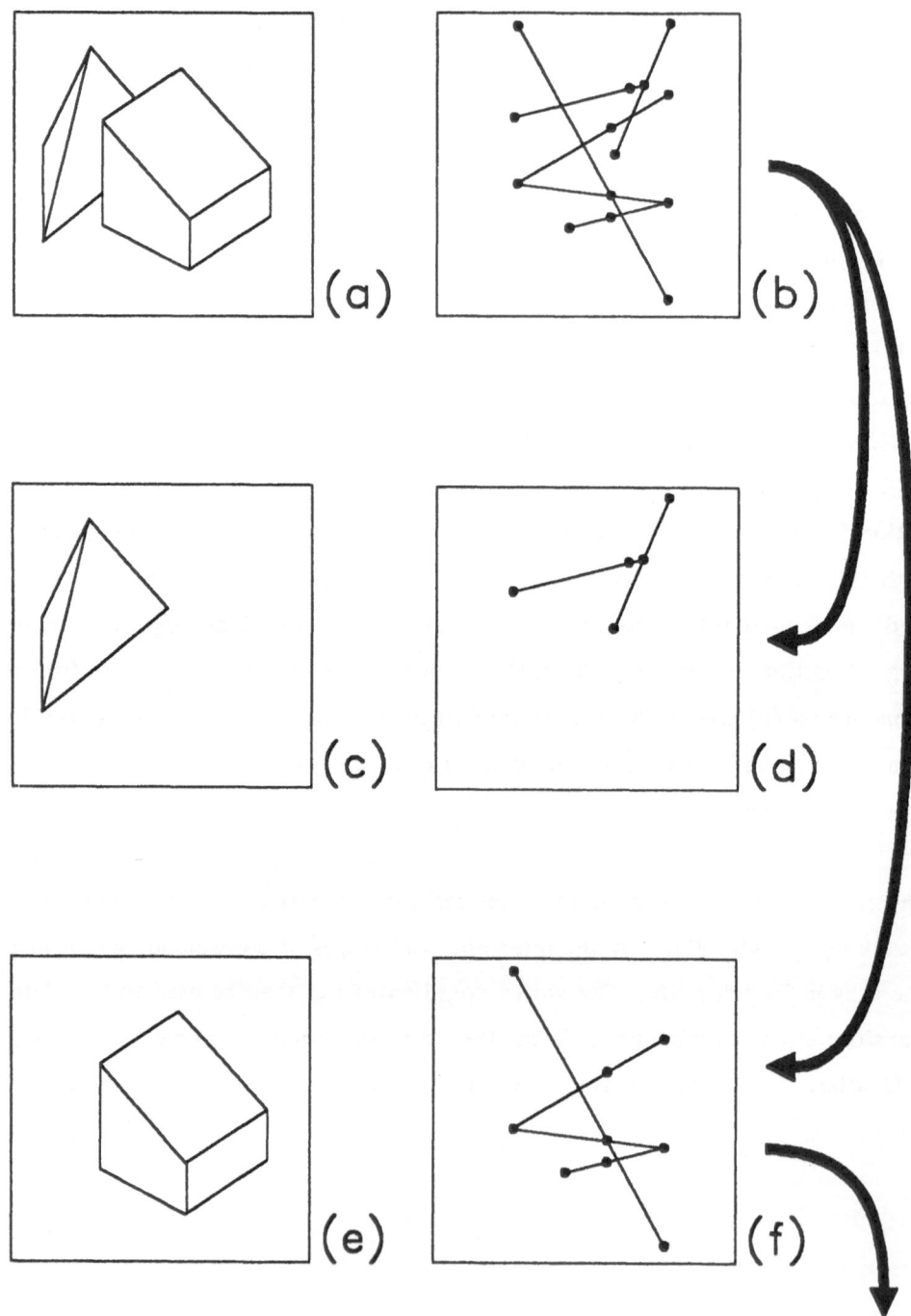

Figure 7. **Decomposition example of polyhedral scene (part 1).** © 1986 IEEE.

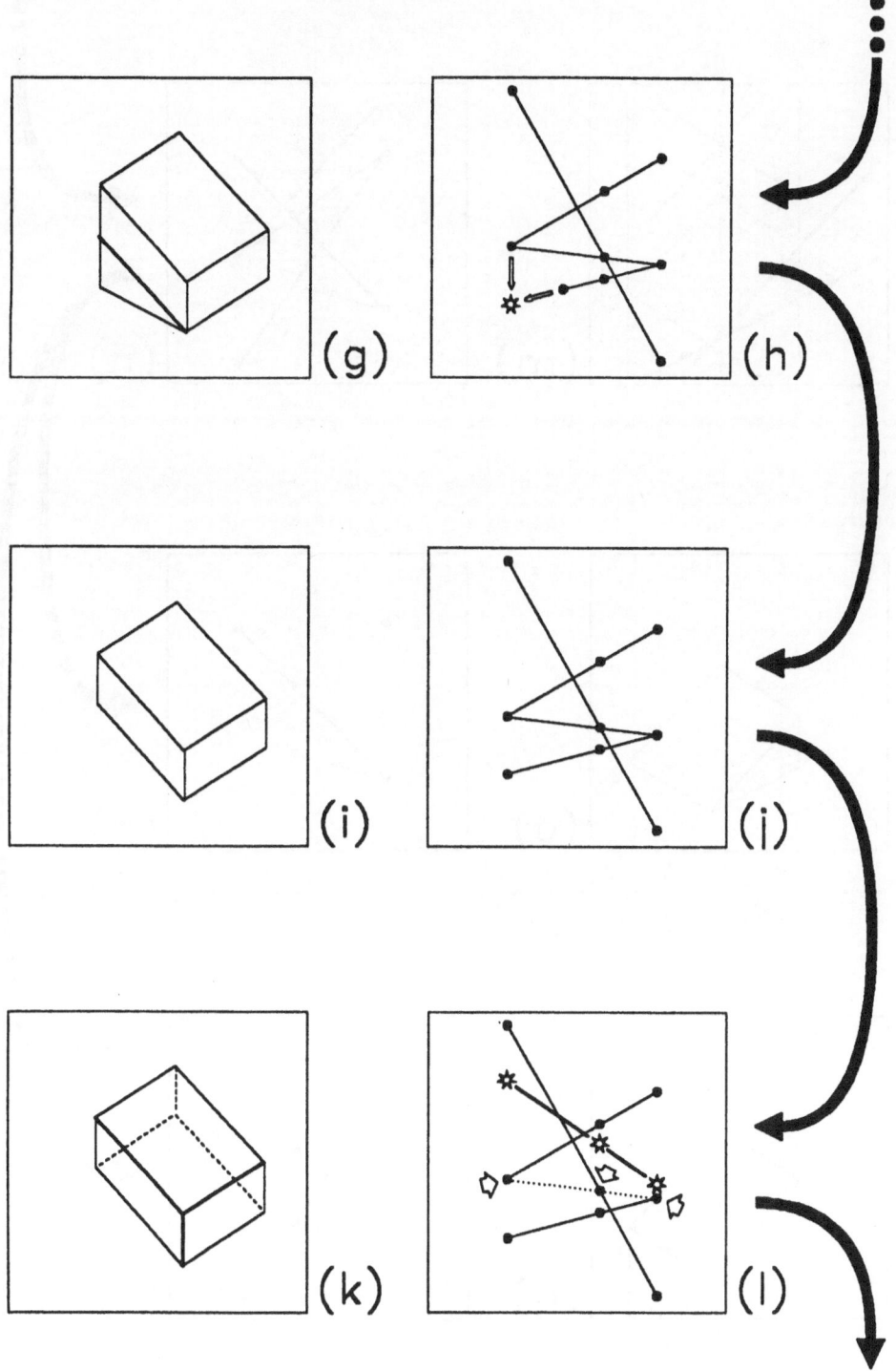

Figure 8. Decomposition example of polyhedral scene (part 2). © 1986 IEEE.

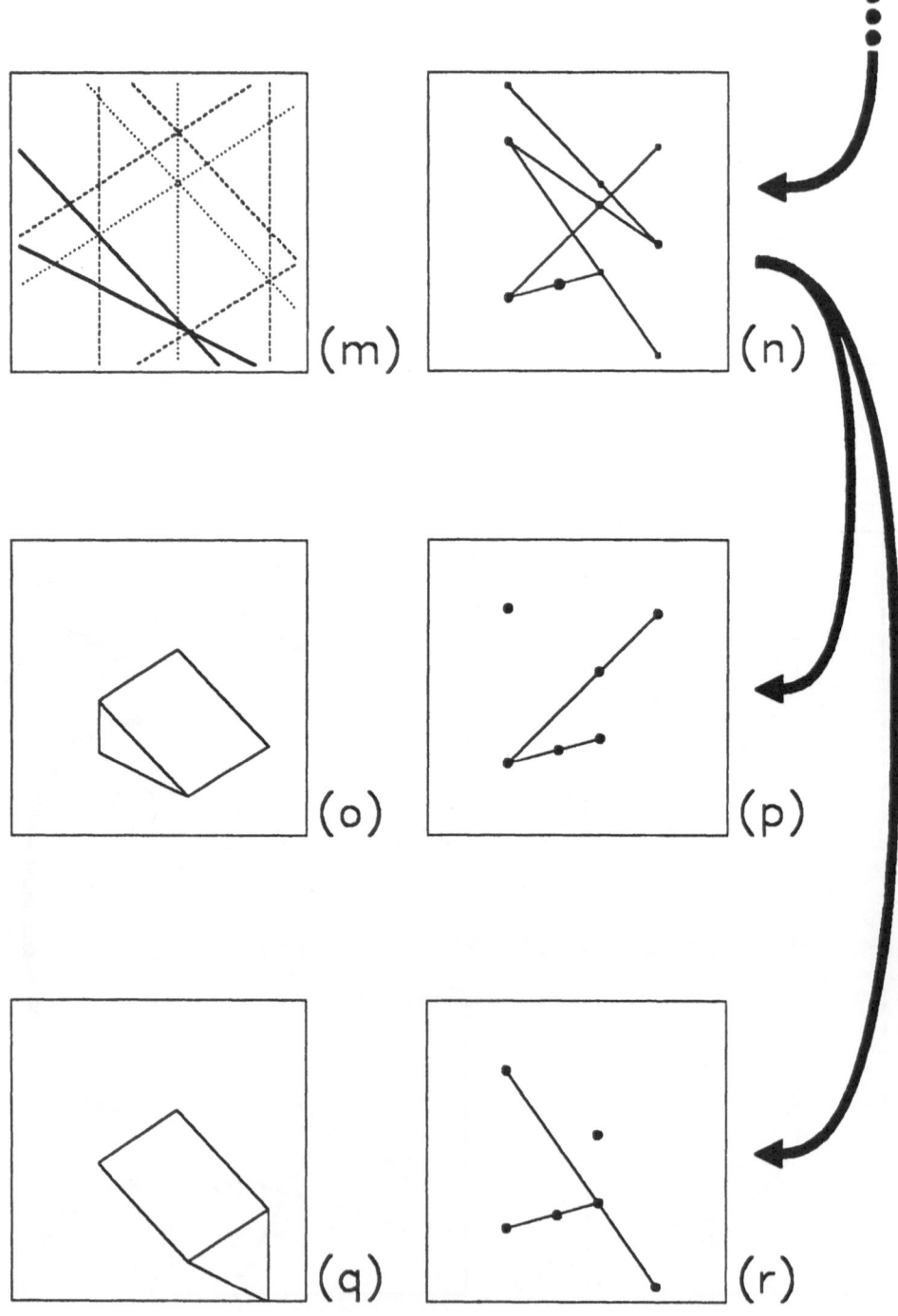

Figure 9. **Decomposition example of polyhedral scene (part 3).** © 1986 IEEE.

Conclusions

We have proposed a new technique to interpret single grey-level images of polyhedral scenes by successive decomposition of their corresponding Hough-space representations. The new approach is attractive as it makes use of the robustness of the Hough transform with respect to image noise and partial occlusions of image edges. As the Hough transform is used in its original two-parameter form, it is still practical to compute. The scene representations in Hough space (Hough nets) are very compact data structures, and accordingly can be analyzed very efficiently. The result of the decomposition process is a complete description of the scene as an assembly of different objects. Each object is represented as one or more object model primitives together with their spatial interrelations in 3d coordinate space. The experience we gained by analyzing model scenes has proven the feasibility of the new method.

However, the problems that may be encountered when processing more complex, real-world scenes are manifold. Too many, too short and fuzzy edges in a noisy environment will make it virtually impossible to identify all desired clusters in the accumulator array. Finite resolution of the camera and perspective distortion by the lens system may prevent from building up the correct Hough nets. Accidental aligning of independent vertices in the 2d projection quite often occurs even for moderate complex scenes.

There are many questions open for further research. As our ultimate goal is to handle real-world objects, such as machine parts, the method has to be extended for inclusion of higher-order analytical shapes. Furthermore, in the long term it would be desirable to have the expert modules generated automatically based on a CAD model library.

References

[BILA86]
H.P. Biland and F.M. Wahl, "Understanding Hough Space for Polyhedral Scene Decomposition", IBM Research Report, RZ 1458, (1986).

[DUDA72]
R.O. Duda and P.E. Hart, "Use of the Hough Transformation to Detect Lines and Curves in Pictures", Commun. ACM, Vol. 15, No. 1, (1972), p. 11 ff.

[HOUG62]
P.V.C. Hough, "Method and Means for Recognizing Complex Patterns", US Patent No. 3,069,654, (1962).

[ROBE65]
L.G. Roberts, "Machine Perception of Three-Dimensional Solids", Optical and Electro-Optical Inf. Proc., MIT press, (1965), pp. 159-197.

[ROSE69]
A. Rosenfeld, chapter 8.4, pp.151: "Coordinate Conversion", in A. Rosenfeld "Picture Processing by Computer", Academic Press, (1969).

[WAHL85]
F.M. Wahl and H.P. Biland, "Identification of Polyhedral Objects in Hough Space", presented at 4th Scandinavian Conference on Image Analysis, Trondheim, (1985).

RUNNING EFFICIENTLY ARC CONSISTENCY

Roger Mohr, Gérald Masini

CRIN Campus Scientifique
Boite Postale 239
54506 Vandoeuvre-lès-Nancy Cédex
France

ABSTRACT

Dealing with several hypothesis for a given fact is a common rule in Artificial Intelligence. Facts are usually interconnected through constraints that restrict the sets of different hypothesis associated with them. Checking the consistency of such a network means removing the hypothesis which make the network inconsistent. Because consistency is a NP-hard problem, people has developed a less strong technique, called arc consistency, that runs in a polynomial time. This paper describes first an optimal arc consistency algorithm, and then explains how this optimal algorithm has been improved for the special constraints introduced by a vision problem.

KEY WORDS

Constraint satisfaction, arc consistency, complexity, relaxation.

NATO ASI Series, Vol. F45
Syntactic and Structural Pattern Recognition
Edited by G. Ferraté et al.
© Springer-Verlag Berlin Heidelberg 1988

1. WHAT IS ARC CONSISTENCY

A constraint network is a general structure used for solving many AI problems. It allows representing a set of objects where each object is associated with several hypothesis. A given hypothesis on a given object introduces some restrictions on the set of hypothesis of each object linked by a constraint to the given object. When the constraints are only binary relations, the resulting structure is a graph the nodes of which are the objects and the edges of which are the constraints. This is the most usual case, but constraints like n-ary relations may be processed too. The problem was studied in [Freuder 78] and [Montanari 74], for instance.

Checking the consistency of such a network is a NP-complete problem. It contains the chromatic number of a graph as a sub-problem. Each set of hypothesis includes the possible colours for the correponding node, and the constraints express the fact that two adjacent nodes must be labeled with two different colours.

This method has been applied by Waltz to label a polyhedron scene [Waltz 75]. In all the examples he described in his paper, Waltz has pointed out that a local criterion was sufficient to find the consistent labelling solutions. This local consistency, called *arc consistency*, checks wether it exists a solution which is locally compatible with all the constraints. Obviously, consistency - which is a global criterion - implies arc consistency, but the reverse is not true. Figure 1 gives a proof of it, in the case of the colouring problem: the network is arc consistent but not consistent, since there are only two colours (*a* and *b*) to label three nodes that must have all different colours. Such an example emphasizes the limits of the theoritical interest of the method. However, in the most part of the real cases, it is not possible to demonstrate the consistency because of the great size of the networks that are involved. Therefore, we have to be satisfied with proving only arc consistency, that allows to prune the initial sets of hypothesis. The remaining hypothesis may be processed afterwards by an exponential algorithm, on condition that their number is sufficiently low.

The ALICE system [Laurière 77] works according to this technique. At each step, local inconsistent hypothesis are discarded by an arc consistency algorithm and then, if some nodes are still associated with several hypothesis, the system restricts a selected set of hypothesis (typically, the smallest one) to a single hypothesis. If an inconsistency is detected during the next step, a backtracking process is triggered. This is the key of the efficiency of the system, which is not known enough in our opinion.

This paper includes two parts: the former presents an optimal algorithm for path consistency and the latter explains how we improved it for a vision project.

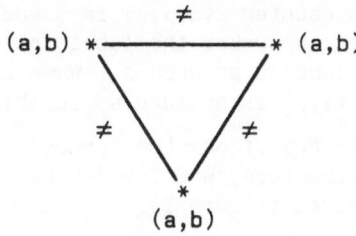

Figure 1: this network is arc consistent but is not consistent.

2. AN OPTIMAL ARC CONSISTENCY ALGORITHM

The most famous arc consistency algorithm has been designed by Waltz [Waltz 75]. It is sometimes called *Waltz filtering* and also discrete relaxation. It was considered up to now as the best algorithm for the problem we are giving our attention to. This algorithm runs in $O(em^3)$ [Mackworth 85], where e is the number of edges and m is the number of hypothesis that can be associated to each node. In a former report [Mohr 86], we have demonstrated that it could be easily improved so as to run in $O(em^2)$. The key idea consists in taking care of avoiding unnecessary work. Following Mackworth's terminology [Mackworth 85], we have named this algorithm *AC-4*.

2.1. The AC-4 algorithm

The constraint satisfaction problem is defined as follows:

- $N = \{i,j,..\}$ is the set of nodes, with $|N| = n$,
- $L = \{a,b,..\}$ is the set of labels (i.e. hypothesis), with $|L| = m$,
- $E = \{(i,j)$ such as (i,j) is an edge in $N \times N\}$, with $|E| = e$,
- $L_i = \{b$ such as $b \in L$ and (i,a) is admissible$\}$,
- R_2 is a binary relation,
  and (i,a)-(j,b) is admissible if $R_2(i,a,j,b)$.

The principle is based on the following assertion: when a label is removed from a node, only labels that may be consequently affected are examined. For this purpose, we have to build a special data structure:

- each label a at node i is associated with the set:
  $S_{i,a} = \{(j,b)$ such as a at node i is consistent with b at node $j\}$,

- for an edge (i,j), a counter $c[i,j,a]$ is associated with each label a at node i, in order to number the labels at node j that are still compatible with the label a at node i. When this counter falls to zero, the labelling (i,a) is no more admissible.

The algorithm runs in two steps: the former is devoted to the construction of the data structure, and the latter takes into account the nodes which have been removed (figure 2).

There is no difficulty in choosing the appropriate data structure for this algorithm. Just notice that the L_i are better implemented through linear lists in step 1, but through a boolean array in step 2.

2.2. Correctness and complexity

Step 1 is a loop on the edges, and, for each edge, a loop examines at most all possible labels of each node of the edge (some labels are already removed at this moment). Therefore, the complexity of step 1 is bounded by $O(em^2)$.

In order to evaluate the complexity of step 2, we have to mention that the initial value of each counter is bounded by m and that there are at most em counters. The inner loop decrements one counter at each step (line 23) and these counters are never negative. So, the complexity of this part is also bound by $O(em^2)$.

Arc consistency runs therefore in em^2 steps in the worst case. Intuitively, since the algorithm has to examine almost all the constraints in some cases and, at this time, has to process m^2 relations for each edge, it is optimal. Several attempts have been made to prove the optimality of Waltz algorithm [Dechter 85], but all of them failed.

So,

AC-4 is optimal.

The correctness proof is easy. By induction, a removed label does not belong to any solution, so any consistent labelling is contained in the result. The result is arc consistent: each remaining label a for each node i and for each edge (i,j) has a positive counter. From the definition, this means that this arc is consistent with a.

3. HOW TO IMPROVE THIS OPTIMAL ALGORITHM

In this section, we shall examine how AC-4 can be improved for some special common relations. These improvements took place in our vision project, called Trident, we shall describe first.

Step 1

```
1      W_list := Empty_list;
2      for (i,a) ∈ NxLᵢ do Sᵢ,ₐ := ∅ ;
3      for (i,j) ∈ E do
4          for a ∈ Lᵢ do
5              begin
6              total := 0;
7              for b ∈ Lⱼ do
8                  if R₂(i,a,j,b) then begin
9                                      total := total + 1;
10                                     Sⱼ,c := Sⱼ,c ∪ {(i,b)}
11                                     end;
12              if total = 0 then begin
13                                  Lᵢ := remove(a, Lᵢ);
14                                  W_list := W_list ∪ {(i,a)}
15                                  end
16                           else c[i,j,a] := total;
17          end
```

At this stage W_list contains all the labels already removed.

Step 2

```
18     while W_list ≠ ∅ do
19         begin
20         (k,c) := pick_one_from (W_list);
21         for (j,b) ∈ Sₖ,c do
22             begin
23             c[j,k,b] := c[j,k,b] - 1;
24             if c[j,k,b] = 0 then
25                             if  b ∈ Lⱼ then
                     /* no more label from k supports label b for j */
26                                             begin
27                                             Lⱼ := Lⱼ - {b};
28                                             W_list := W_list ∪ {(j,b)}
29                                             end
30             end
31         end
```


Figure 2: algorithm AC-4.

3.1. The Trident vision system

The purpose of this project is to experiment perception strategies in three dimensional vision [Masini 85]. The system is built according to an architecture which is now quite usual. The a priori knowledge about the world is stored in a *model*, and the knowledge about the scene is extracted from the image by some perception tools. Then, an interpretation module matches the image features with a subset of the model features, in order to find the structure and to label the objects composing the scene to be recognized. Features represent elementary shapes, that is to say elements of surfaces in our case.

The system works on range images that are synthesized or taken from a laser range finder. A segmentation process associated with a 3D Hough transform [Muller 84] is able to extract planar and quadratic surfaces from this input.

3.1.1. The model

We have taken our inspiration from the Acronym model [Bro-81], but primitive shapes we use are surfaces. We have not either attempted to conceive a model as universal as the Acronym one, for we are convinced that this kind of model is too complex to be efficiently used for describing real scenes and to be efficiently used by the interpretation level.

The description of a scene consists in a set of rules giving the structure of the objects and their relationships. Each rule describes an object by means of unions of subshapes (figure 3). Elementary shapes, called *primitives*, are pieces of planar surfaces or simple volumes like cones, cylinders, spheres and ellispsoids.

Parameters and characteristics of an object are represented by attributes of its own. Some of them, like length, take their values in a continuous domain, and other ones, like colour, take their values in finite discrete sets. A set of constraints is associated to each rule. They are linear equations or inequations and allow to bound the values of the attributes. For instance, they express that a house must have a stated size, that the roof elevation must be greater than the wall elevation and so on (figure 3). The whole scene is described in an absolute 3D system of coordinates. The relationships between objects are described by predefined relations, like *put_upon* (figure 3), that are translated automatically into constraints on the coordinates of the concerned objects. A same object may be described by several rules, introducing alternatives on the shape and the structure of the object.

The formalism of the model is rather complicated and its manipulation requires a great care. Consequently, a specialized module verifies the consistency of the model, that is to say it makes sure that an

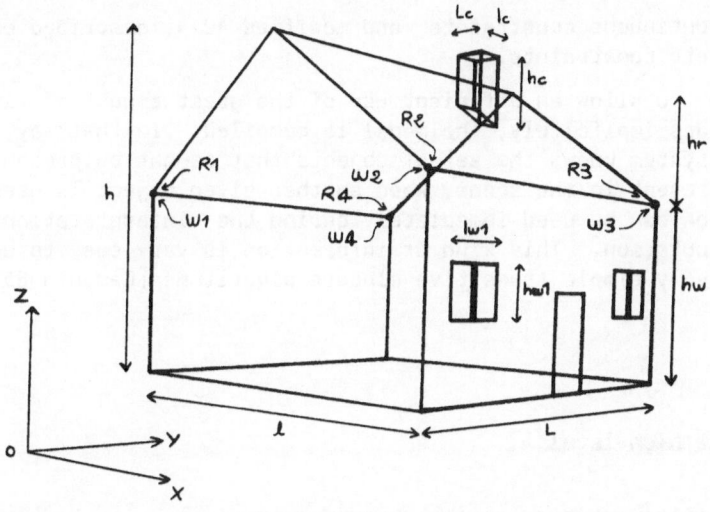

rule # 5:

house (L, l, h) → **walls** (Lw, lw, hw, colour_w, W1, W2, W3, W4)
roof (Lr, lr, hr, colour_r, R1, R2, R3, R4)
chimney (Lc, lc, hc, colour_c, C1, C2, C3, C4)
door (ld, hd, colour_d)
window_1 (lw1, hw1)
window_2 (lw2, hw2)

constraints:
h = hw + hr
L = Lw
l = lw
L = Lr
l = lr
Lc = 0.75
lc = 0.50
hc = 1.00
...
$W1_z = R1_z$ ⎤
$W1_x \geq R1_x$ ⎟
$W1_y \geq R1_y$ ⎬ put_upon(roof, walls)
... ⎟
$W4_y \leq R4_y$ ⎦
...

Figure 3: a rule describing a house.

instance of the model can be built, with all its constraints satisfied
[Camonin 85]. The whole model constitutes an and/or graph. A node *and*
is labeled with a rule identifier. Its sons are nodes *or*, labeled with
the subshapes of the object described by the rule. A node *or* is labeled
with an object name, and its sons are nodes *and*, giving the possible
rules describing the object. Building an instance of the model means
extracting a path from the graph. During this process, the consistency
of the constraints is verified by using Shostak's algorithm [Shostak 77]

for the continuous constraints, and modified AC-4, described below, for the discrete constraints.

Then, to allow an efficient use of the great amount of information it includes implicitely, the model is compiled. In that way, for example, the system knows the set of objects that cannot be present or that must be present in the scene, when another given object is present. The information can be used immediately during the interpretation, without extra computation. This kind of information is very easy to compute and requires only simple transitive closure algorithms [Camonin 85].

3.1.2. The high level

The interpretation process has been inspired by syntactic analysis techniques, and it mixes both of the bottom-up and top-down approaches. The result is a tree describing hierarchically the scene and the different objects that have been recognized. A shape is represented by a node the sons of which are its subshapes.

The process itself is very loose in order to be able to deal with the problem of occluded objects. Its role consists in managing and synchronizing the operations required by the construction of the tree. These operations are not executed according to a predefined and invariable scenario. Their logical sequence depends on the quality of the knowledge acquired by the system.

The process consists essentially in a large loop. At each step, the system chooses between different actions:

- starting a new partial recognition from a well-identified primitive, i.e. starting a new bottom-up analysis,

- extending a tree in a top-down way, i.e. inferring the subshapes of an identified object represented by a leaf of the tree,

- extending a tree in a bottom-up way, i.e. inferring an object from one of its identified subshapes represented by the root of the tree,

- merging two partial disjoined analysis, that is to say merging their trees,

- locating a given primitive surface in the image of the scene,

- checking the consistency of all the partial information gathered in the different partial analysis till now,

- backtracking, when the information collected by the system is no longer consistent.

The process stops when the greatest amount of image features has been labeled. They are represented by the leaves of the tree. A node of the tree without label is considered as representing an occluded part.

The selection of actions to be performed is guided by a strategy designed to reduce the combinatory explosion. Priority is given to deterministic actions, for instance top-down or bottom-up operations on the tree, when only one rule is admissible to describe the object represented by a given node. Contextual actions come afterwards. For instance, each time any action brings some new information, this one is propagated immediately through the system. Backtracking belongs also to this category: it is triggered as soon as an inconsistency is detected by any other action.

Another important principle is to arbitrarily select hypothesis, that is to say labels for shapes, as late as possible. To this effect, when a deterministic working is no longer practicable, a new analysis is started from an image feature that has a high confidence rate. We expect that the information discovered by the new analysis, combined with the information given by the other analysis and by the compilation of the model, will allow to solve the indeterminism. So, several analysis run in parallel, building firm basis for the final tree.

When this technique cannot be applied or when a deterministic working cannot be restored, heuristics [Laurière 77] are used in order to minimize the number of solutions to be explored, and so the number of backtrackings.

3.2. Constraint propagation

Let us now have a closer look on the constraints and the propagation algorithm as they are implemented in the Trident system. All along its running, the interpretation process builds a network of hypothesis about the identification of the objects extracted from the image. This network - as usual in AI - will grow exponentially if it is not pruned. For this reason, the constraints introduced by the model are used to get more precise information on the attributes of each predicted shape and to verify if their values are compatible with each other. Inconsistent hypothesis are then removed.

Dealing with attributes taking their values in a continuous domain, like in the Acronym system, led us to heavy computation. So, we left off the idea and converted all the attribute domains into discrete ordered sets. For instance, the domain of an attribute like *height*), that was taking previously real values, has been sliced into a range of small intervals, according to a predefined step.

There are four kinds of constraints:

- $A_i = A_j$,
- $A_i \neq A_j$,

- $A_i < A_j$,
- $A_i \in \{v_1, v_2, \dots v_k\}$.

The last constraint explicitly states that A_i is restricted to a given set of values.

During the interpretation, the consistency is checked regularly. Each node (i.e. hypothesis) that has no more label (i.e. value) for one of its attributes must be removed. From there, several other hypothesis hierarchically linked to the removed node may also disappear. This is a key for controlling the growth of such a distributed recognition process.

3.3. The equality constraints

In order to reduce the number of nodes in the network, all nodes linked through equality constraints are grouped in an equivalence class. For instance, the following set of constraints:

$A_1 \leq B$ $\quad A_1 \in \{1, 2, 3, 4, 5\}$
$A_2 = A_1$ $\quad A_2 \in \{3, 4, 5, 6\}$
$A_1 \neq B$

becomes:

$A_{12} < B$
$A_{12} \in \{3, 4, 5\}$

where A_{12} is the representation of the equivalence class $\{A_1, A_2\}$.

The equivalence nodes are easily implemented by the Union-Find data structure [Horowitz 78].

3.4. The differ relation

Let us consider the behaviour of AC-4 for the relation $\neq$. For each labelling (b,i) and for each edge (i,j), the value of the counter $c[i,j,b]$ gives exactly the number of labels for j which are different from b. This counter will become null only when node j has b as single remaining label. So, AC-4 wastes time in decreasing the counter, generally initialized with a big value, while the important fact is the moment when it comes to zero, to tell that b has to be removed from node i.

The improvement is obvious: without changing the spirit of AC-4, it is better to introduce in this case a counter for node j which counts how many labels are still associated with it. When the value of this counter reaches 1 with b as remaining label, all the nodes directly connected with node j are told to remove label b from their domain. This event occurs only once for each node i. So, if d_i is the degree of node i, it takes at worst Σd_i = em steps for the whole graph.

This result can be illustrated by an example easy to understand: the graph is built upon n nodes numbered from 1 to n, and each node i is associated with the set {1, 2,... i} of possible labels. In addition, the graph is complete: all nodes are connected to all nodes, see figure 4 for n=4. In this case, the modified version of AC-4 runs in O(em) but original AC-4 runs in $O(em^2)$.

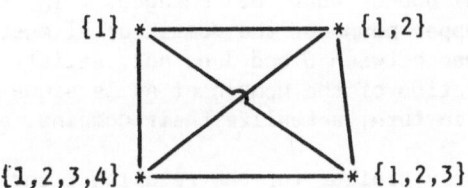

Figure 4: the worst case for AC-4 with the differ relation.

Sets of labels are finite ordered ones, and an order relation is defined for each type of values. The inequality relations introduce special case constraints: they deal only with the upper and lower bounds of the domain of each node.

Let AI and AJ be two nodes,
let i (resp. j) be the lower bound of the domain of AI (resp. AJ),
let I (resp. J) be the upper bound of the domain of AI (resp. AJ).

The processing of this type of constraint is once more very simple: we have only to consider all the different relative positions of the domains and deduce the new bounds of each one. For instance, let us consider the relation AI ≤ AJ. We shall not show all the possible cases, but just one example for each of the three groups of cases.

```
---[----]----{-----}---
 j    J    i     I
```

Case1: it is the simplest one: J < i and so the constraint is never satisfied. The network is not consistent.

```
---{----}----[-----]---
 i   I    j    J
```

<u>Case2</u>: this time, the constraint is always satisfied because I < j.
Consequently, it can be removed from the network and the
corresponding edge is deleted. The number of relations to process decreases, involving a gain in the computational time.

```
---{----[-----]----}---
 i   j    J    I
```

<u>Case3</u>: one or several bounds must be changed.   In this particuliar
case, the upper bound of the domain of AI must be shifted to J,
for the values between J and I do not satisfy the constraint.
The modification of the domain of AI is signaled to the related
nodes that, in turn, actualize their domains, and so on...

The cases are very similar for the relation < and need not to be
discussed.

The implementation of this strategy requires a list of the nodes
that are linked by the ≤ or < relations, keeping only the extrema values
of the domain of each label of each node.

For discussing the improvement we obtained, let us suppose that
only ≤ relations are used in the network. Each time an extremum of node
i is modified, i propagates the information to its adjacent nodes.   If
d_i is the degree of node i, the whole process takes d_i steps. The
total number of steps is so at most $m \Sigma d_i$, and the algorithm runs in
$O(em)$.

Figure 5 gives an example for which the modified AC-4 runs in
$O(em)$, while the initial algorithm runs in $O(em^2)$: e nodes are connected
through a cycle by the ≤ relation, excepting two nodes that are connected by a < relation. Each node has m labels: $\{1,2,... m\}$. Obviously, this system is inconsistent and the modified version of AC-4 will
discover it by running m times the cycle. The initial version of AC-4
will also run m times the cycle, but will perform an average $O(m)$ work
at each step.

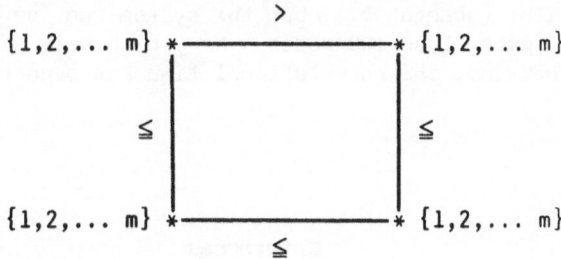

Figure 5: an example where modified AC-4 runs in O(em), with e=4.

So,
   for =, ≠ and < relations, modified AC-4 runs in O(em).

4. DISCUSSION AND CONCLUSION

This paper emphasizes that the key for building efficient programs is to perform action only when necessary. This idea is not new and has been already fully exploited in the ALICE system [Laurière 77]. For this purpose, actions are activated by events and an appropriate data structure links together events to actions.

Applying this idea to the arc consistency problem led us immediately to an optimal algorithm which is easy to program. Compared to the best known results, its complexity is reduced by a factor m where m is the number of hypothesis that can label a node. The same approach [Mohr 86] was also used for improving the path consistency algorithm given in [Mackworth 85]. In this case however, the data structure may be too large for an effective use.

For special kinds of relations, like ≠ or inequalities, the previous result has been reduced once more by a factor m. Here again the same paradigm has been applied, introducing inexpensive new data structures.

We developped this arc consistency algorithm for our 3D vision project. The nodes represent the different matchings between the shapes extracted from the scene and the shapes of the model. The labels are the possible values for the attributes of each shape. The model introduces constraints on these attributes. At each step, the system checks the consistency and removes inconsistent matchings. This approach requires that attributes have finite domains. Like Brooks [Brooks 81], we worked previously with real numbers and linear inequalities using

Shostak algorithm [Shostak 77], but the system ran several hours when dealing with more than 100 nodes. Now, with only simple inequalities and discrete domains, the computational time has been reduced to a few seconds.

References

[Brooks 81]
R. Brooks, *Symbolic reasonning among 3D models and 2D images*, Ph. D. thesis, Standford University, 1981

[Camonin 85]
M. Camonin, G. Masini, *Conception et compilation du modèle dans le système de vision Trident*, Actes Congrès Cognitiva, Paris, 1985, Tome 1, pp 131-136

[Dechter 85]
R. Dechter, J. Pearl, *A problem simplification approach that generates heuristics for constraint-satisfaction problems*, Technical Report UCLA, 1985 (to appear in Machine Intelligence)

[Freuder 78]
E.C. Freuder, *Synthesizing constraint expressions*, Communications of the ACM 21, 1978, pp58-966

[Horowitz 78]
E. Horowitz, S. Sahni, *Fundamentals of computer algorithms*, Springer-Verlag, New York, 1978

[Laurière 77]
J.L. Laurière, *A language and a program for stating and solving combinatorial problems*, Artificial Intelligence 10, 1977, pp 29-127

[Mackworth 77]
A.K. Mackworth, *Consistency in networks of relations*, Artificial Intelligence 8, 1977, pp9-118

[Mackworth 85]
A.K. Mackwoth, E.C. Freuder, *The complexity of some polynomial network consistency algorithms for the contraint satisfaction problems*, Artificial Intelligence 25, 1985, pp 65-74

[Masini 85]
G. Masini, R. Mohr, E. Thirion, *Stratégie de perception pour un modèle hiérarchique*, Actes 5ème Congrès AFCET Reconnaissance des Formes et Intelligence Artificielle, Grenoble, 1985, Vol. 2, pp 631-640

[Mohr 86]
R. Mohr, T.C. Henderson, *Arc and path consistency revisited*, Artificial Intelligence, Vol. 28, No. 2, 1986, pp 225-233

[Montanari 74]
U. **Montanari**, *Networks of constraints: fundamental properties and application to pictures*, Information Sciences 7, 1974, pp5-132

[Muller 84]
Y. **Muller**, R. **Mohr**, *Planes and quadrics detection using Hough transform*, Proc. 7th International Joint Conference on Pattern Recognition, Montreal, 1984, Vol. 2, pp 1101-1103

[Shostak 77]
R. **Shostak**, *On the sup-inf method for proving Presburger formulas*, Journal of the ACM 24, No. 4, 1977, pp 529-543

[Waltz 75]
D. **Waltz**, *Understanding line drawings of scenes with shadows*, in "The Psychology of computer vision", P.H. Winston editor, McGraw-Hill, New York, 1975

SMITH: AN EFFICIENT MODEL-BASED TWO DIMENSIONAL SHAPE MATCHING TECHNIQUE[1]

R. Mehrotra(*), W.I. Grosky(**)

(*) Computer Science Dept.
Univ. of South Florida
Tampa, Florida 33620
(**)Computer Science Dept.
Wayne State University
Detroit, Michigan 48202 U.S.A.

ABSTRACT

Previous research in model-based 2−D object recognition has compared the scene representation either with each model in a database of models or with each of a collection of features belonging to a database of models until the scene was completely analyzed. In other words, these techniques were model-driven and lacked a data-driven indexing mechanism for model retrieval. In this paper, we develop such an indexing mechanism. This mechanism is part of an overall scheme called **SMITH** (Shape Matching Utilizing Indexed Hypothesis Generation and Testing) for 2-D object recognition which is based on a dynamic programming implementation of attributed string matching. It is computationally efficient and works effectively for both non-occluded and occluded shapes. Another advantage of our technique is that models may be inserted or deleted with relatively little cost.

1. INTRODUCTION

With increasing emphasis on automated manufacturing, there has been a growing interest in interpreting images containing 2−D objects. A number of approaches have been proposed for these types of scenes. Some of these approaches can handle only simple scenes in which touching and occlusion are not allowed, whereas others are capable of handling more complex scenes. Most of the current 2−D object recognition systems are model-based. In such a system, the representation of each of the known set of objects is precompiled and stored in a database of **models** and later used to recognize each instance of an object in the image. Thus, there are two phases in the overall functioning of such a model-based system: the **training phase** to learn the representations of the known objects and use the learned representations to develop a recognition strategy, and the **recognition phase** to produce the required description of the unknown scenes.

Most of the current model-based recognition techniques employ the information conveyed by the object boundary [AyF86, BhF84, Dav79, GrL86, Pav78, Per79, TsY85]. In these techniques, the boundary or the approximated boundary of an object is represented by a set consisting of certain primitive structural units. This set and the information derivable from it form the basis of both model building and the scene analysis process. Perkins [Per79] represents the object boundary as an ordered set of straight lines and circular arcs, called **concurves**. The recognition process is based upon cross-correlating the curvature as a function of the curve length between the scene representation and the database model. This approach is

---

[1]This research has been partially supported by the Institute for Manufacturing Research at Wayne State University.

NATO ASI Series, Vol. F45
Syntactic and Structural Pattern Recognition
Edited by G. Ferraté et al.
© Springer-Verlag Berlin Heidelberg 1988

capable of handling complex scenes but involves expensive preprocessing, and is sensitive to scaling. Davis [Dav79] uses an ordered set of corners to represent an object or a scene, whereas Bhanu and Faugeras [BhF84] use an ordered set of line segments for this purpose. The recognition processes of both of these systems are based on the use of relaxation, and hence are computationally very expensive if implemented in a sequential manner.

The local-feature-focus method of Bolles and Cain [BoÇ82] can analyze more complex scenes. Their approach is based on the assumption that the objects to be recognized contain clusters of local features that can uniquely identify and orient the object. The recognition process tries to find clusters of local features in the scene whose relative configurations are consistent with those of the model. The method used to search for consistent clusters utilizes the detection of maximal cliques in a graph, which is computationally expensive. Recently, Turney, Mudge, and Volz [TMV85] proposed an efficient approach to recognizing partially visible objects. Their method is based on a saliency weighted Hough transform. The object boundary is divided into overlapping segments called **subtemplates**. A vector pointing to a standard position in the object is associated with each subtemplate, the latter of which is assigned a weight based on its saliency. To recognize an object, each of its subtemplates is matched against the edges in the scene. In case of a successful match, the accumulator at the position pointed to by the given subtemplate vector is incremented by the subtemplate weight. The location with the largest value above a given threshold is considered to be the location of the object. A weakness of this approach is that the complex saliency calculations have to be redone when models are inserted or deleted. Knoll and Jain [KnJ86] utilize the ideas of the local-feature-focus method as well as saliency in order to develop an approach called **feature-indexing**. They assume that generating hypotheses is less costly than verifying them, and hence generate as many hypotheses as possible. They choose features in a more formal way than previous approaches, which leads to an increase in power and efficiency. Specifically, they show that the recognition time grows as the square root of the size of the model set. However, changing the model set would still entail expensive recalculations.

Another approach to the $2-D$ object recognition problem is to treat an object boundary as a string of primitive structural units and then utilize some form of string matching [BaJ75, LoW75, LuF78, WaF74] for analyzing the scenes. These techniques are commonly used in the areas of signal processing and speech recognition [AnG83, BaJ75, MyR81a, MyR81b, SaC78]. Recently, similar techniques have been proposed for the shape recognition problem. Grosky and Lu [GrL86] have used chain-code based strings to represent shape. Shape recognition in their approach is based upon finding a generalization of the longest common substring between the strings derived from the model and the unknown shape. Since only the discrete symbols of the chain-coding scheme are used, these strings are usually very large, and the matching process is thus computationally expensive. Tsai and Yu [TsY85] have represented shape by strings of attributed line segments. This shape recognition process has employed a modified dynamic programming based string matching technique. The inclusion of attributes results in the reduction of the size of the strings needed to represent the shape boundary, and hence has increased the speed of the recognition process. Both of these systems are not capable of handling occluded objects or complex scenes, however.

The basic problem with all of these systems is that they are primarily model-driven. In order to analyze an unknown scene, the representation of the scene is compared either with each model in a database of models or with each of a collection of features belonging to a database of models until the scene is completely analyzed. There is not any way of hypothesizing the presence or absence of one or more objects in the scene without such an exhaustive comparison. In other words, there is a lack of a **data-driven** indexing mechanism for model retrieval. We view the recognition phase of scene analysis as a two step process. The first step is to hypothesize the presence of one or more objects in the scene and to retrieve **only** these models from the database. The second step uses this set of models for further analysis of the scene. This is a data-driven as well as model-driven manifestation of the classical **hypothesize-and-test paradigm.** A sophisticated indexing mechanism thus has the possibility of speeding up the recognition phase. Such an indexing mechanism is also very useful for query processing in a generalized image database management system [MeG85]. The $2-D$ object recognition technique described in this paper, called **SMITH** (Shape Matching Utilizing Indexed Hypothesis Generation and Testing), is based on such an iconic indexing mechanism. The entire approach uses a dynamic programming implementation of attributed string matching. This method is computationally efficient and works effectively for both occluded and non-occluded shapes. Another advantage of our technique is that models may be inserted or deleted with relatively little cost.

Our paper is organized as follows. In Section 2 we discuss our scheme of model and index construction, while Section 3 shows how our index can be modified. Section 4 concerns our approach to indexed hypothesis generation and verification. Some experiments are discussed in Section 5. Finally, in Section 6 we present our conclusions.

2. MODEL BUILDING AND INDEX DESIGN

The training phase of our system includes model building as well as index design and modification. Both the model and scene representation building process are the same. These processes involve the following sequence of operations:

1. We acquire a good contrast image of the object or scene. The images used in the examples provided in this paper were taken by an inexpensive vidicon camera. These images are 128 x 128 and consist of 256 gray-levels.

2. We threshold the image to obtain a binary image.

3. We trace the boundary [DuH73] to obtain an ordered sequence of boundary points. In the examples discussed, the starting point of a sequence is found by a raster scan of the given image from bottom to top, and the boundary is traced in a clockwise direction. To reduce the preprocessing time, one may retain only the first of every 3 or 5 points.

4. We obtain the high curvature points. Curve splitting [Pav77] can also be performed. This provides a better approximation to the object boundary in case curve segments exist. We used the curvature computation method proposed by Rosenfeld and Johnston [RoJ73] and then retained all the points whose curvature was locally maximum.

5. We connect the ordered set of high curvature points by straight-line segments to obtain the polygonal approximation of the object or scene boundary.

6. We obtain an ordered set of feature vectors. Each vector is a quadruple $<A_V, D_V, X_V, Y_V>$, where A_V is the internal angle at the vertex point V, D_V is its distance from the next clockwise vertex point, and (X_V, Y_V) are the coordinates of V.

As mentioned previously, index construction and modification is also a part of the system's training. The index is initially constructed for a group of models. Insertions and deletions of models are handled by modifying the index. This process uses a few selected subsets of contiguous vectors, called **privileged strings**, of the feature vector set obtained in Step 6 above. A distance measure, called the **edit distance**, is also employed. The edit distance between two strings is the cost of converting one string to the other. This distance function is described in the Appendix.

Let the cost of converting a string s_1 to a string s_2 be denoted by C_{s_1,s_2}. As mentioned in the Appendix, this cost is not symmetric. The index design process then involves the following sequence of steps:

1. We select a few privileged strings of fixed length from each model. In our implementation, the size of each such string is 5. The first privileged string of a model always starts from the sharpest corner, as we feel that such corners are less sensitive to noise. We then find other locations in the given models, if any, where a similar string (one whose edit distance is less than a given threshold value) exists. We associate with each string a list of such locations. A maximum of 5 privileged strings from each model is then chosen. Let the collection of privileged strings from all models be $S = \{s_1,...,s_n\}$.

2. We compute C_{s_i,s_j} and C_{s_j,s_i} for every pair $<s_i,s_j>$, where $s_i,s_j \epsilon S$ and $s_i \neq s_j$. The measure of similarity of the 2 strings, E_{s_i,s_j}, is then taken as the average of these 2 costs, in order for this measure to be symmetric. As shown in [SaC78], the performance of matching systems is improved if one uses a symmetric cost function as opposed to a non-symmetric one.

3. We form clusters of similar privileged strings and find representatives for each of these clusters. In succeeding steps, only these cluster representatives, called **reference strings**, will be used. We would like to have the string corresponding to the center of the cluster represented by S as the reference string of S. This reference string may not belong to S. To find the center, one may have to solve an optimization problem which is computationally expensive. This will be the case if we take as the reference string that string whose sum of distances to the other members of the cluster is minimal. We, however, do the following,

    a. We select privileged string $s \epsilon S$, such that $\max(\{E_{s,s_k}|s_k \epsilon S\})$ is minimum. This privileged string is treated as the **reference string** for the collection S.

    b. We split the collection S into 2 comparably sized collections S_1 and S_2. In our implementation, we sort the privileged strings in the collection by increasing order of distance from the reference string and then the midpoint is used as the collection split point.

    c. We find the reference strings for the 2 collections S_1 and S_2 as done in Step 3a. Call them r_1 and r_2, respectively.

    d. We adjust S_1 and S_2 such that for all privileged strings $s_i \epsilon S_1$ we have that $E_{s_i,r_1} \leq E_{s_i,r_2}$, and for all privileged strings $s_j \epsilon S_2$ we have that $E_{s_j,r_1} > E_{s_j,r_2}$.

        A. We transfer a privileged string $s_i \epsilon S_1$ to S_2 if $E_{s_i,r_1} > E_{s_i,r_2}$.

B. We transfer a privileged string $s_j \epsilon S_2$ to S_1 if $E_{s_j, r_1} \leq E_{s_j, r_2}$.

f. We recursively apply Step 3 to the collections S_1 and S_2. Termination occurs when the size of a collection becomes 1.

The hierarchy generated by this process is used as the iconic indexing mechanism for model retrieval. The reason for the binary split as against any higher-order split is explained by the fact that the search time for an n string k-ary index tree is $O(m \cdot k \cdot \log_k n) = O(m \cdot k \cdot \log_2 n / \log_2 k)$, where m is the cost of computing the similarity measure between 2 strings. Since $k > \log_2 k$, the best choice for sequential processing is when $k = 2$. To allow more concurrency, a choice of $k > 2$ could be taken. It is easily shown that this index building mechanism ensures that the privileged strings in a particular collection are more similar to the reference string of that collection than the reference string of the neighboring collection.

3. MODEL RETRIEVAL AND VERIFICATION

Once the indexing mechanism is developed, the model retrieval process is very simple. It starts with the development of the scene representation. Then a privileged string is extracted from this representation. Let the level of the root of the index tree be 0. The model retrieval process starts from level 1 of the index tree. A privileged string, p, of the scene is compared to the 2 reference strings at level 1. $E_{r_1, p}$ and $E_{r_2, p}$ are computed. If $E_{r_1, p} \leq E_{r_2, p}$, the subtree having root r_2 is eliminated from further consideration and the search proceeds to the subtree having root r_1, while if $E_{r_1, p} > E_{r_2, p}$, the search proceeds to the subtree having root r_2. The search continues until a leaf node is reached. If the edit distance is less than a given threshold then the privileged string, s, represented by that leaf node is taken to be most similar to p, and the models associated with s are retrieved.

The privileged strings extracted from the scene are used for model retrieval. The indexing provides an efficient mechanism to find where and in which models a given privileged string of the scene exists. Thus, one privileged string of the scene generates one or more hypotheses. The next step is to verify the presence of these hypothesized models. The matching of privileged strings can find two or more corresponding points. These points can be used to compute transformation parameters. Then, all points of the hypothesized model are transformed. If the transformed model points match more than a given threshold number of scene points, the model is assumed to be found, otherwise the given hypothesis is rejected. The process of hypothesis/verification continues until the number of scene points not yet matched is less than another given threshold. If no acceptable hypotheses can be found, then the presence of an unknown object is declared.

One of the following two methods can be employed to compute the transformation parameters:

1. Let a pair of model points and the corresponding scene points be known. The scaling k, the rotation θ, the translations T_x and T_y can then be computed as follows,

$$k = L_s / L_m,$$

$$\theta = A_s - A_m,$$

$$T_x = X_s - k(X_m \cos\theta - Y_m \sin\theta),$$

$$T_y = Y_s - k(X_m \sin\theta + Y_m \cos\theta),$$

where L_s is the distance between the two scene points, L_m is the distance between the two corresponding model points, A_s is the orientation of the segment formed by joining the two scene points, A_m is the orientation of the segment formed by joining the two corresponding model points, and (X_s, Y_s) and (X_m, Y_m) are the midpoints of the scene and model segments, respectively. The transformation parameters can be computed using two or more pairs of matching points, and the average value of the transformation parameters can be used for verification.

2. Let $\{(X_{s_1}, Y_{s_1}), ..., (X_{s_n}, Y_{s_n})\}$ be the vector of scene points and $\{(X_{m_1}, Y_{m_1}), ..., (X_{m_n}, Y_{m_n})\}$ the corresponding vector of model points obtained by the privileged string matching process. Then, the transformation vector $T = (k\cos\theta, k\sin\theta, T_x, T_y)^T$ can be computed as follows. Let M be the matrix

$$\begin{bmatrix} X_{m_1} & -Y_{m_1} & 1 & 0 \\ Y_{m_1} & X_{m_1} & 0 & 1 \\ \cdot & & \cdot & \cdot \\ \cdot & & \cdot & \cdot \\ \cdot & & \cdot & \cdot \\ X_{m_n} & -Y_{m_n} & 1 & 0 \\ Y_{m_n} & X_{m_n} & 0 & 1 \end{bmatrix}$$

and B the column vector

$$\begin{bmatrix} X_{s_1} \\ Y_{s_1} \\ \cdot \\ \cdot \\ \cdot \\ X_{s_n} \\ Y_{s_n} \end{bmatrix}$$

Then, the error $E = MT - B$. The transformation, T, that minimizes the square of the error can be computed using the pseudo-inverse method [DuH73]. We get that $T = (M^T M)^{-1} M^T B$. The greater the number of matching points, the better the estimate of T, but the more computationally expensive the estimation process becomes.

4. INDEX MODIFICATION

Our scheme efficiently supports the insertion and deletion of models. This is handled by a simple index tree modification.

To delete a model, one must search the index tree for each privileged string from the given model. If the associated list of model locations in the leaf node to which we arrive due to the search contains more than one model, then the given model is deleted from this list. Otherwise, there are 2 cases. In the first case, the privileged string represented by the leaf node and its parent are not the same. We then tag the leaf node as **deleted**. This node is then only used for index purposes and not for hypothesis generation. In the second case, the privileged string represented by the leaf node and its parent are the same. We then delete the leaf node and replace its parent by its sibling, thus reducing the height of the tree.

To insert a model, one must insert in the index tree each privileged string from the given model. One first conducts a search for the closest match to the given string. If whatever leaf node one reaches is marked as **deleted**, then this node is reclaimed and the new string is inserted into it. Otherwise, we have 2 cases. In the first case, a match is found whose measure of similarity is below some threshold value. The new string is then added to the list of model locations associated with this node. Otherwise, a new level in the tree is created, where one of the sons is a copy of the given father tree node and the other corresponds to the new string.

As is clear from this index modification technique, one could start with just a single model and insert other models as the need arises.

The index tree may be periodically reconstructed using only the untagged leaf nodes. We are currently studying efficient ways of maintaining optimally balanced index trees.

5. EXPERIMENTS

We have tried our techniques on numerous simple and complex real scenes. In one of these experiments, our models consist of 11 flat shapes, some of which are shown in Figures 1-3. Each figure shows first, the thresholded binary image, and second, the polygonal approximation. We note here that our display has introduced some distortion into these images, as the x and y scalings are not equal. See Table 1 for a listing of the 9 privileged strings, 3 strings/model, their corresponding models, and where they are located in the polygonal approximation. Note that each segment is of length 5. Figure 4 displays the index tree for this data.

Figures 5-6 consist of two scenes on which we tried our techniques, with the results shown in Tables 2-3. Note that each image is taken to be in the third quadrant of a standard Cartesian coordinate system and that a positive angle of rotation is assumed to be clockwise from the positive X-axis.

6. CONCLUSIONS

Previous approaches to the problem of 2-D shape matching have been model-driven. Our technique, on the other hand, employs a data-driven indexing mechanism for model retrieval (hypothesis generation) as well as the standard model-driven approach to hypothesis verification. We have demonstrated the efficacy of using all the noise insensitive model features to construct an index into the set of models, rather than going the computationally expensive route of choosing those features which differentiate one model from another. This latter approach is very sensitive to model insertions and deletions, while our approach handles them easily.

We have also presented a computationally efficient matching scheme which is robust. It relies on well-studied techniques whose time-complexity is quite good. We are currently applying these techniques to other vision problems such as object tracking.

For the future, we hope to apply this combination of efficient matching and model indexing to range images.

APPENDIX

The quantity C_{s_1,s_2} is defined as the cost of converting string s_1 to string s_2, where s_1 is the representation of an unknown shape/segment and s_2 is the representation of a model or model segment. It is defined as $d[|s_1|,|s_2|]$, where $d[i,j]$ is the minimum cost of converting the first i features of s_1 to the first j features of s_2. In our representation, these features are the corners in the polygonal approximation of the given object boundary.

To convert the shape represented by s_1 to the model represented by s_2, the following 4 operations are used:

1. **Deletion** of corners belonging to s_1

2. **Insertion** of corners belonging to s_2 into s_1

3. **Changing** corners belonging to s_1 to the corresponding corners of s_2

4. **Smoothing** one or more corners of s_1 and s_2

Smoothing is introduced in order to eliminate noisy corners and hence to decrease the sensitivity of our system with respect to noise. Tsai and Yu [TsY85] introduced a similar operation called **merge** for analyzing simple 2-D scenes.

The quantity $d(|s_1|,|s_2|)$ may be computed using a dynamic programming technique. The recursive relations are as follows. We define $d[0,0] = 0$, $d[0,i] = d[0,i-1] + 1$, $d[i,0] = d[i-1,0] + 1$, and $d[i,j] = \min(d[i,j-1] +$ cost of inserting the jth corner of s_2, $d[i-1,j] +$ cost of deleting the ith corner of s_1, $d[i-1,j-1] +$ cost of changing the ith corner of s_1 to the jth corner of s_2, $\min_{2 \leq m \leq i, 2 \leq n \leq j}(d[i-m,j-n] +$ cost of smoothing the corners between the $(i-m+1)^{st}$ to the ith corner of $s_1 +$ cost of smoothing the corners between the $(j-n+1)^{st}$ to the jth corner of $s_2 +$ cost of changing the $(i-m+1)^{st}$ corner of s_1 to the $(j-n+1)^{st}$ corner of $s_2 +$ cost of changing the ith corner of s_1 to the jth corner of s_2)). Notice that the cost of the **change** operation is included in the cost of the **smoothing** operation (when $m=n=1$), and hence can be eliminated. As is standard in dynamic programming algorithms, the shortest $2-D$ path from $d[0,0]$ to $d[|s_1|,|s_2|]$ gives the corresponding points.

The deletion and insertion costs are constants k_d and k_i, respectively.

The cost of changing a corner c_1 with angle θ_{c_1} and distance from the next point d_{c_1} to a corner c_2 with angle θ_{c_2} and distance from the next point d_{c_2} is taken to be,

$$p \frac{\left|\theta_{c_1} - \theta_{c_2}\right|}{\theta_{max}} + q \frac{\left|d_{c_1} - d_{c_2}\right|}{d_{max}} \tag{1}$$

where θ_{max} is the maximum chargeable angle difference, d_{max} is the maximum chargeable distance difference, and p and q are given weights. In our implementation, $\theta_{max} = 20°$, $d_{max} = 0.4 \cdot d_{c_2}$, p=0.7 (an emphasis on angle), and q=0.3. We chose $d_{max} = c \cdot d_{c_2}$, and thus we have that $C_{s_1,s_2} \neq C_{s_2,s_1}$.

The cost of smoothing a corner is

$$a \frac{\left|180 - \theta_c\right|}{\phi_{max}} + b \frac{\left|d_1 - d_2 - d\right|}{e_{max}} \tag{2}$$

where ϕ_{max} is the maximum smoothable angle and e_{max} is the maximum allowable distance, a and b are constants, and we refer to Figure 7. Here, the $(m-1)^{st}$ corner is smoothed by joining the $(m-2)^{nd}$ corner to the mth corner by a straight line. In our implementation, $\hat{\theta}_{max} = 20°$, $\hat{e}_{max} = 0.15 \cdot d$, a=0.7 (an emphasis on angle), and b=0.3.

As is clear from (1), the smaller the angle and length difference, the lower the change cost. When these differences are greater than θ_{max} and d_{max}, respectively, a higher change cost results. Similarly, (2) indicates that the sharper the corner, the higher the smoothing cost. Also, the distance component of the smoothing cost takes care of the height of the corner.

Notice that $C_{s_1,s_2} = 0$ if $s_1 = s_2$ and increases with the difference in the two strings.

In case more than one corner is smoothed, θ_{c_2} is taken to be equal to the angle of that original corner which differs most from $180°$, and d_2 is taken to be the sum of the distances associated with the smoothed corners.

Table 1
List of Privileged Segments

Privileged String	Model	Starting Vertex in the Polygonal Approximation
7	3	17
8	3	5
9	3	10
13	5	5
14	5	9
15	5	4
25	9	10
26	9	1
27	9	4

Table 2
Unknown Scene 1

Hyp.	Starting Scene Vertex in the Polygonal Approximation	Hypothesized Privileged String	Result A-Accept R-Reject	Transformation			
				Scaling	Rotation	Translation	
						X	Y
1	18	27	R				
2	6	13	A/Model 5	0.94	8.13°	4.67	−11.00
3	15	26	A/Model 9	1.08	−182.12°	149.06	128.02

Table 3
Unknown Scene 2

Hyp.	Starting Scene Vertex in the Polygonal Approximation	Hypothesized Privileged String	Result A-Accept R-Reject	Transformation			
				Scaling	Rotation	Translation	
						X	Y
1	6	27	A/Model 9	1.14	−78.47°	−54.81	−137.33
2	22	13	A/Model 5	1.14	15.25°	−1.50	−22.50
3	35	7	A/Model 3	0.91	−164.06°	115.96	129.34

 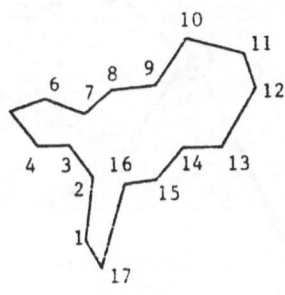

Figure 1 - Thresholded Binary Image and Polygonal
Approximation of Model 3

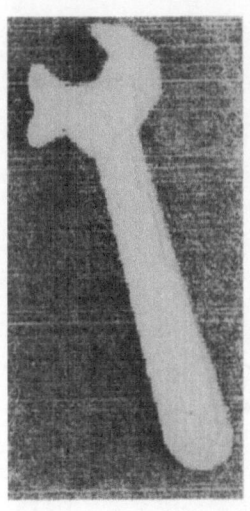 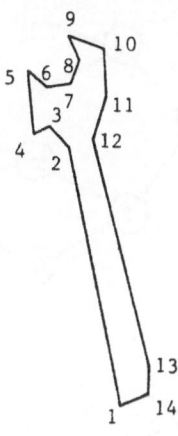

Figure 2 - Thresholded Binary Image and Polygonal
Approximation of Model 5

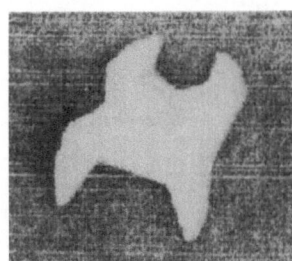

Figure 3 - Thresholded Binary Image and Polygonal
Approximation of Model 9

244

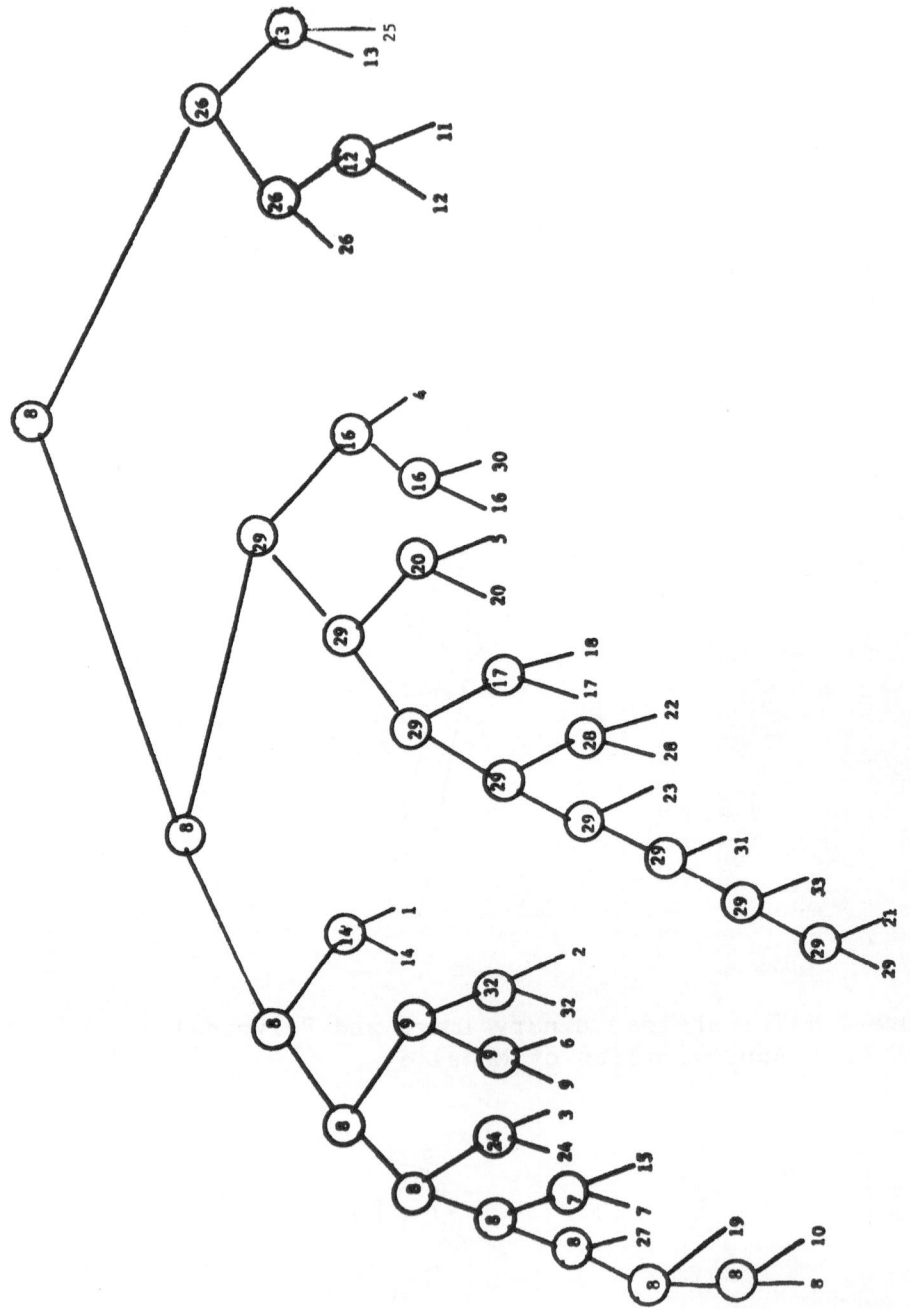

Figure 4 – 33 Feature Index Tree

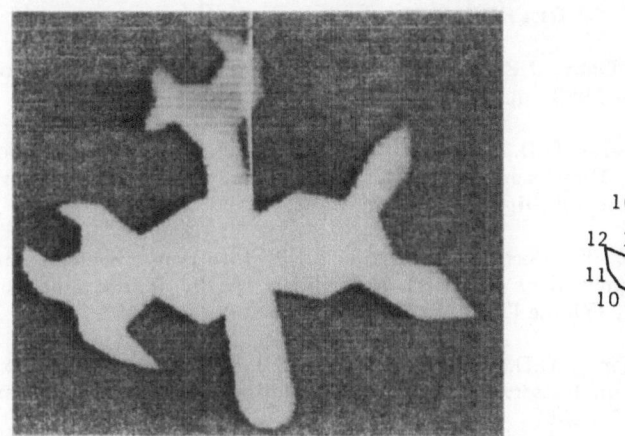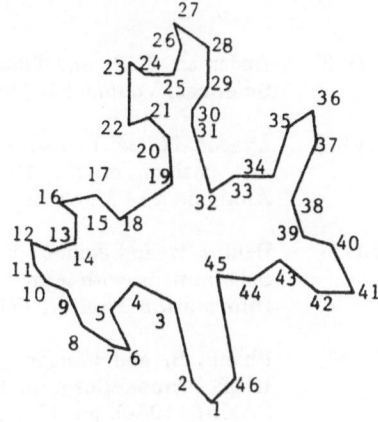

Figure 5 - Thresholded Binary Image and Polygonal
Approximation of Scene 1

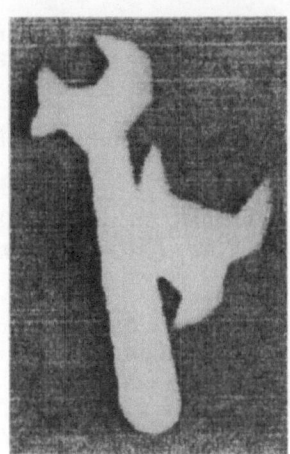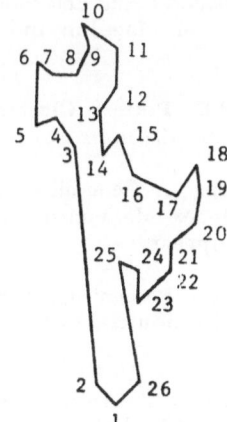

Figure 6 - Thresholded Binary Image and Polygonal
Approximation of Scene 2

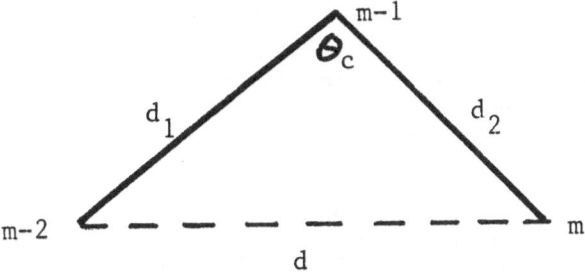

Figure 7 - Cost of Smoothing a Corner

REFERENCES

[AnG83]   Anderson, K.R. and Gaby, J.E., 'Dynamic Waveform Matching,' **Information Sciences**, Volume 31 (1983), pp. 221–242

[AyF86]   Ayache, N. and Faugeras, O.D., 'HYPER – A New Approach for the Recognition and Position of Two-Dimensional Objects,' **IEEE Transactions on Pattern Analysis and Machine Intelligence**, Volume PAMI-8 (1986), pp. 44–54

[BaJ75]   Bahl, L.R. and Jelinek, F., 'Decoding for Channels with Insertions, Deletions, and Substitutions with Applications to Speech Recognition,' **IEEE Transactions on Information Theory**, Volume IT-21 (1975), pp. 404–411

[BhF84]   Bhanu, B. and Faugeras, O.D., 'Shape Matching of Two-Dimensional Objects,' **IEEE Transactions on Pattern Analysis and Machine Intelligence**, Volume PAMI-6 (1984), pp. 137–156

[BoC82]   Bolles, R.C. and Cain, R.A., 'Recognizing and Locating Partially Visible Objects: The Local-Feature-Focus Method,' **International Journal of Robotics Research**, Volume 1 (1982), pp. 57–82

[Dav79]   Davis, L.S., 'Shape Matching Using Relaxation Techniques,' **IEEE Transactions on Pattern Analysis and Machine Intelligence**, Volume PAMI-1 (1979), pp. 60–72

[DuH73]   Duda, R.O. and Hart, P.E., **Pattern Classification and Scene Analysis**, John Wiley and Sons, New York, 1973

[GrL86]   Grosky, W.I. and Lu, Y., 'A Generalized Pattern Matching Approach to the Design of a Pictorial Index Mechanism,' **Computer Vision, Graphics, and Image Processing**, To Appear

[KnJ86]   Knoll, T.F. and Jain, R.C., 'Recognizing Partially Visible Objects Using Feature Indexed Hypotheses', **IEEE Journal of Robotics and Automation**, Volume RA-2 (1986), pp. 3–13

[LoW75]   Lowrance, R. and Wagner, R.A., 'An Extension of the String-to-String Correction Problem,' **Journal of the ACM**, Volume 22 (1975), pp. 177–183

[LuF78]   Lu, S.Y. and Fu, K.S., 'A Sentence-to-Sentence Clustering Procedure for Pattern Analysis,' **IEEE Transactions on Systems, Man, and Cybernetics**, Volume SMC-8 (1978), pp. 381–389

[MeG85]   Mehrotra, R. and Grosky, W.I., 'REMINDS: A Relational Model-Based Integrated Image and Text Database Management System,' **Proceedings of the IEEE Computer Society Workshop on Computer Architecture for Pattern Analysis and Image Database Management**, Miami, Florida, November 1985, pp. 348–354

[MyR81a] Myers, C.S. and Rabiner, L.R., 'A Level Building Dynamic Time Warping Algorithm for Connected Word Recognition,', **IEEE Transactions on Acoustics, Speech, and Signal Processing**, Volume ASSP-29 (1981), pp. 284–297

[MyR81b]   Myers, C.S. and Rabiner, L.R., 'Connected Digit Recognition Using a Level Building DTW Algorithm,' **IEEE Transactions on Acoustics, Speech, and Signal Processing**, Volume ASSP-29 (1981), pp. 351–363

[Pav77]   Pavlidis, T., **Structural Pattern Recognition**, Springer-Verlag, New York, 1977

[Pav78]   Pavlidis, T., 'A Review of Algorithms for Shape Analysis,' **Computer Graphics and Image Processing**, Volume 7 (1978), pp. 243–258

[Per79]   Perkins, W.A., 'A Model-Based Vision System for Industrial Parts,' **IEEE Transactions on Computers**, Volume C-27 (1979), pp. 126–143

[RoJ73]   Rosenfeld, A. and Johnston, E., 'Angle Detection on Digital Curves,' **IEEE Transactions on Computers**, Volume C-22 (1973), pp. 875–878

[SaC78]   Sakoe, H. and Chiba, S., 'Dynamic Programming Optimization for Spoken Word Recognition,' **IEEE Transactions on Acoustics, Speech, and Signal Processing**, Volume ASSP-26 (1978), pp. 43–49

[TMV85]   Turney, J.L., Mudge, T.N., and Volz, R.A., 'Recognizing Partially Occluded Parts,' **IEEE Transactions on Pattern Analysis and Machine Intelligence**, Volume PAMI-7 (1985), pp. 410–421

[TsY85]   Tsai, W.H. and Yu, S.S., 'Attributed String Matching with Merging for Shape Recognition,' **IEEE Transactions on Pattern Analysis and Machine Intelligence**, Volume PAMI-7 (1985), pp. 453–462

[WaF74]   Wagner, R.A. and Fischer, M.J., 'The String-to-String Correction Problem,' **Journal of the ACM**, Volume 21 (1975), pp. 168–173

TRAINING AND MODEL GENERATION FOR A SYNTACTIC CURVE NETWORK PARSER

J.R. Stenstrom

General Electric Company
Research and Development Center
KW C625 One River Road
Schenectady, NY 12345
U.S.A.

1. Introduction

This report describes a syntactic pattern recognition system which permits automatic model generation as well as efficient and effective model matching. The system uses a perspective invariant boundary curve segmentation that is uniform for all shapes. Boundary curves are categorized by attributes that vary gently with changes in viewing perspective. This enables automatic construction of the essentially finite state models used to model object shapes. The process is completed by a fast parser for pattern grammars made possible using an extension of the finite state machine. The system represents an effective compromise between a 2D matcher and a more general but computationally expensive 3D matcher, using the characteristic view/stable position paradigm as [1].

The system was first reported in [2] where the emphasis was principally on the syntactic matcher. Invariance of the curve segmentation with respect to non-trivial imaging perspective is presented in [3]. This report will concentrate on the syntactic models, their automatic synthesis, and their use in matching.

Robert Ledley's work, [4-6] for example, marks beginning of syntactic pattern recognition. Ledley begins by identification of the chromosome boundary. Each boundary point is considered as the center of a short curve. Through a simple approximation, the angle of curvature is estimated at each edge point. Peaks in curvature as the boundary contour is traversed become the boundary-segment curve centers. The curvature as well as the curve length, a tangent length, and the distance between adjacent curve segment centers permit the categorization of these boundary curves as one of a few basic types. These boundary-segments are combined according to a formal grammar to identify submedian and telocentric chromosomes. Thomason and Gonzalez [7] have extended the Ledley chromosome grammar for the standard substitution, insertion, and deletion errors.

Another important syntactic classification procedure is from Pavlidis and Ali [8], (see also [9]). The idea applies when rotation and scale are known. Each segment of a boundary, as within a particular section of the image, is described by a few parameters. These parameters are made into a short string within the window. The outline of the shape is the concatenation of these shorter strings. There are a large number of individual classifications made and the resulting string is usually long. A grammar necessary to accomodate variations in shape becomes large. However, such a grammar defines a regular language (see [10,11]). A regular language can always be put into a form where it is very easy to determine membership. Also, the automatic generation of a regular grammar is straight-forward, even from a very large number of classified examples.

Different applications have seemed to require different choices in boundary curve primitives. This human involvement effectively blocked any hopes of a generally applicable system. You and Fu [12-14] devised a general description for any planar curve based upon four purely numerical attributes. These four attributes are similar to those underlying the curve classification of Ledley. The usage of these attributes for curve classification depended upon being able to determine closed

NATO ASI Series, Vol. F45
Syntactic and Structural Pattern Recognition
Edited by G. Ferraté et al.
© Springer-Verlag Berlin Heidelberg 1988

object boundaries. Their error-correcting matching techniques also effectively precluded consideration of complex networks of classified curves.

Davis and Henderson [15] suggest a stratified context-free grammar where symbols have both attributes and attaching points (see also [16 to 19]). The authors have developed a grammar for airplane shapes which fills 28 typed pages. Davis and Henderson make the telling observation that "It is doubtful, at this time, that such grammars could be inferred automatically and without human intervention...". One problem cited with grammar synthesis relates to structuring the nonterminals to "correspond to meaningful, rather than arbitrary, pieces of the shape." They also specifically mention the difficulty in inferring semantic rules for attributes involved in productions. Davis and Henderson recognized the need to "allow *all* plausible descriptions or labels" to each primitive.

In this report a method is developed where each shape primitive is classified with a single label at model building time. When input patterns are considered, each primitive is allowed to have all plausible labels. It will come clear that alternate segmentations, in fact entire alternate models, can be directly incorporated into simpler grammars.

The system itself may be viewed as an extension of the work of You and Fu. You and Fu had proposed a curve classification scheme with applicability to general shapes. Their technique overcame this most critical obstacle. It remained to be shown how the curve classification could be applied to the more difficult problem of recognizing shapes not arbitrarily isolated from their background.

2. Lines, Angles, and Curves

Consider figure 1 and a path $l_1 l_2 l_3$. Refer to an angle such as that made by l_3, to the left of the continuation of l_2 (the dotted line) as an *inward* or *a –type angle* or simply an *a* . Refer to an angle into the other half plane, to the right of the continuation, such as that made by l_4 as an *outward* or *b –type angle* or simply a *b* . The terms outward and inward are based upon the convention of a counter-clockwise traversal of an object boundary contour. Lines are labeled $l_1 l_2 l_3 \cdots l_n$ in counter-clockwise traversal. Consider a-type angles to have positive sign and b-type angles to have negative sign with magnitudes determined as pictured in figure 2.

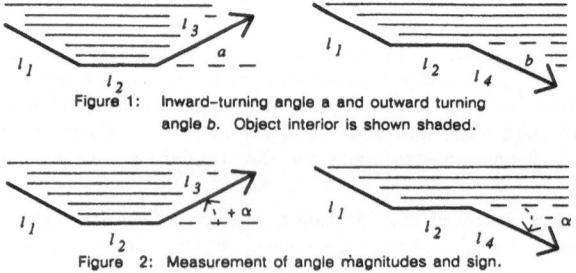

Figure 1: Inward–turning angle a and outward turning
angle *b*. Object interior is shown shaded.

Figure 2: Measurement of angle magnitudes and sign.

A curve $A = l (al)^+$ or $B = l (bl)^+$ will be called *pseudo –convex* (see figures 3a thru 3d). The term pseudo-convex was chosen to reflect the notion that curves such as those pictured in figure 3c and 3d are not the border of any convex body.

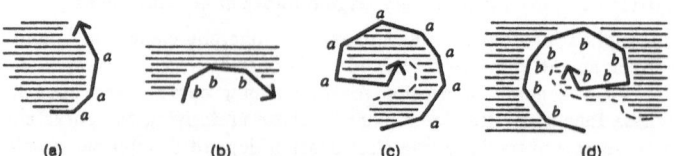

Figure 3: Examples of pseudo-convex curves. (a) and (c) are inward–turning or A–type
curves. (b) and (d) are outward–turning or B–type curves. Shaded areas
correspond to the inside of objects or the convention of a counter-clockwise
traversal of the boundary contour. For help in visualization, the boundary
contour is continued with broken lines in (c) and (d).

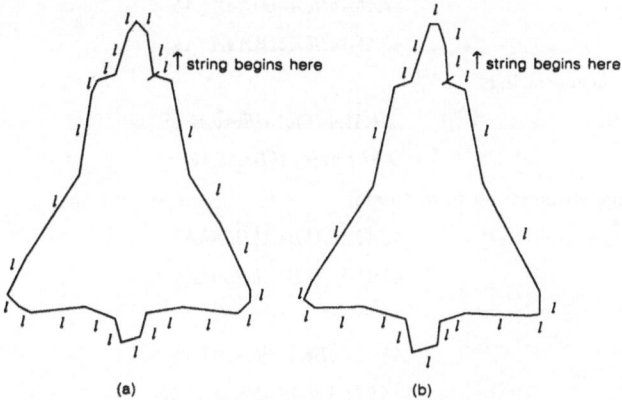

Figure 4: Two line–fitting results for aircraft shape.

3. Rewriting Line-Angle Strings as Symbols in the Curve Alphabet

Consider the shapes in figure 4a and 4b. The figures show two different segmentations for boundaries representing the same aircraft shape at the same scale. In terms of the line-angle alphabet the boundary of 4a is

$$lalalalalblalalblalalalalblblalalblblalalalalblalb$$

and for 4b is

$$lalalalblalblalalalblblalalblblalalalblalb.$$

It is well known that the variations involved in the digitization of an image prevent the line and angle string representation from usefully modelling the object. We seek to translate such strings over the line-angle alphabet into strings over the curve alphabet. The translated strings will not only be shorter but also will create meaningful equivalence classes for strings over the line-angle alphabet. Such strings may be expected to be both representative of the object and robust enough to occur across variations in digitization.

Recall that a pseudo-convex curve was defined to be a sequence of line segments and the connecting angles where either all connecting angles are a, or all are b. The form for an A is $l(al)^*$. Similarly, the form for a B is $l(bl)^*$. A *rewriting* of a string over the line-angle alphabet:

translates some substrings $l(al)^*$ to A,
translates some substrings $l(bl)^*$ to B, and
translates some angles a and b to the same symbols in the curve alphabet

where

each symbol in the string over the line-angle alphabet is used in exactly one translation. That is, each line-angle symbol must be used and there can be no overlapping use of the line-angle symbols.

The symbols a and b from the curve alphabet represent those angles at which subsequent curves connect and are called *connecting angles*.

Several rewritings of the string from 4a are:

$$AaAaBaAbAaAaBbAaBbAaAbAb$$

$$AaAaBaAbAaAaBbAaBbAbAb$$

$$\cdots$$

$$AbAbAbBaAbBaAaAbAb$$

$$AbAbAaBbAaBbAaAbAb$$

or either of the shorter strings

$$AbAbAaBbAaBbAbAb$$

$$AbAbAbBaAbBaAbAb.$$

Several rewritings of the string from 4b are:

$$AaAbAbAbBaAbBaAbAb$$

$$AbAbAaAbBaAbBaAbAb$$

$$\cdots$$

$$AbAbAaBbAaBbAaAbAb$$

$$AaAbAbAaBbAaBbAaAbAb$$

or either of the shorter strings

$$AbAbAaBbAaBbAbAb$$

$$AbAbAbBaAbBaAbAb.$$

The final string is the same as the final string of example 4a.

The shapes of figures 4a and 4b are derived from the same image and differ because of varying the parameters involved in the line fitting and merging. For both figure 4a and 4b, a starting point near the nose of each airplane was used. If different starting points in the line-angle string were used, even more rewritings would be produced. Two different images would be expected to produce even greater variation because of differences in illumination, different orientations, chance digitization effects, and variations in scale. Even with two shapes from the same image, there is little correspondence between the two line-angle strings or the various curve strings representing the shapes.

Yet there is a procedure that will select exactly a decomposition of each line-angle string that insures rewriting to consistent curve alphabet symbols for both examples. The next section develops a uniform method to select the connecting angles that join the pseudo-convex curves.

4. Developing Canonical Strings Over the Curve Alphabet

Define the cyclic equivalence class of the string s over the line-angle alphabet, $cycle\,(s)$:

$$\{s' : s' = s_2 s_1 \text{ where } s = s_1 s_2, \text{ and } s_1, s_2 \in (la \cup lb)^* \}.$$

Similarly, define the cyclic equivalence class of the string S over the curve alphabet, $Cycle\,(S)$:

$$\{S' : S' = S_2 S_1 \text{ where } S = S_1 S_2, \text{ and } S_1, S_2 \in (Ab \cup Aa \cup Ba \cup Bb)^* \}.$$

Let s be any string of lines and connecting angles chosen from $cycle\,(s)$ representing an object contour. The plan is to develop a rewriting rule that will automatically translate from s to a string S over the curve alphabet. Let s and s' be two strings where s and s' rewrite to strings S and S'. It would be ideal to have the property that if $s' \in cycle\,(s)$, then $S' \in Cycle\,(S)$. That would mean that a cyclic permutation of the rewriting would result, independent of the starting vertex.

Only a weaker starting point invariance result may be established. A *normal form* will be defined to restrict the cyclic rotations of a string. The set $\eta(s)$ will represent the normal form cyclic rotations of s. It is always true that $s \in cycle\,(s)$ but it may *not* be the case that $s \in \eta(s)$.

Define a string s' in $cycle\,(s)$ be in *normal form* when s' is either:

1) $(lb \cup la)^* lalalb$,
2) $(lb \cup la)^* lblbla$,
3) $(la)^*$, or
4) $(lalb)^+$.

Such an s' will be called alternatively a *normal form rotation* for s, a *normal form* for s, or simply a *normal form*.

It should be clear that each string over the line-angle alphabet (excluding $(lb)^+$) is either in normal form or can be transformed to normal form by a simple rotation. Moreover, if a string has either a third or fourth normal form, it has only third or fourth normal forms respectively.

The goal is to usefully rewrite normal form strings over the line-angle alphabet to a cyclic equivalence class of strings over the curve alphabet. But any partition of a string over the line-angle alphabet into parts such that each part corresponds to a symbol in the curve alphabet is a valid rewriting. It is essential to identify particular connecting angles from the line-angle strings to become the connecting angles in a curve rewriting.

A *designated* angle is either the first angle of the string or an angle whose immediately previous angle is of the same type. A *continuable* angle is either a designated angle or a nondesignated angle whose last previous designated angle is of the same type. A *breakable* angle is one that is not continuable. Thus, for example, the leftmost angle in any normal form is continuable and the first different angle to the right is breakable. The final angles in either a first or second normal form are breakable. For strings in third normal form there can be no breakable angles. For any substring $ala(la)^* b$ (or $blb(lb)^* a$) the single b (or a) is a breakpoint for a curve ending at the b (or a).

These breakable angles in a normal form turn out to be precisely the right choice of connecting angles to allow different normal forms for the same string to rewrite to curve strings within the same cyclic equivalence class. This is because the *same* breakable angles are preserved under any normal form rotation of a string (the proof is presented in [20]). This means that for a polygonal shape, as represented by *any* normal form line-angle string, certain angles take on the distinguished role of connecting pseudo-convex curves. Thus breakable angles are a consistent feature upon which to synchronize our curve generation from lines and connecting angles. Define σ to rewrite a line angle string in normal form grouping all continuable angles as interior angles in pseudo-convex curves A or B and all breakable angles as the connecting angles a or b in the curve alphabet. It is shown in [20] that σ fully rewrites any normal form. These breakable angles lead to the proof of the main result of this section:

The Cyclic Invariance Theorem
For any string s in normal form,

$$Cycle\,(\sigma(s)) = \sigma(\eta(s)).$$

The proof itself is presented in [20].

The cyclic invariance theorem shows that for any given normal form string of lines and angles, a curve string is obtained that is unique to within a cyclic permutation.

5. Invariance of Curve Segmentation Under Perspective

An important result of the present research concerns the curve segmentation. Perspective images of planar polygons are considered. It is shown in [3] that under perspective transformations resulting from ordinary imaging, the image of a breakable angle is a breakable angle and the image of a continuable angle is a continuable angle -- unless the entire image is reduced to a single line segment.

6. Curve Categorization

This section considers a particular categorization procedure based upon four attributes of plane curves. These attributes, developed by You and Fu [12-14], have been found particularly effective for categorizing pseudo-convex curves. You and Fu represent each curve by a feature vector (V, L, A, S) combining the four attributes.

The first curve attribute V is the vector at the origin that has correct magnitude and direction to connect the first and final points in a curve, see figure 5a. The attribute L is the total length of the curve, figure 5b. The attribute A is the sum of the outer angles of a sequence of vectors, figure 5c. The attribute S is a measure of angular change as a function of curve length, see [12]. For a symmetric pseudo-convex curve, the S attribute is zero. If there is greater angular

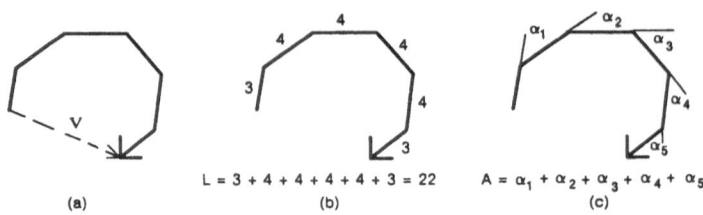

$$L = 3 + 4 + 4 + 4 + 4 + 3 = 22 \qquad A = \alpha_1 + \alpha_2 + \alpha_3 + \alpha_4 + \alpha_5$$

(a)   (b)   (c)

Figure 5: (a) Curve Attribute V, the vector between starting and ending points.
            (b) Curve Attribute L, lengths are given near line segments.
            (c) Curve Attribute A.

change per unit length in the first part of the curve, the attribute takes on a positive value. If the angular change occurs mostly in the later part of the curve, S is negative.

You and Fu give simple computational procedures for finding these attributes in the case of a sequence of vectors approximating a smooth underlying curve.

You and Fu theorem C from reference [12]: $M = (v_1, v_2, \ldots v_m)$ is a vector chain. Let M_i denote $(v_1, \ldots v_i)$, i.e. $M_m = M$, a_i is the angle between v_i and v_{i+1}, $D(v_i) = (V_i, L_i, 0, 0)$, $1 \leq i \leq m$.
Then $D(M_j) = (V_{Mj}, L_{Mj}, A_{Mj}, S_{Mj})$, $1 \leq j \leq m$, where:

$$V_{Mj} = V_{M(j-1)} + V_j = \sum_{i=1}^{j} V_i$$

$$L_{Mj} = L_{M(j-1)} + L_j = \sum_{i=1}^{j} L_i$$

$$A_{Mj} = A_{M(j-1)} + a_{j-1} = \sum_{i=1}^{j-1} a_i$$

$$S_{Mj} = G_{Mj} - \frac{1}{2} A_{Mj} L_{Mj}$$

$$G_{Mj} = G_{M(j-1)} + A_{Mj} L_j = \sum_{i=1}^{j} A_{Mi} L_i = \sum_{i=1}^{j} \sum_{l=1}^{i-1} a_l L_i.$$

An important feature of the particular attributes developed by You and Fu is that they permit a simple computation of the attributes of a curve made up of two smaller curves. Assume curves with attribute vectors (V_1, L_1, A_1, S_1) and (V_2, L_2, A_2, S_2) are being combined at an angle a. Then the attribute vector for the combined curve is (V, L, A, S) where:

$$V = V_1 + V_2$$

$$L = L_1 + L_2$$

$$A = A_1 + a + A_2, \text{ and}$$

$$S = S_1 + S_2 + \frac{1}{2} [(A_1 + a) L_2 - (A_2 + a) L_1].$$

You and Fu offer a normalization that reduces an object made up of a sequence of the attributed curves to be rotation, translation, and scale invariant (without the effects of digitization or noise) in images:

$$N(V, L, A, S) = (\frac{|V|}{L}, \frac{L}{L_{total}}, A, \frac{S}{L})$$

where L_{total} might be chosen as the total object perimeter, the sum of the individual L. Since the object's total perimeter is not available until a full match to a curve sequence is determined, a different transformation must be used for curve networks. The transformation continues to make use of the curve's length to normalize two other attributes and so remains rotation and translation invariant:

$$T(V,L,A,S) = (\left\lfloor 100\frac{\mid V \mid}{L} \right\rfloor , \lfloor L \rfloor , A , \left\lfloor 100\frac{S}{L} \right\rfloor)$$

Multiplication by 100 and flooring makes the first transformed attribute an integer between zero and 100. The length is taken as an integer number of pixels. The final symmetry attribute is also chosen to be an integer. This final transformation is the one actually used in this implementation. Consistent scaling will be enforced in the matching phase.

7. From Curve Attributes to Categorized Curves

Consideration has been given to dividing an objects boundary contour into a sequence of pseudo-convex curves. Several curve attributes have been presented which might be used to classify the pseudo-convex curves. The transition from curve attributes to classified curves is quite simple.

First cutoffs are computed for curve attributes. An automatic procedure makes a histogram of each of the curve attributes independently and arrives at fixed percentile breaks for each attribute. These percentile ranges form categories for each attribute. The curves themselves are categorized as the cartesian product of the separate attribute categories.

As with any categorization scheme, curves near attribute category boundaries are subject to misidentification with repeated samples. To prevent this most basic sort of misidentification, a more tolerant technique must be used when a trial string is compared to a model string. When an attribute value is near a category boundary, the attribute is freely considered to be *either* category. When classified based upon the three curve attributes, curves in a trial string may be determined to be in 1, 2, 4, or as many as 8 categories. String matching is made reliable in this way, even for distorted boundary curves.

8. Perspective Effects on Curve Attributes.

The recognition or matching decision is based upon a string of categorized curves. For a curve model to be valid over a useful range of viewing positions, it is necessary that the curve categorization vary as little as possible with changes in perspective. In this section a simple experiment is performed to establish that the curve classification remains stable under significant changes in viewing perspective.

Four perspective transformations were considered. Refer to figure 6a-6d. The first transformation 6a is the identity transformation. The second 6b is merely a scale reduction. The third 6c is a considerable shear and results in looking from head on at an angle of about one-half radian. The fourth 6d is at a severe angle, almost one radian.

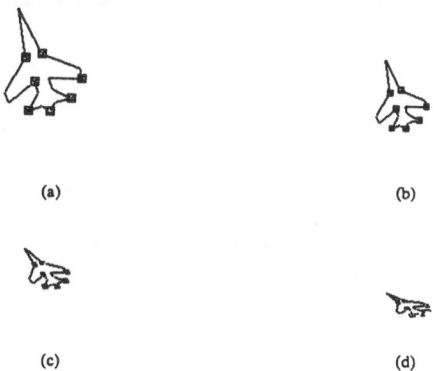

(a)

(b)

(c)

(d)

Figure 6: A shape under four different viewing perspectives.

The model shape under consideration has seven pseudo-convex curve primitives. The primary curve classification is the one used ordinarily in model building. The experiment considers primitive classification exactly as it is done in the matching. Under each of the four perspective transformations, each of the seven curves was categorized to include the original primary curve classification -- with one exception. One curve, under the most severe perspective, included only a neighboring classification to the primary curve classification. Even in the case of the most severe perspective, there was a matching substring of length six out of a maximum of seven.

9. The Scaled Finite State Machine

Just as it is sometimes convenient to attach values or attributes to a grammar it is useful to do similar computations with finite state machines. It is an important problem in syntactic pattern recognition to insure conformity of the sizes of component features found to be in the proper configuration by a shape grammar.

As a practical example consider figure 7. The model for a timing light is shown in the upper left corner. At this time consider the contour to be made up of only two types of curves. For actual matching purposes curves are categorized with greater specificity using the attributes and categorization of sections 6 and 7. For the present example it is clearer to use only two types of curves. The lengths of the various curves are represented as curve attributes. The model string may be written as $A_{294}B_{72}A_{93}B_{30}$ where A_{294} is the curve at the top of the timing-light.

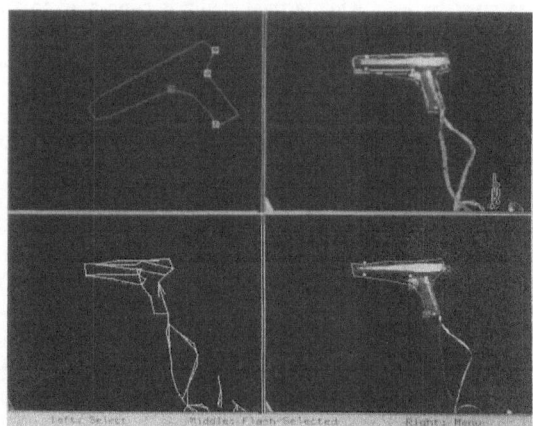

Figure 7: Timing light example. Model in upper left, image with edges overlaid upper right, line–angle network lower left, match discovered in network overlaid on test image lower right.

The problem is to recognize strings $ABAB$ with appropriate length attribute combinations. Each corresponding curve in the recognized string must have approximately the same portion of the total sum of lengths as the curve in the model string. Assume we wish to test whether a *trial string*, such as $A_{201}B_{39}A_{55}B_{25}$, conforms to the model.

The correspondence of each curve in a trial string to a curve in the model string permits an estimate of the overall perimeter for the shape represented knowing only the length of the curve. Let the length of the curve in the trial string be l, the length of the corresponding curve in the model be m, and the total model perimeter be M. Then the expected size of the completed trial perimeter would be:

$$M\left(\frac{l}{m}\right).$$

The length conformity is checked using a set of overall *scales* or ranges for the total perimeter of the shape. For simplicity, assume we are interested in only four scales: 0 to 150, 151 to 300, 301 to 450, and 450 or more. It is useful to associate a bit-vector with several scales. Let the scale 150 or less be associated with bit position 0. Similarly, let scales 151 to 300, 301 to 450, and 450 or more be associated with bit positions 1, 2, and 3 respectively.

Typically, approximate lengths developed over several stages of preprocessing are used to derive the estimate of total shape perimeter. The ranges for total perimeter used in the four scales are broad. Nevertheless, there are cases where the implied total shape perimeter might fall near the boundary of one of the above ranges. If reliable scale matching is to be performed, it is essential to allow overlapping ranges of perimeters to determine scale. The table below gives one possible choice for range overlap in determining scale. Also shown is the corresponding bit-vector representation for the range.

Range	Bit-Vector	Associated Scale
0 to 100	0001	scale 0
101 to 200	0011	scales 0 or 1
201 to 250	0010	scale 1
251 to 350	0110	scales 1 or 2
351 to 400	0100	scale 2
401 to 500	1100	scales 2 or 3
501 or more	1000	scale 3

Matching of each of the four trial curves, in order, to the four model curves implies four bit-vectors. If the four bit-vectors share a one in any common bit position, the trial string is recognized as an instance of the model string.

The scaled finite state machine (scaled-FSM) associates a bit-vector or *scale–vector* B_i to each state S_i. The scaled-FSM also associates an update function
$$f_{A:i \to j}(l)$$
to each transition under letter A with associated length attribute l from state S_i and S_j. Initially, the scale-vector for the start state is all ones and the scale-vector for every other state is all zeros. The update rule for transition
$$(A_l, S_i, S_j)$$
on symbol A of length l from state S_i to S_j with the scale-vectors B_i and B_j respectively is
$$B_j \leftarrow B_j \ or(B_i \ and f_{A:i \to j}(l)).$$

The lengths of the letters of an input string accepted by the scaled finite state machine are called *conformable* if the scale-vector associated with any final state is not all zeros at the end of the computation.

A complete generalization of the scaled finite state machine to an attributed finite state machine is possible. By attaching general attributes to states and associating suitable update functions to transitions it is possible to mirror exactly the computations involved in an attribute grammar. The construction is tedious and represents no gain over the attributed grammar for our purposes.

Consider the steps in recognizing the trial string $A_{201}B_{39}A_{55}B_{25}$ conforms to the model string $A_{294}B_{72}A_{93}B_{30}$.

Represent the model string by the scaled-FSM given in figure 8a. It will be helpful to consider one bits in the scale-vector of state S_0 as meaning that a partial recognition of a substring of the corresponding scale has been discovered in the trial string.

Define the functions $f_{A:i \to j}(l)$ for each transition to return the bit vector corresponding to the range selected based upon the result of $\frac{l}{m}M$ in the table above.

Consider the scaled-FSM given in figure 8b. This is the result of updating the state S_1 based upon the first symbol A_{201} of the trial string. Since
$$\frac{201}{294} \times 489 = 334,$$
by the table above
$$f_{A:0 \to 1}(201) = 0110.$$
There is a length one substring in the trial string that is consistent with the model string at a perimeter of between 151 and 450.

Advancing by another input symbol B_{39} leads to figure 8c. Since

$$\frac{39}{72} \times 489 = 265$$

the scale transition function is

$$f_{B:1 \to 2}(39) = 0110.$$

There is a length two substring in the trial string that is consistent with the model string at a perimeter of between 151 and 450.

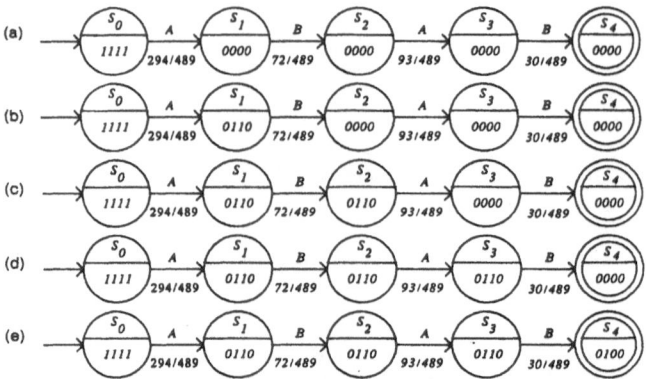

Figure 8: Scaled-FSM:

    (a) Initial Machine

    (b) Scaled-FSM after transition on A_{201}.

    (c) Scaled-FSM after transition on $A_{201} B_{39}$.

    (d) Scaled-FSM after transition on $A_{201} B_{39} A_{55}$.

    (e) Final Scaled-FSM after transition on $A_{201} B_{39} A_{55} B_{25}$.

Using the next symbol of the trial string A_{55} leads to figure 8d. Once again compute

$$\frac{55}{93} \times 489 = 289$$

or that $f_{A:2 \to 3}(55) = 0110$. There is a length three substring in the trial string that is consistent with the model string at a perimeter of between 151 and 450.

The result of the computation for the trial string is given in figure 8e. The correspondence of the symbol B_{25} from the trial string to the symbol B_{30} of the model leads to the estimate

$$\frac{25}{30} \times 489 = 407$$

for total object perimeter. Thus, the scale-vector update based upon the table above is

$$f_{B:3 \to 4}(25) = 1100.$$

There is a length three substring in the trial string that leads to a perimeter of between 151 and 450. There is also a curve symbol that advances from state S_3 to S_4. That curve symbol, by correspondence to the model string, leads to a perimeter size of greater than or equal to 301. Thus, it is easily concluded that the trial string may be recognized as the model at a perimeter of between 301 and 450. Once the scale-vector is non-zero for the final state, the final state has been reached at a conformable size. The correctly matched curve string of conformable size is displayed in the lower-right of figure 7.

As each transition of length l is taken from state S_i with scale-vector B_i to S_j with scale-vector B_j we update B_j using the bitwise logical operators. The computational overhead compared to the simplest implementation of a finite-state machine is a table lookup followed by several fixed-size logical bit operations. Typically eight bits or fifteen ranges are used. By increasing the number the ranges we can achieve greater specificity of the object perimeter.

10. Model Creation

At the present time models are generated in two steps. First a digital image is processed up through the stages of a smoothing, boundary point detection, fitting lines to the boundary points, and coalescing nearby junctions. The model is selected as a sequence of pseudo-convex curves. Each curve is in turn selected as a number of sequential lines. In the present implementation the connecting angle is determined as the angle between the last line segment of one curve primitive and the first line segment of the next.

Once the curve primitives and connecting angles are selected it is a simple matter to convert them into efficient syntactic models. The syntactic processes of the previous section are *fully automated* and directly produce the syntactic model.

Every string over the curve-angle alphabet representing an object shape is converted into a scaled-FSM representing either the string or, at the users option, the cycle of that string (see section 4). The cycle construction creates a final scaled-FSM with no more than $n\left(\frac{n}{2}\right)+2$ states for a model string of length n. Figure 9 shows the cycle scaled-FSM corresponding to figure 8.

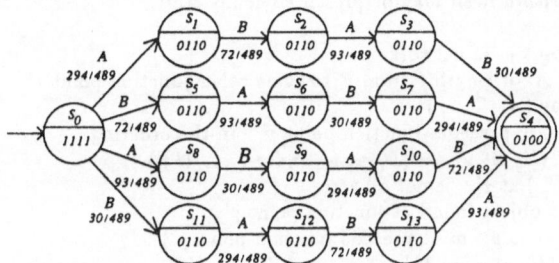

Figure 9: Cycle Scaled-FSM of Scaled-FSM of figure 9.

It is a simple matter to perform the operations of union and intersection on the scaled finite state machines. Thus, the processes of merging multiple models for the same shape (as from two different views) is fully automated. Also, there is an automatic check to assure that two different model classes have an empty intersection.

11. Building the Curve Network

Each vertex joining two or more line segments is considered to determine which segments make a breakable angle with another as defined in section 4. The curve primitives are grown only from these breakable line segments, since the segmentation of the model boundaries occurs only at breakable vertices. No vertex ever occurs in a single curve more than once. Such a non-simple curve would not occur in the segmentation of an object boundary. Because of irregularities in digitization, curve growth must continue across some junctions that are not strictly convex. At the same time a shorter curve should be produced ending at each such non-convex angle when that too would be a reasonable interpretation of the line segment sequence.

A number of particular criteria are used to determine when a curve may be grown across an oppositely-signed angle. Other heuristics limit such things as total curve length, sizes of connecting angles, and total angular change.

12. Parsing Networks of Scaled Curves

In some applications it is possible to isolate an object from its surroundings. The image becomes a binary image, object pixels with one value, everything else with the other. Without the strictest control on the imaging situation, object isolation is not a manageable problem. Rather than having an object border which can be brought into simple correspondence with a string, there are typically a number of edge points meandering along various contours and intersecting at various junctions in the image. In this section consideration is given to the problem of how we might process such a *network* of curves, classified as terminals or *primitives* in formal grammatical models.

The plan is to maintain a scaled-FSM at each junction of curve intersection in the network. If a particular bit is set in the scale-vector for a particular state at a given junction, there is a path to that junction that would cause a scaled finite state machine to reach that state and scale.

Having all zeros in a state for a given junction means that no path of conformable scale leading to that junction and state has been located.

Besides considering whether each state and scale can be reached at a given junction from the start state, consider also whether any final state can be reached from a given state and scale at that junction. This requires the use of two bit-vectors as scale-vectors for each state. Call the vector that tracks the ability to reach "down" to a given state and scale from the start state the D vector. The vector U will track whether there is a path of consistent scale that will lead back "up" to a final state from a given state. A complete conformable path is discovered when a given scale bit is set for both the vector D and the vector U for any state.

At the start of the algorithm the scale-vector D for the initial state and the scale-vector U for each final state are set to all ones at each junction of the network. Every other bit is set to zero. No bit is ever changed from one to zero in the course of operation of the algorithm.

The entire algorithm itself may be presented compactly:

for each junction point p in network
  for each primitive A of length l from p to some other junction point p'
   for each transition (A_l, S_i, S_j)
    if junction point p precedes junction point p' in the ordering
      D_j at p' ← $(D_j$ at p' or $(D_i$ at p and $f_{A:i \to j}(l)))$
     if $(D_j$ and $U_j)$ at p' is not all zero
      then report object found at junction point p'
    else {junction point p' must precede junction point p }
      U_i at p ← $(U_i$ at p or $(U_j$ at p' and $f_{A:i \to j}(l)))$.
     if $(D_i$ and $U_i)$ at p is not all zero
      then report object found at junction point p

Consider an example to clarify the algorithm. Once again, refer to the simple shape illustrated in figure 7. The timing-light has four curve segments represented by the string $A_{294}B_{72}A_{93}B_{30}$ as before.

Recall that the shape corresponds to the string scaled-FSM of figure 8 or the cycle scaled-FSM of figure 9. It will be simpler to consider only the string model of figure 8. Figure 10a represents a simplified subset of the line network pictured in figure 7. The figure also shows part of an initial state scale-vector configuration. Notice that all the scale bits corresponding to initial and final states are set to one at every junction and that no other scale bits are set. Notice too that the points are identified in raster order P_0, P_1, P_2, and P_3.

This example will show the actual process by which node updates reach the conclusion of a timing-light present at node P_3. This corresponds to the trial string $A_{201}B_{39}A_{55}B_{25}$ being identified as conforming to the model string exactly as in section 9.

The process considers curve junction nodes in the raster order presented above, beginning with P_0. There are two phases to each junction update. First every curve primitive that begins at that junction is used to advance the D-vector states and scales. There is only one such curve primitive of interest to the example, the primitive A_{201} from P_0 to P_2. Recall the scaled-FSM update made use of bit-vector

$$f_{A:0 \to 1}(201) = 0110$$

on the transition from state S_0 to state S_1. The corresponding update for the network scaled-FSM is

$$D_1 \leftarrow D_1 \text{ or } (D_0 \text{ and } f_{A:0 \to 1}(201))$$

or, substituting the values,

$$D_1 = 0110 \leftarrow 0000 \text{ or } (1111 \text{ and } 0110).$$

This update takes the D-vector field for state S_0 of P_0 and uses it in an *and* operation with the result of $f_{A:0 \to 1}(201) = 0110$. The result, 0110, is used in a boolean *or* along with the value of the D-vector field for state S_1 at P_2, 0000. The result of the *or*, 0110, is redeposited in the D-vector field for state S_1 at P_2. Figure 10b shows, at P_2, the result of this transition in the scaled-FSM network.

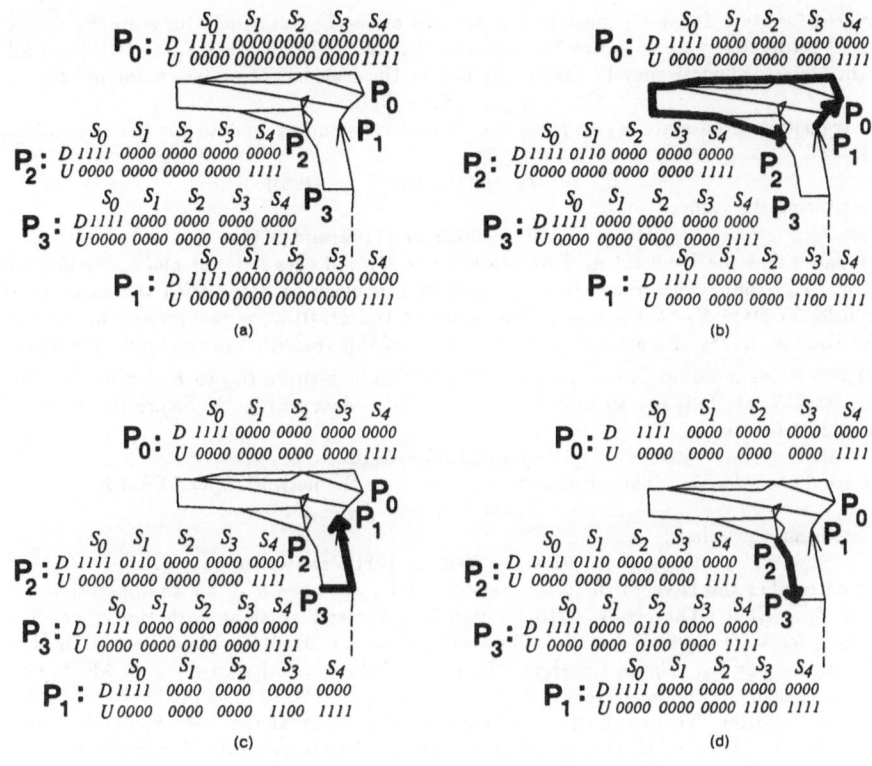

Figure 10: Network example:

(a) Initial Configuration.

(b) After A_{201} from P_0 to P_2 and B_{25} from P_1 to P_0.

(c) After A_{55} from P_3 to P_1.

(d) After B_{39} from P_2 to P_3. At P_3 ; S_2 : $0110 \cdot 0100 = 0100 \neq 0000$

The other phase to the junction update is used to advance the U-vector states and scales. The D-vector tracks the progress forward through the scaled-FSM, tracking the ability to reach a state and scale by a substring from the start state. In a completely analogous way, the U-vector tracks the progress "backward" through the scaled-FSM, tracking the ability to reach from a state and scale by a substring to a final state.

There is only one useful curve primitive of interest to the U-vector update at P_0, B_{25}. This curve goes from P_1 to P_0. The scaled-FSM update of section 9 made use of

$$f_{B:3\to4}(25) = 1100$$

on the transition from state S_3 to state S_4. The corresponding update for the network scaled-FSM is

$$U_3 \leftarrow U_3 \text{ or } (U_4 \text{ and } f_{B:3\to4}(25)).$$

or, substituting the values,

$$U_3 = 1100 \leftarrow 0000 \text{ or } (1111 \text{ and } 1100).$$

This update takes the U-vector field for state S_4 at P_0 and uses it in an *and* operation with the result of $f_{B:3\to4}(25)$. The result, 1100, is used in a boolean *or* along with the value of the U-vector field for state S_3 at P_1, 0000. The result of the *or*, 1100, is redeposited in the U-vector field for state S_3 at P_1. Figure 10b shows, at P_1, the result of this transition in the scaled-FSM network.

The next curve vertex in raster order, P_1, has only one curve of interest to this example. A_{55} took the scaled-FSM from state S_2 to state S_3. In the curve network A_{55} goes from curve vertex P_3 to P_1. For the transition to do something useful in the curve network either the D-vector field for state S_2 at P_3 must be not all zero or the U-vector field for state S_3 at P_1 must be not all zero. Referring to figure 16, only the U-vector field for state S_3 at P_1 is not all zero. Then the U-vector update may be taken. As before, the scaled-FSM update made use of

$$f_{A:2\rightarrow3}(55) = 0110$$

on the transition from state S_2 to state S_3. The corresponding update for the network scaled-FSM is

$$U_2 \leftarrow U_2 \; or \; (U_3 \; and \; f_{A:2\rightarrow3}(55)).$$

or, substituting the values,

$$U_2 \; = \; 0100 \leftarrow 0000 \; or \; (1100 \; and \; 0110).$$

This update takes the U-vector field for state S_3 at P_1 and uses it in an and operation with the result of $f_{A:2\rightarrow3}(55)$. The result, 0100, is used in a boolean or along with the value of the U-vector field for state S_2 at P_3, 0000. The result of the or, 0100, is redeposited in the U-vector field for state S_2 at P_3. Figure 10c shows the result of this transition in the scaled-FSM network.

There is one more update of interest. At P_2 there is a curve B_{39} to P_3. Since the D-vector field for state S_1 at P_2 is not all zero, there is a useful update to P_3. As before this leads to scale vector update function

$$f_{B:1\rightarrow2}(39) = 0110$$

from state S_1 to state S_2. The corresponding update for the network scaled-FSM is

$$D_2 \leftarrow D_2 \; or \; (D_1 \; and \; f_{B:1\rightarrow2}(39))$$

or, substituting the values,

$$D_2 \; = \; 0110 \leftarrow 0000 \; or \; (0110 \; and \; 0110).$$

This update takes the D-vector field for state S_1 of P_2 and uses it in an and operation with the result of $f_{B:1\rightarrow2}(39)$. The result, 0110, is used in a boolean or along with the value of the D-vector field for state S_2 at P_3, 0000. The result of the or, 0110, is redeposited in the D-vector field for state S_2 at P_3. Figure 10d shows the result of this transition in the scaled-FSM network.

Now consider both the D-vector field for state S_2 at P_3, 0110, and the U-vector field for state S_2 at P_3, 0100. There is a one in a common bit position at the state S_2 at P_3. Similar to the case with the scaled-FSM, this means that there is a length two substring in the trial network that ends at P_3 that would advance the scaled-FSM from state S_0 to S_2 that conforms with the model occurring at a scale of between 151 and 450. There is also a length two substring in the trial network that would advance the scaled-FSM from state S_2 to S_4 beginning at P_3. Exactly as before, it is easily concluded that there is an instance of a length four string in the trial network that may be recognized as the model at a perimeter of between 301 and 450. Once the bit-vectors have a one bit in the same position in both the D-vector field and the U-vector field for state, a string has been found in the trial network at a conformable size. The string of conformable size is displayed in the lower-right of figure 7.

13. Merging

Using primitive merging in the matching process allows fewer curves to be created from the line-angle network. Other research has independently verified the utility of merging when dealing with problems of inconsistent segmentation, see Tsai and Yu [21].

The merging step takes place during the final phase of matching. First each curve primitive at a point is used directly. Next the curve is considered for a possible merging. The curve may be merged with any other suitable connecting angle and curve primitive at the junction where the first ends. This may be done symbolically by specifying special production rules for merging. This merging may be done numerically as well. The curve attributes have simple rules allowing computation from the two adjacent curves. There is reason to prefer the numerical merging. It requires no special operator training phase, identifying curve-angle-curve triples and naming the curve that results from the combination.

14. A Polygonal Study.

Five simple polygonal shapes were considered for purposes of illustration. The idealized decomposition of these shapes into the component curves is pictured in figure 11.

Figure 12 is a table of the actual values sorted by attribute. It should be apparent that subdividing the transformed length attribute l at 5, 12, and 50 yields an effective separation of the

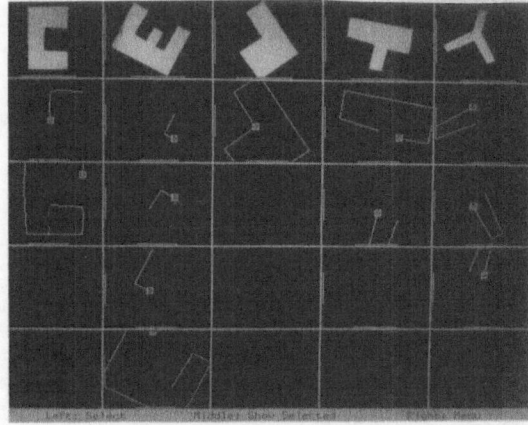

Figure 11: Decomposition of five polygonal shapes into pseudo–convex curves.

Connecting Angle	L	A	S
−1.54	0	−1.65	−32
−1.49	9	−1.55	−32
−1.49	15	−1.55	−27
−1.46	17	1.52	−18
−1.46	19	3.0	−8
−1.43	19	3.0	−6
−1.42	20	3.0	−4
−0.72	25	3.05	−2
−0.62	72	6.16	−1
1.51	73	7.7	11
1.54	75	7.82	13
1.58	77	7.82	14

Figure 12: Sorted values for connecting angle and curve attributes.

values into four well-divided classes. Similarly an effective grouping for the transformed angle attribute a may be accomplished to by splitting values at 0.0, 2.25, 4.5, and 6.8 yielding five classes. A reasonable set of values for the transformed symmetry attribute is -10, 1, and 5 giving four categories once again. Dividing the values for the connecting angles at -1.0, 0.0, and 1.0 gives a reasonable separation of the connecting angles into four groups.

For each of the groupings number the classes from zero to three or four as appropriate. Refer to the connecting angles as ap0 to ap3. The curve primitive number for a curve primitive p is based upon:

$$s + a n_s + l \; n_s \, n_a$$

where s, a, and l are the number of the category for each of p's attributes S, A, and L and n_s is the number of the s categories, in this case four, and n_a is the number of the a categories, here five. In this way each curve primitive may be classified as being one of curve primitives 0 through 79.

The scaled finite state machine for the "E" shape is given in figure 13.

A second image is provided for each shape at a different scale and in a different orientation. An edge detector is used and line-fitting is performed exactly as in the first image. Breakable line segments are found at junction points and curves are formed starting with those segments. Using this simple formulation each shape was found without exception and with no false matches.

The string of line segments and connecting angles forming curve primitives found to match the shape "E" are highlighted in figure 14.

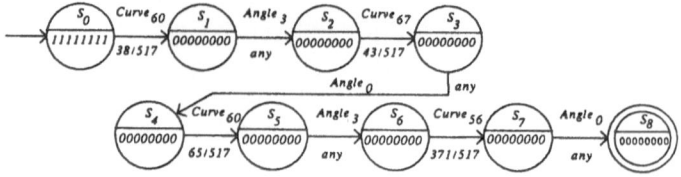

Figure 13: Scaled–FSM for "E" shape polygon.

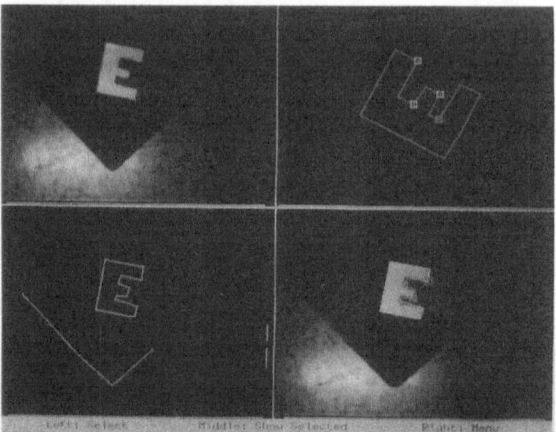

Figure 14: Illustration of Match to "E" shape. Test image upper left. Model "E" shape upper right. Edges from test image lower left. Match discovered overlaid on test image lower right.

It is interesting to consider the elapsed times (on a Symbolics 3600 Lisp Machine) for different steps in the processing of this example:

syntactic model synthesis - 20 seconds, only once per type of problem
smoothing, edge detection - 90 seconds
line fitting and linking - 120 seconds
breakable angle determination - < 1 second
curve primitive construction - 1.5 seconds
parsing - 3.45 seconds

It is clear that the time involved for the initial processing dominates. Fortunately, these operations may sometimes be done quickly in special hardware. Both the curve construction and curve parsing are perfectly parallel operations and could be well-suited to parallel computation.

15. Two Aircraft Shapes

More complicated shapes are provided by two airplanes images. Models were formed of the shapes in figure 15a and 15b. These images were matched to find the corresponding airplane shape in figure 16a and 16b. The shapes have relatively long curve sequences, length 6 and 7. With angles considered they have representative string lengths of 12 and 14. With such long strings avoidance of misclassification is easier. The scaling condition forces corresponding training and input curves to represent roughly the same proportion of the object contour. Even without using the scaling criterion, or considering the connecting angles, the longer number of classified curves gives a more unique match than with shorter models.

Figure 15: Two aircraft shape model segmentations.

16. A Curve Network Example

As a final example, return to figure 7. There are 72 line segments forming a highly interconnected network. This example serves to demonstrate the need for and efficacy of the various curve formation heuristics. If curves are formed only at breakable junctions, are allowed to end only at breakable junctions, and are allowed to turn in only a single direction there were 712 curve primitives generated. If curves are grown from every line at each junction are allowed to end anywhere provided that they are allowed to turn in only a single direction and that at least one angle occurs there are 1203 curves.

At the other extreme, if curves may be grown from any line junction and pass through angles of either type as long as no more than four wrongly turning angles are passed and the net angle change is in the proper direction there are 41210 curves. If five wrongly turning angles may be passed the number reaches 73942. Restricting the curves to starting and ending at breakable angles reduces the numbers to 24666 and 43558 curves respectively.

The preprocessing steps result in fairly long lines when possible. Using 0.3 radians, about 17 degrees, as a limit for the size of a wrong turning angle to be freely allowed it was not necessary to allow any larger angles to be included. Using that value along with the fairly liberal constraints on curve growth only 2282 curves were grown. The actual heuristics limited the length of the curve to 300 pixels in length, the total angular change to 7.0 radians (more than a full circle), the largest angle used in a curve to 3.0 radians, and the second largest and all smaller angles to 2.0 radians. Assuming that the imaging geometry permitted knowing the expected size of the largest curve to be 200 units and that the angular change was restricted to a smaller value of 5.0 radians the number of curves was reduced to 768.

Without any parallel computation but using the most judicious heuristics, the curves were constructed and the actual string matching time to the model of the timing light was about 15 seconds each. Using the less judicious heuristics the curves were constructed in about 30 seconds. The actual time for matching the strings of curve primitives to the model of the timing light was also about 30 seconds. It should be noted that thousands of string sequences were evaluated in that time and many of those matches were trying to *simultaneously* match to a number of other models. In fact, it is interesting that a simple backtracking tracer sometimes takes about six times longer to find just the single match at a point, even knowing 1) that a specific match occurred there, 2) exactly what state the match occurred on, and 3) the actual object scale.

17. Discussion

A hierarchical decomposition of the general contour matching problem has been developed. One key element in the process is a robust boundary segmentation. This segmentation has the advantage of perspective invariance [2]. At the same time, there is a simple test for possible curve segmentation points in a line-angle network. A formal automata style model, the scaled-FSM, has been developed for an important subclass of attributed grammar systems. This essentially finite-state model permits fast contour matching while enforcing a consistent scale on the component parts.

The hierarchical decomposition permits a generic training phase since each object shape is modeled by a uniform segmentation of pseudo-convex curves. Automatic grammar synthesis is provided in way both conceptually simple and computationally manageable. The scaled-FSM model makes simple the cycle, union, and intersection operations. These operations are useful in generalizing, combining, and testing multiple syntactic models.

The application of edge detectors to grey-level image data yields a network of lines and angles. As a result of the segmentation and the scaled-FSM, it becomes feasible to process such networks. This avoids an intractable problem. It is no longer necessary to arbitrarily divide the image into object and background.

The process is completed by a flexible curve primitive categorization. During the matching phase, individual curve primitives are freely counted as being in any category to which they are sufficiently close. Only when such curve primitives occur, connected with the proper angles, in the correct order, and at a uniform size, is a match produced.

18. References

[1]   Chakravarty, I., Ph.D. Thesis, Rensselaer Polytechnic Institute, Troy, New York, 1982.

[2]   Stenstrom, J. R., "Syntactic Pattern Recognition for Robot Vision", IEEE First Int. Conference on Robotics, Atlanta, 1984.

[3]   Stenstrom, J. R., "An Improved Segmentation for a Syntactic Curve Network Parser", Eighth Int. Conf. on Patt. Recog., Paris, 1986.

[4]   Ledley R. S., "Automatic Pattern Recognition for Clinical Medicine", Proc. IEEE, vol. 57, no. 11, 1969.

[5] Ledley, R.S. "High Speed Automatic Analysis of Biomedical Pictures", Science, vol. 146, 1965.

[6] Ledley, R. S. "Practical Problems in the Use of Computers in Medical Diagnosis", Proc IEEE, vol. 57, 1969.

[7] Thomason, M.G. and R.C. Gonzalez, "Syntactic Recognition of Imperfectly Specified Patterns", IEEE Transactions on Computers, January, 1975.

[8] Pavlidis, T. and F. Ali, "A Hierarchical Syntactic Shape Analyzer", IEEE Tr. Patt. An. and Mach. Intell, vol PAMI-1#1, 1979.

[9] Pavlidis, T., *Structural Pattern Recognition* , New York: Springer-Verlag, 1977.

[10] Hopcroft, J. E. and J. D. Ullman, *Formal Languages and Their Relation to Automata* , Reading, Massachusetts: Addison-Wesley, 1969.

[11] McNaughton, R., *Elementary Computability* , *Formal Languages* , *and Automata* , Englewood Cliffs, New Jersey: Prentice-Hall, 1982.

[12] You, K.C. and K.S. Fu, "Syntactic Shape Recognition Using Attributed Grammars", TR-EE 78-38, Purdue University, September 1978.

[13] You, K.C. and K.S. Fu, "A Syntactic Approach to Shape Recognition Using Attributed Grammars", IEEE Transactions on Systems, Man, and Cybernetics, Vol. SMC-9, No. 6, 1979.

[14] You, K.C. and K.S. Fu, "Distorted Shape Recognition Using Attributed Grammars and Error-Correcting Techniques", Computer Graphics and Image Processing, Vol 13, 1-16 1980.

[15] Davis, L.S. and T.C. Henderson, "Hierarchical Constraint Processes for Shape Analysis", IEEE Tr. Patt. An. and Mach. Intell, vol PAMI-3#3, 1981.

[16] Pfaltz, J.L. and A. Rosenfeld, "Web Grammars", Proc. 1st Int. Joint Conf. on Art. Intell., Washington, D.C., 1969.

[17] Feder, J., "Plex Languages", *Information Sciences* , vol 3, 1971.

[18] Rosenfeld, A. and D.L. Milgram, "Web Automata and Web Grammars", *Machine Intelligence* 7, Edinburgh U. Press, 1972.

[19] Brayer, J.M. and K.S. Fu, "The Derivation Diagram of a Web Grammar and Its Application to Scene Analysis", Joint Workshop on Pattern Recognition and Artificial Intelligence, IEEE, 1976.

[20] Stenstrom, J. R., Ph.D. Thesis, Rensselaer Polytechnic Institute, Troy, New York (in preparation).

[21] Tsai, W.H. and S.S. Yu, "Attributed String Matching With Merging", Seventh Int. Conf. on Patt. Recog., Montreal, 1984.

VI.　APPLICATIONS II

KNOWLEDGE-BASED COMPUTER RECOGNITION OF SPEECH

R. de Mori

McGill University
School of Computer Science
805 Sherbrooke St. W.
Montréal, Quebec H3A 2K6
Canada

Abstract

Shape recognition by fast syntactic methods is possible when there exists a natural linear (one dimensional) order on component shapes. This may not be available for *structural* shape descriptions taking the form of unordered, variable-length sets of simpler shapes. In this case, it is tempting to fall back on slower exhaustive correlation, graph matching, and relaxation methods. However, if the structural shapes are themselves *simple*, it is possible to apply multi-dimensional search techniques for asymptotically fast feature identification. I exploit the fact that many simple shape types may be parameterized as points in low-dimensional spaces where distance models dissimilarity. During training, shapes are clustered heuristically within each class, then among all classes, giving a small set of characteristic shape distributions. Each os these is then associated with a binary feature variable taking the value one when any input shape falls within the distribution. This mapping from a structural description into a bit-vector is an example of a *feature identification* method. Selecting such a mapping is slow and heúristic, but fully automated, applicable uniformly to many shape types, and controlled by only a few natural statistical parameters. A mapping, once selected, can be applied quickly using kD-trees. Large-scale statistically-significant trials have shown the technique to be superior to simpler fixed mappings, in an OCR context.

ACKNOWLEDGEMENTS

This research was supported by the Natural Sciences and Engineering Research Council of Canada with grant no. A2439.

NATO ASI Series, Vol. F45
Syntactic and Structural Pattern Recognition
Edited by G. Ferraté et al.
© Springer-Verlag Berlin Heidelberg 1988

1. INTRODUCTION

At present, a number of scientists and engineers seem to be quite interested in doing research in the area of speech recognition by computer. Different workers in the field have different approaches, and might even describe their motivations for doing speech recognition research somewhat differently. A very common position, for example, is that the main goal of speech recognition research is to develop techniques and systems for speech input to machines. If we consider machines which are real computers, rather than mere automatic dictation devices, this makes speech recognition an instance of the general problem of designing a convenient and pleasant human-computer interface, the ultimate goal being the ability to talk to computers in much the same way we now talk to fellow human beings. Indeed, if both speech recognition and general machine intelligence make sufficient progress in our lifetimes, we could conceivably encounter computers that not only listen but also reply sensibly.

Figure 1 shows the essential transformations of information involved in speech communication. Initially, a sentence generator produces an abstract representation S of the sentence to be transmitted. This abstract representation S is then converted by the speaker into a sequence of sets of discrete articulatory commands which drive the vocal tract actuators, producing a continuously time-varying pressure signal $x_1(t)$.

The signal $x_1(t)$ is transmitted through a noisy acoustic channel, resulting in a different signal $x(t)$ which is perceived by the listener.

The listener transforms the signal $x(t)$ into an acoustic pattern Ω.

The purpose of this transformation is to have a representation of the spoken message which better exhibits the linguistic features than does the signal itself. Furthermore, an attractive hypothesis about the usefulness of this transformation is that it could allow the application to speech patterns of the same interpretation strategies that are used for vision.

The acoustic pattern is then interpreted by a recognition/understanding system which first transforms the acoustic information in Ω into an abstract, linguistic representation Λ.

Unfortunately, Λ is not S. Rather, it may be a continuous string or a lattice of characters or of word hypotheses which has to be interpreted in order to produce $\hat{S}$, the recognition system's final interpretation of $x(t)$. In a satisfactory recognition system, the self-correcting mechanisms of perception function well enough to produce an $\hat{S}$ which is S most of the time.

Some speech recognition workers have chosen to apply the discipline of information theory to the construction of recognizers. That theory as originally conceive is a mathematical theory designed to measure the amount of information necessary to reduce the receiver's doubt concerning given alternatives. The contrasting approach taken here is to model both the source and the sentence interpreter as rule-based systems, where the rules encode the a priori knowledge we have about human speech generation and understanding.

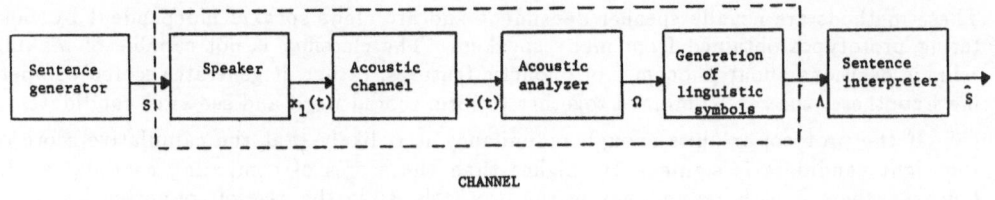

CHANNEL

```
S:        sentence representation
x₁(t):    generated signal
x(t):     received signal
Ω:        acoustic pattern
Λ:        lattice of hypotheses composed of linguistic symbols
Ŝ:        interpretation
```

Figure 1: The Speech Communication Channel

There are many dimensions affecting the feasibility and performance of a speech recognition system.

One of the main features affecting the complexity of a speech recognition task is whether the speech is connected or is spoken one word at a time. In the latter case, the system complexity mainly depends on the range of vocabulary. Even the recognition of isolated words is a difficult task because some acoustic ambiguity may be present in the spoken word to be recognized. This ambiguity may depend on the speaker, his personality, his dialect or his speaking rate.

Some of the difficulties may be removed by asking the speaker to be cooperative, pronouncing the words carefully, with sufficiently long pauses between them. But even when speakers are cooperative, there may be ambiguities in the recognition system due to the limitations of the acoustic analyzer and to our imperfect knowledge of the features to be used and the way they have to be extracted. Fortunately, contextual (inter- and intra-syllabic) information can be used to resolve such ambiguities.

The reasons for the inherent difficulty of computer recognition of continuous speech are, at present, well understood. In continuous speech it is difficult to determine where one word ends and another begins. Moreover, co-articulation and other effects lead to a far greater context-dependent variability in the acoustic characteristics of sounds and words. For these reasons continuous speech recognition systems do not achieve the very high performance of isolated word recognition systems. We suggest that very high performance continuous speech recognition could conceivably come, in the long run, as rule-based computer recognition systems mature into deep knowledge-based systems.

Knowledge-based speech recognition systems require mechanisms to pay attention to distinctive acoustic information, detailed phonetic features, and the like. These mechanisms can perhaps be justified by considering alternate recognizers which use little or no speech knowledge.

For example, many workers use a recognition model based on feature extraction and classification. With such an approach, the same set of features are extracted at fixed time intervals (typically every 10 msecs.) and classification is based on distances between feature patterns and prototypes [LEVINSON 81] or likelihoods computed from a Markov model of a source of symbols generated by matching centisecond speech patterns and prototypes [BAHL 83].

These methods are usually speaker dependent and are made speaker independent by clustering prototypes obtained from many speakers. The classifier is not capable of making reliable decisions about phonemes of phonetic features; rather, it generates scored competing hypotheses that are combined together to form scored word and sentence candidates.

If the protocol exhibits enough redundancy it is likely that the cumulative score of the right candidate is significantly higher than the scores of competing candidates. If, however, there is little redundancy in the protocols, as in the case of connected letters or digits or in the case of a large lexicon, then it is important that ambiguities at the phonetic level are resolved before hypotheses are generated. Examples of these difficulties have been reported in the recent literature [BAHL 84, RABINER 84]. For example, in the case of connected letters, in order to distinguish between /p/ and /t/ the place of articulation is the only distinctive feature, and its detection may require the execution of special sensory procedures on a limited portion of the signal with the time resolution finer than 10 msec.

Some people have argued that speech recognition will require fundamental progress in machine intelligence to deal with problems of context-dependence and disambiguation; we make the more modest claim that knowledge-based recognition systems need only increase their speech competence to improve their recognition performance.

We have referred to the fact that most of the systems proposed so far take advantage of the redundancy of the protocols they use. The most difficult and unsolved problems arise when tasks have little redundancy or when speaker independence is required for complex tasks. Examples of such complex tasks are the recognition of letters and digits (isolated or connected) and the recognition of large vocabularies. For such problems, it seems reasonable to extract a large variety of acoustic properties.

2. A MODEL FOR COMPUTER PERCEPTION

Drawing on important computer science and machine intelligence motivations, a system has been implemented in which both the extraction of acoustic properties and the generation of syllabic hypotheses result from the collaboration of several distinct processes. This cooperation of computational activities has been conceived using the paradigm of an Expert System Society [MINSKY 75, ERMAN 80].

Each expert is associated with a Long Term Memory (LTM) containing the specific expert's knowledge and a Short Term Memory (STM) where data interpretations are written.

Experts are computing agents which execute reasoning programs using structural and procedural knowledge. The knowledge of each expert is expressed by a set of plans, some of which can be executed in parallel. Communication between cooperating tasks is performed by message passing.

Interpretations of the speech waveform are generated by an Expert System Society. Its structure is shown in Figure 2.

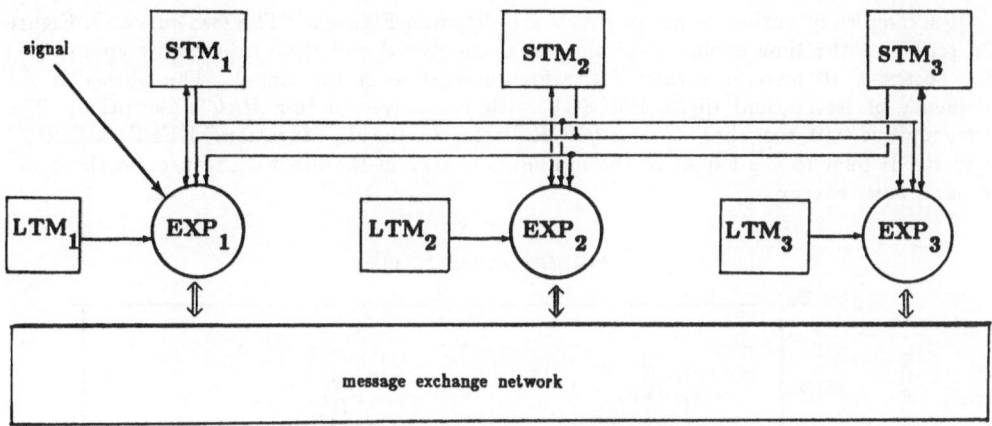

Figure 2: Expert System Society

EXP_1 is the Acoustic Expert (AE). It has the task of sampling and quantizing the signal, performing various types of signal transformations, and extracting and describing acoustic cues. The term acoustic cues will be used for indicating spectral or signal properties describing aspects that are relevant for hypothesizing phonetic features. Examples of acoustic cues are formant loci, characteristics of burst spectra like compactness, and peaks and valleys of signal energy.

EXP_1 can perform, for example, an analysis based on Linear Prediction Coefficients (LPC) for segments labelled with vocalic hypotheses in order to find formant loci capable of describing the place and manner of articulation. EXP_1 can also perform a broad-band spectral analysis based on the Fast Fourier Transformation (FFT) when hypotheses of nonsonorant continuant sounds have been made. Like other experts, EXP_1 may carry out both spontaneous data-driven activities and expectation-driven activities arising out of requests issued by other experts.

Requests and control messages are exchanged among experts through the message exchange network shown in Figure 2. Data, cues, descriptions and hypotheses are written by an expert into its own Short Term Memory (STM). Only the expert which owns the STM can write into it, but any expert can read any STM.

EXP_2 is the Phonetic and Syllabic Expert (PSE). It translates descriptions of acoustic cues into phonetic feature hypotheses. These features describe the manner and the place of articulation of each segment of the spoken utterance. This translation may involve the extraction of new acoustic cues by asking EXP_{1k} to execute sensory procedures.

There are some acoustic cues, like peaks and valleys of time evolutions of energies in fixed bands of the signal, that can be extracted by context-independent algorithms. These acoustic cues will be called Primary Acoustic Cues (PAC) and the phonetic features related to them will be called Primary Phonetic Features (PPT). A definition of the primary cues and features used in the system described here is given in Tables I, II and III. These algorithms for extracting PACs generate descriptions of a time interval of the signal without being constrained by contextual information extracted from adjacent segments.

Examples of various types of PACs are shown in Figure 3. The two curves in Figure 3a represent the time evolution of the signal energy (–) and the zero-crossing counts (---) in successive 10 msec intervals of the first derivative of the signal. The phrase is the sequence of letters and digits E3KBCV with the corresponding PAC description. The time unit is 0.01 sec. LONG and SHORT refer to the dip duration. DEEP, MEDIUM and HIGH refer to the height of the minimum energy in the dip with respect to the background noise energy.

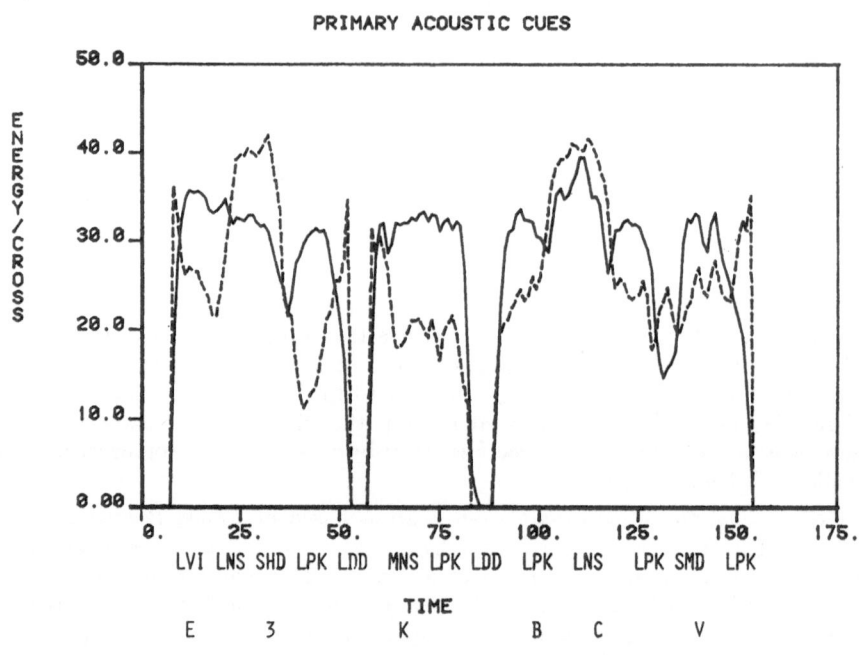

Figure 3: A Concatenation of letters and digits

Legend
(––) Total energy
(--) zero crossing counts

Other functions of EXP_2 include those of segmenting the speech signal into Pseudo Syllabic Segments (PSS) and of checking or evaluating phonetic hypotheses. How exactly hypotheses are formulated and then tested, whether the hypotheses are highly specific or more open and less specific, and how central the asymmetry between positive evidence and negative evidence is to perception, are all deep questions both for the psychology and philosophy of perception and for the design of computer perception systems; we have identified these questions as subjects for research but do not claim to possess definitive answers to any of them. The current system implementation implicitly models some of these aspects in a tentative fashion.

In EXP_2, the activity of generating PPFs is data-driven, while the activities of extracting other phonetic features expectation-driven. Expectations may arise from a strategy inside EXP_2 or they can be requests transmitted by EXP_3. This distribution of expectation sources in the computer recognition system has a direct analogy to the fact that, in humans, perception is distributed throughout large regions of the brain; we see again the distinction between distributed sensors and distributed intelligence.

EXP_3 is the Lexical Expert (LE) that generates lexical hypotheses based on prosodic features, phonetic hypotheses, and syntactic and semantic constraints.

3. ACOUSTIC PROPERTIES, PHONETIC FEATURES, AND PLANS

This section describes how relations of phonetic features to acoustic properties are embedded in <u>plans</u> which are executed by the expert system.

With this approach a phoneme PH_i is expressed by a set of phonetic features, i.e.

$$PH_i = (pf_{i1}, pf_{i2}, \ldots, pf_{ij}, \ldots, pf_{iJ}) \tag{1}$$

Each phonetic feature ph_z is represented by a relation R_z to a set properties ap_z, i.e.

$$pf_z = R_z(ap_{z1}, ap_{z2}, \ldots, ap_{zk}, \ldots, ap_{zK}) \tag{2}$$

For example, the phoneme /p/ is represented as follows:

Table I
Primary Acoustic Cues

Symbol	Attributes	Description
LPK	tb,te,ml,zx,	long peak of total energy (TE)
SPK	"	short peak of TE
MPK	"	peak of TE of medium duration
LOWP	"	low energy peak of TE
LNS	tb,te,zx	long nonsonorant tract
MNS	"	medium nonsonorant tract
LVI	tb,te,ml,zx	long vocalic tract adjacent to a LNS or a MNS in a TE peak
MVI	tb,te,ml,zx	medium vocalic tract adjacent to a LNS or a MNS in a TE peak
LDD	emin,tb,te,zx	long deep dip of total energy
SDD	"	short deep dip of total energy
LMD	"	long dip of total energy with medium depth
SMD	"	short dip of total energy with medium depth
LHD	"	long non-deep dip of total energy
SHD	"	short non-deep dip of total energy

Attribute description

Attribute	Description
tb	time of beginning
te	time of end
ml	maximum signal energy in the peak
emin	minimum total energy in a dip
zx	maximum zero-crossing density of the signal derivative in the tract

/p/ = (nonsonorant-interrupted-consonant,tense,labial) = (nit,labial).

The phonetic feature 'labial' in the context of 'nit' features is represented by the following relation R_k:

(**relation** R_k)

(**left-side**   (3)
(**feature** (labial))
       (**feature context** (nit))
       (**temporal context** (**followed-by** font vowel)))
(**right-side**
(suprasegmental and time-domain properties)
(formant-transition properties)
(burst-spectra properties)))

The rule for 'labial' takes into account different types of contextual dependencies. One contextual dependency is represented by the other features that appear with 'labial' in a plosive phoneme. The other contextual dependencies are represented by the class of phonemes that can follow or precede the plosive phoneme under consideration. Relations are used by plans executed by the expert system.

In many cases, acoustic property extraction is context dependent; for example, we may be forced to impose precedence relations on the extraction processes. For more on the latter, we refer to [DEMICHELIS 83] and [KOPEC 84].

A plan is a sequence of items. Each item may contain a precondition expression for applying rules of the type R_k, operators containing sensory procedures for extracting the properties used by R_k, and an algorithm for evaluating the evidence of the hypothesis generated by R_k. An example of a plan applying rules like R_k will be given later in this section. In practice, a plan is a sequence of operators. Each operator is associated with a precondition and an action.

The Acoustic Expert is capable of extracting acoustic cues based on the requests it receives from the Phonetic and Syllabic Expert.

An example of such a request is the following:

$$RQ := \text{plosive-place-of-articulation } (t_1,t_2,t_3,\text{ctx})$$

where t_1 is the time of the beginning of the plosive silence or buzz-bar, t_2 is the time of the beginning of the burst, t_3 is the time of the beginning of the voice onset, and ctx is the context in which the plosive sound is hypothesized. For example ctx can be (front-vowel, F1, F2, F3). The time instants t_1, t_2 and t_3 are attributes of the primary acoustic cues which generated a plosive hypothesis.

When the Acoustic Expert receives a request for the extraction of the cues which generated a plosive hypothesis.

When the Acoustic Expert receives a request for the extraction of the cues for the place of articulation of plosive sounds it creates a process "plosive-place-of-articulation (t_1,t_2,t_3,ctx)" which executes a plan whose details are given in the following: Algorithm for process : plosive-place-of-articulation (t_1,t_2,t_3,ctx)

   begin
       find-burst-spectra (t_2,t_3,bs);

Table II
Rule-set (1)

No.	Precondition				Confusion set
1	LNS	MVI	SHD	LPK	3
2	MNS	MVI	SHD	LPK	3
	.				
	.				
8	SDD	LOWP	SMD	LPK	v
	.				
	.				
11	LDD	SPK	LHD	LPK	k
	.				
	.				
	.				
	.				
41	SDD	LPK			b, d, v
	.				
	.				
	.				
45	LDD	(LPK+LVI)			p, e, b, d, k, 3 v
	.				
	.				
	.				
74	LDD	MNS	MVI		k, g, t
	.				
	.				
	.				
90	SHD	MPK	LPK		v

```
burst-evidence := f1(bs);
compact-evidence := f2(bs);
diffuse-falling-ev := f3(bs);
diffuse-rising-ev := f4(bs);
track-formants-in-the-plosive-transitions (for,ctx,t₃);
compute-pseudo-loci (for,pl);
for-labial-evidence := R_{lp} (pl,ctx);
for-alveolar-evidence := R_{lp} (pl,ctx);
ev(labial) := R_{lp} (for-labial-evidence, diffuse-falling-ev, burst-evidence, ctx);
ev(alveolar:) := R_{lp} (for-alveolar-evidence, diffuse-rising-ev, burst-evidence, ctx);
ev(palatal) := R_{lp} (for-palatal-evidence, burst-evidence,compact-evidence, ctx);
send (PSE(ev(labial, ev(alveolar), ev(palatal), plosive(t₁,t₂,t₃)))
end
```

PSE collects all the evidences about syllabic hypotheses in a time interval (t_a, t_b) and compares them according to the following algorithm.

"PSE evaluator of hypotheses (t_a, t_b)"

```
    begin
        repeat
            receive (evidences,t_a,t_b);
            compose (evidences,ctx,t_a,t_b)
        until all-requests-satisfied;
        decide-hypotheses-to-be-kept (evidences,t_a,t_b,syll-hyp)
    end
```

4. DETAILED DESCRIPTION OF A PLAN

The speech signal is first analyzed on the basis of loudness, zero-crossing rates and broad-band energies as described in [DE MORI 85]. The result of this analysis is a string of symbols and attributes. Symbols belong to an alphabet of Primary Acoustic Cues (PAC).

The use of coarse acoustic properties (PAC) for segmenting continuous speech into acoustic segments has been described in a previous paper [DE MORI 85]. The segmentation algorithm based on an attributed grammar has been simplified, in such a way that a preliminary segmentation is performed first based only on signal energy deep dips and friction intervals. Each *acoustic segment* obtained in this way may contain one or more syllables but it never contains less than one vowel. Each acoustic segment could be further described following the scheme shown in Figure 4.

Figure 4 shows the general scheme of an action hierarchy and the planning system that generates its compiled version. Each expansion of an action represents a plan refinement. Each refinement is a sub-plan consisting of a sequence of operators. Preconditions for refinement are generated and updated by a Learning System [DE MORI 84].

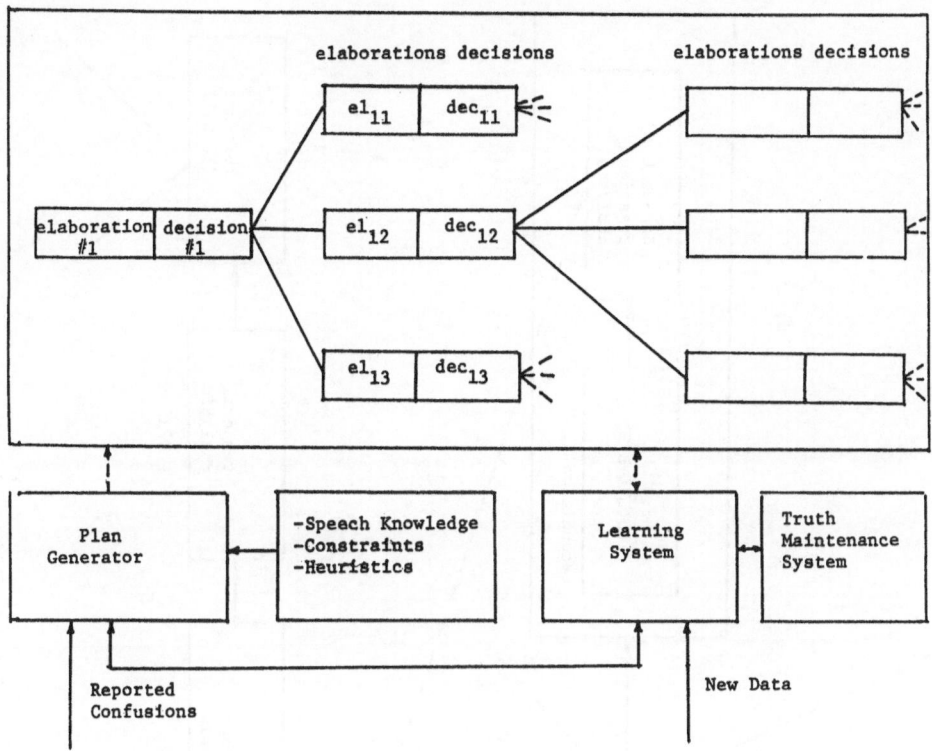

Figure 4: Scheme for Generating Network of Actions in Compiled Form

An action hierarchy is applied for interpreting the consonantal segment of every syllable.

Figure 5 shows the more abstract part of the action hierarchy. PAC extraction and segmentation is a preliminary spontaneous activity that is followed by a decision phase where rule-set(1) is applied. Rule-set(1) is made of preconditions and actions. Preconditions are sequences of PACs. Actions describe what acoustic properties are worth to be extracted given the suprasegmental morphology described by the PAC expression. Rule set (1) is represented in Table II.

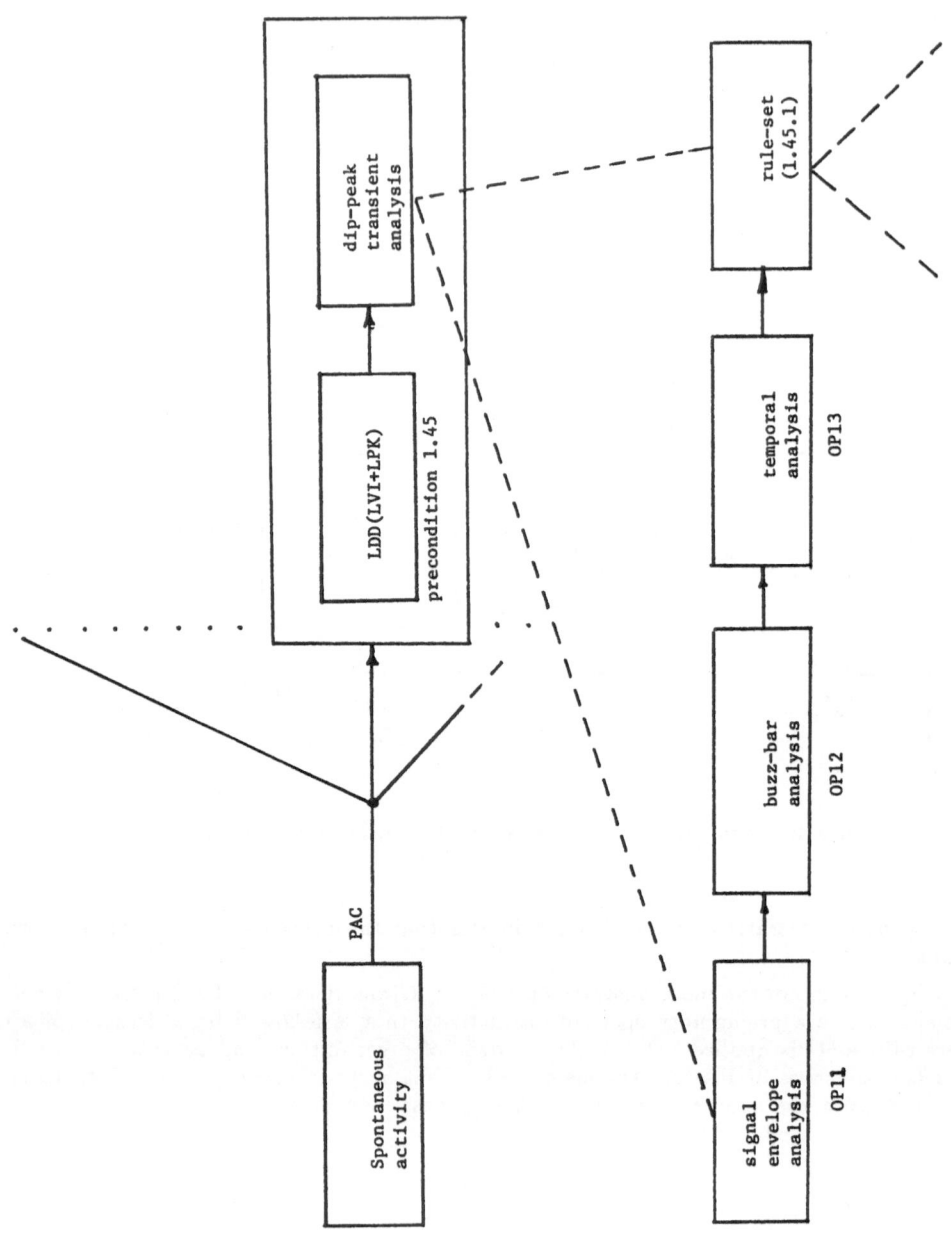

Figure 5: Example of Network Actions

For the sake of simplicity only the action corresponding to rule (1.45.1) is shown in Figure 5. The precondition for this rule is (LDD) (LPK + LVI). They sumbol + represents here logical disjunction. This PAC morphology may correspond to a vowel after a pause, to the consonants |v| or |g| or to plosive consonants for which burst is not evident in the loudness curve. This precondition represents a speaking mode that is rather frequent for |b| and is one of the most difficult to analyze. The action to be executed in this case is the analysis of the (deep dip) | (peak) transition. Useful operators are those that detect voicing, the existence of burst and some temporal relations among the onset of the energies in various bands because delays in these onset time are cues for the presence of plosive sounds.

The approach is that of extracting a *redundant set* of cues so that the final decision about the acoustic properties that describes them is reliable.

According to Figure 5, the activity "dip peak analysis" is refined by a more detailed sub-plan which is produced by the planning system introduced in the previous Section. As broad-band energies have to be analyzed at the end of the buzz-bar and when the signal envelope starts rising, the chaining order shown in Figure 5 is obtained.

If there is enough evidence for burst and if there is still a need of discriminating among classes, then a rule of rule-set (1.45) will invoke the execution of a burst analysis action

Op11 in Figure 5 produces an envelope description by analyzing the signal amplitude before and after preemphasis. Envelope samples are obtained every msec by taking the absolute value of the difference between the absolute maximum and the absolute minimum of the signal in a 3 msec interval. The envelope description is based on the following alphabets:

A111 =
{SHORT–STEP(ST1),
LONG–STEP(ST2),
NO–STEP(ST0)} (5)

A112 =
{HIGH LOW FREQUENCY ENERGY(BZ1),
ABSENCE OF BUZZ INDICATOR IN THE ENVELOPE (BZ0)}

A113 =
{POSSIBLE BURST (PB1),
ABSENCE OF BURST EVIDENCE (PB0)}

A114 =
{STRONG BURST EVIDENCE (BU1),
NO STRONG BURST EVIDENCE (BU0)}.

The description produced by Op11 as well as those produced by Op12 and Op13 are attached to the DDN describing the segment under analysis.

Because of the shape of the signal envelope, a description (ST0, BZ1, PB1, BU0) is obtained.

Op12 describes the buzz-bar by analyzing the shape of the time waveform and of the spectra before the voice onset. The alphabets of the description it produces are:

BZA1 = {BI1,BI2,BI3,BI4,BI5}

for the time waveform and

$$BZA2 = \{BR1,BR2,BR3,BR4,BR5\}$$

for the spectra.

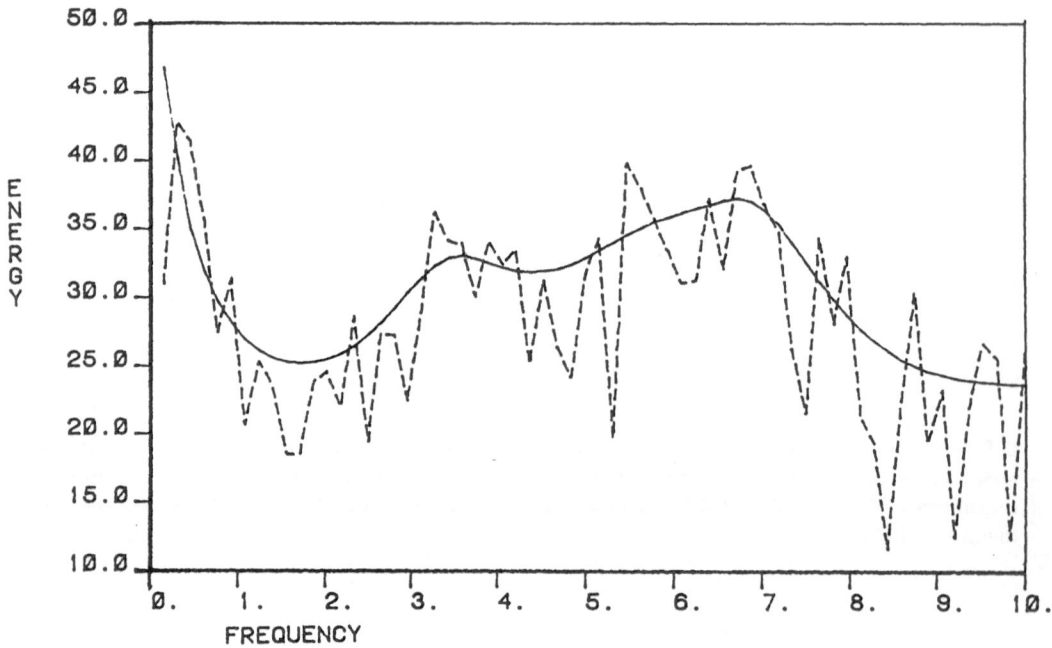

Figure 6:

Figure 6 shows the LPC (continuous line) and the FFT (dashed line) spectra computed before the TE peak onset. The consistent low frequency peak in the FFT spectrum is the cue for buzz-bar. The low frequency oscillations in the waveform are other buzz cues.

BI1 and BR1 mean no buzz and the other symbols describe degrees of buzz-bar evidence: (BI1,BR2: little evidence; BI5,BR5: strong evidence).

Based on the waveform of Figure 6 and the spectra shown in Figure 6 the segment is described as (BI3,BR5).

Op13 analyzes temporal events of the energy in some frequency bands at the voice onset. These events are related to voice onset time. They are:

NP : number of short peaks in the time evolution of the energy in the 2–4 kHz band,

RE : a bar of energy at low frequency before the peak

DL : the delay between the onset of low and high frequency energies,

ZQ : the duration of the largest zero-crossing interval of the signal at the onset,

ZR : the number of zero-crossing counts in the largest sequence of successive zero-crossing intervals with duration less than 0.5 msecs.

DR : minimum value in a dip of ratio between low and high frequency energies.

Table III

rule number	precondition	letter
1	BZ0 ST0 RE0 NP0 NP1	DR3 DR4 DL1 DL2 ZQ2 ZQ4 BI1 BI2 BR2 BR2	p
2	PB0 ST1 BZ1 BU0 NP0 NP1	DR3 DR4 DL1 DL2 ZQ3 ZQ4 BI3 BI4 BR3 BR5	d
3	BI0 STBI1 BZ1 BU0 NP0 NP1	DR3 DR4 DL1 DL2 ZQ3 ZQ4 BI3 BI4 BR3 BR5	b
4	PB0 ST1 BZ0 BU0 BU1 RE0 RE1 NP0 NP1	DR4 DR4 DL1 DL2 ZQ4 ZQ5 BI1 BI4 BR3 BR3	p
5	PB0 ST0 BU0 RE0 NP0 NP1	DR3 DR4 DL1 DL2 ZQ3 ZQ5 BI1 BI1 BR2 BR4	b
6	PB0 ST0 BI1 BZ0 NP0 NP1	DR3 DR4 DL1 DL2 ZQ2 ZQ4 BI1 BI2 BR2 BR3	e
7	PB1 ST0 BI1 BZ0 RE0 NP0 NP1	DR2 DR4 DL1 DL2 ZQ2 ZQ5 BI2 BI2 BR4 BR4	p
8	PB1 ST0 BZ0 BU0 RE0 NP1	DR2 DR3 DL2 DL2 ZQ5 ZQ5 BI4 BI4 BR4 BR5	p
9	PB0 ST0 BZ1 BU RE0 NP0	DR4 DR4 DL1 DL1 ZQ1 ZQ1 BI1 BI1 BR2 BR2	e
10	PB0 ST0 BZ1 BU0 RE0 NP1	DR2 DR2 DL4 DL4 ZQ4 ZQ4 BI2 BI2 BR3 BR3	3
11	PB0 ST1 BZ1 BU0 RE0 NP0	DR3 DR4 DL4 DL4 ZQ1 ZQ3 BI4 BI4 BR3 BR3	v
12	PB1 ST0 BZ0 BU0 RE0 NP1	DR2 DR3 DL2 DL2 ZQ1 ZQ3 BI1 BI1 BR1 BR2	k
13	PB1 ST0 BZ0 BU0 RE0 NP1 NP2	DR2 DR2 DL2 DL2 ZQ1 ZQ3 BI1 BI1 BR1 BR3	d
14	PB1 ST0 BZ1 BU0 RE0 NP1	DR3 DR3 DL1 DL1 ZQ4 ZQ4 BI1 BI1 BR2 BR2	p
15	PB0 ST0 BU0 RE0 NP0 NP1	DR3 DR4 DL1 DL2 ZQ2 ZQ4 BI1 BI1 BR2 BR2	v
16	PB0 ST0 RZ1 RE0 NP0	DR4 DR4 DL1 DL1 ZQ1 ZQ1 BI1 BI1 BR2 BR2	e

The first two cues represent morphologies described by the alphabets: {NP0,...,NP5},{RE0,RE1}. Intervals of parameters are coded with symbols {DL1,...,DL5}. {ZQ1,...,ZQ5}, {DR1,...,DR5}. From a dip-peak transition, a vector of values of (NP,RE,DL,ZQ,ZR,DR) is extracted and represented by a conjunction of symbols according to the intervals the parameters fall into. Intervals are determined after clustering vectors corresponding to the same letter and intersecting clusters of different letters.

The description obtained by OP11, OP12 and OP13 for the |b| is the following:

Descr. 1.45 : {PB1 ST0 BZ1 BU0 RE0 NP1 DR4 DL2 B13 BR5}

This description is compared with the premises of the rule set given in Table III. The letter candidate with the highest SM with Descr. 1.45 is considered along with the letter candidates with SM close to this maximum. All these letters constitute an active confusion set. If the active confusion sets are found containing more than one element, then an action has to be introduced by the planning system. If a refinement already exists, then a detailed subplan will be executed.

For our example the following similarity measures were found:

Letter	SM	Rule component
d	1.000	(2)
b	1.000	(3)
p	0.888	(7)
p	0.824	(8)
p	0.812	(1)
p	0.771	(14)
k	0.757	(12)
e	0.729	(6)
b	0.669	(4)
3	0.656	(10)
d	0.556	(13)

In our example the active confusion set contains two letters [b, d] and the refinement is based on a more detailed analysis of the burst and transition of spectral lines. The refined action contains the calculation of burst spectra and the description of spectral profiles as well as other properties like, center of gravity, frequency above which there is 70% of the spectral energy and other parameters whose details are omitted for the sake of brevity.

In particular example under analysis, the system has tracked the time evolution of the major energy concentration in the frequency band of the second formant. Based on a rule on the curve slope it has decided that the letter was |b|.

Notice that the same hypotheses, i.e., a letter of the E1 set can be hypothesized following different paths in the network of actions. The path followed in generating the hypotheses is remembered by adding a suffix to the hypothesized letter. In the way, the hypothesization of B through the n-th possible path to B will be described a B_n.

5. EXPERIMENTAL RESULTS

The proposed approach has been tested in a multi-speaker environment. An initial learning phase was performed on a corpus of 1000 connected pronounciations of symbols of the E1 set in strings EGP3V and KCBTD of five symbols each. The strings were pronounced by five male speakers, and five female speakers. Each speaker pronounced them ten times.

Knowledge acquisition consisted in plan refinement as well as in precondition learning.

After the first 10 speakers were analyzed and knowledge was determined and updated, an overall error rate of 3% was achieved in the learning set consisting of 1000 samples.

The rules of the planning system do not necessarily produce a single hypothesis, although a-priori probabilities can be collected in the case of multiple hypotheses and used for forcing the system to generate the most plausible hypothesis according to some decision criterion.

This possibility was not used in the experiment described in this Section.

Failure to generate the expected hypothesis is considered an error, while the generation of the expected hypothesis together with other candidates is considered an ambiguity.

In the learning set, 9% of the cases resulted in ambiguities. The average numberr of ambiguous hypotheses generated in this 9% was 2.2.

The experiments continued by presenting to the system data collected from successive sets containing always new speakers. Each set was made of 10 speakers randomly chosen among a large student with predominance of Francophones and Anglophones. Each speaker in each set was asked to pronounce two sequences of elements of the E1 set. Sequences were generated at random by a computer program with the only constraint that |e| could appear only at the beginning of the sequence. This fact as well as the number of letters of each sequence were not known to the recognition program.

Performances were computed before refinement so that the system knowledge was used for recognizing phrases of speakers that were not previously analyzed by the system. Nevertheless, after the analysis of each set of speakers, plan and precondition refinements were performed.

After 5 sets were analyzed containing 100 data from 10 speakers each, all the 1500 utterances corresponding to the first 15 sets were processed again for recognition and statistics of rule application were collected in order to prepare a stochastic model.

Inductive learning of rules is supposed to characterize different speaking modes through chains of property descriptions for short time intervals whose duration typically varies between 10 and 50 msecs. Some properties may have a larger scope both backwards and forwards.

Variations of parameter values and minor distortions of the inferred rules can be characterized by stochastic networks.

A-priori probabilities of rule application have been inferred using data from the first 60 speakers.

For each PAC precondition, other parameters and morphologies are extracted and expressions are derived. If there is enough data available for a given precondition, then statistics of the rules applied at the second level of the action hierarchy can be collected.

Having captured a model of the structure of speaking modes, it is possible now to perform a stochastic generalization by using the knowledge obtained with inductive learning for setting initial conditions.

Initial conditions can be set as follows.
If a transition is associated with a single symbol, then a probability distribution on the symbols of the same alphabet is set with a high value assigned only to the single symbol.

For example, ST1 will generate the distribution

$$\{(ST0, .01); (ST1, .98); (ST2, .01)\}$$

If a transition is associated with n symbols of a vocabulary of N symbols, then a probability equal to .01 is assigned to the (N - n) symbols not appear in the transition and a probability of $(1 - .01(N-n))/n$ is associated to the others.

For example, NP0,1 will enerate the distribution:

$$\{(NP0, .48); (NP1, .48); (NP2, .01); (NP3, .01); (NP4, .01); (NP5, .01)\}$$

If a transition is associated with n symbols of a vocabulary of N symbols, then a probability equal to .01 is assigned to the (N - n) symbols not appear in the transition and a probability of (1 - .01(N-n))/n is associated to the others.

For example, NP0,1 will enerate the distribution:

$$\{(NP0, .48); (NP1, .48); (NP2, .01); (NP3, .01); (NP4, .01); (NP5, .01)\}$$

If a transition is associated to a "don't care" symbol x, then all the symbols of the alphabet will appear with equal probability.

For example, x on RE will generate the distribution

$$\{(RE0, .5); (RE1, .5)\}$$

Parameter intervals can be seen as 3σ values of gaussian distributions with mean just in the middle of the interval.

Forward-backward algorithms can be applied for refining the statistics of these stochastic automata.

Stochastic generalizations were performed only for those subnetworks for which there were enough data available. In the other cases, a-priori probabilities of rule application were used for disambiguating multiple generation of letter hypotheses.

The confusions reported in the following refer to a test set containing 500 data from 50 new speakers.

Errors are now due to two causes, namely fault in generating the right hypothesis or fault of selecting the right hypothesis when disambiguation is performed using stochastic networks.

A total of 23 errors due to segmentation and 58 errors due to misclassification have been found making the overall performance of the system in a multispeaker environment equal to 84%.

6. CONCLUSIONS

The idea of using a number of phonetically significant properties in a recognition system based on the planning paradigm appears very promising. The analysis of the behavior of each plan and of the errors generated by their application suggests the actions that have to be taken in order to improve recognition accuracy.

Rule 45 is the most frequently used and the one responsible for most of the errors. In particular, in 14 confusions of |b|, |d|, |p| and |3| with |e| are due to the fact that the system has not been able to correctly detect and process the transient at the vowel onset. Transients in such a case are difficult to analyze and the system tends to see the cues of |e| rather than those of the consonant that is supposed to preceed it.

Planning and inductive learning are still interactive and involve human activities, but they are aimed towards characterization of speaking modes and speech styles.

Different speaking modes and styles can be well characterized using descriptions of acoustic properties that represent different phonetic events.

Finer variations of characteristic descriptions of acoustic properties can be represented and learned using stochastic methods. Letters exhibit little redundance because they are very short sounds. This justifies the introduction of a relatively large number of acoustic properties for characterizing a short time segment. Redundancy helps in reducing ambiguities.

Using different descriptions for the same letter allows one to recognize pronouncia-tions of rare speaking modes provided that descriptions do not interfere with similar descriptions of other letters in other speaking modes.

Criticial problems, are the distinction |d| vs |b| due to the fact that when the burst cues for |d| are not very evident, the system tends to apply the rule that a weak burst is more likely for |b| than for |d|. Other errors are due to misrecognition of voicing (confu-sion |d|, |t|) and to the incorrect classification of frication noise (confusion |3|, |c|). A few errors may be due to a wrong pronounciation of the speakers. In fact, perceptual tests were not performed in order to assess if the speakers really pronounced what they saw on the screen.

The use of suprasegmental features described by PACs is very effective for locating signal segments wehre properties have to be extracted even if a few problems remain to be fixed. The research will continue with the analysis of letters and digits including dip-thongs and also towards the characterization of rare speaking modes of speakers with different mother tongue.

REFERENCES

Bahl, L.R., Jenlinek, F., Mercer, R.L., A Maximum Likelihood Approach to Continuous Speech Recognition, IEEE Trans. on Pattern Analysis and Machine Intelligence, Vol. PAMI-5, No. 2, pp. 179-190, March 1983.

Bahl, L.R., Das, S.K., de Souza, P.V., Jelinek, F., Katz, S., Mercer, R.L., Picheny, M.A., Some Experiments with Large-Vocabulary Isolated Word Sentence Recognition, Proc. of the IEEE Conference on Aoustics, Speech, and Signal Processing, San Diego, CA, pp. 2651-2653, March 1984.

Church, K.W., Phrase-Structure Parsing: A Method for Taking Advantage of Allophonic Constraints, MIT/LCS/TR-296, Cambridge, MA, January 13, 1983. (MIT Ph.D. thesis)

Demichelis, P., De Mori, R., Laface, P. and O'Kane, M., Computer Recognition of Plosive Sounds Using Contextual Information, IEEE Trans. on Acoustics, Speech, and Signal Processing, Vol. ASSP-31, No. 2, pp. 359-377, April 1983.

De Mori, R., Giordana, A., Laface, P., Saitta, L., An Expert System for Interpreting Speech Patterns, Proc. of the AAAI-82, pp. 107-110, 1982.

De Mori, R., Computer Models of Speech Using Fuzzy Algorithms, Plenum Press, New York, NY, 1983.

De Mori, R. and Gilloux, M., Inductive Learning of Phonetic Rules for Automatic Speech Recognition, Proc. of the CSCSI-84, London, Ontario, pp. 103-106, May 1984.

De Mori, R., Laface, P., and Mong, Y., Parallel Algorithms for Syllable Recognition in Continuous Speech, IEEE Trans. on Pattern Analysis and Machine Intelligence, Vol. PAMI-6, pp. 56-69, January 1985.

Doyle, J., A Truth Maintenance System, Artificial Intelligence, Vol. 12, No. 3, pp. 231-272, 1979.

Erman, L.D., Hayes-Roth, F., Lesser, V.R., Reddy, D.R., The HEARSAY-II Speech-Understanding System: Integrating Knowledge to Resolve Uncertainty, Computing Surveys, Vol. 12, No. 2, pp. 213-253, June 1980.

Kopec, G.E., Voiceless Stop Consonant Identification Using LPC Spectra, Proc. of the IEEE Conference on Acoustics, Speech, and Signal Processing, San Diego, CA, pp. 4211-4214, March 1984.

Levinson, S., Rabiner, L.R., Isolated and Connected Word Recognition: Theory and Selected Applications, IEEE Trans. on Communications, Vol. COM-29, No. 5, pp. 621-659, May 1981.

McCarthy, J., Some Expert Systems Need Common Sense, in The Computer Culture, H. Pagels, ed., Annals of the New York Academy of Sciences, Vol. 426, (1984).

Michalski, R.S., A Theory and Methodology of Inductive Learning, in Machine Learning: An Artificial Intelligence Approach, Tioga Publishing Company, Palo Alto, CA, pp. 83-134, 1983.

Minsky, M., A Framework for Representing Knowledge, in The Psychology of Computer Vision, P. Winston, ed., McGraw-Hill, New York, NY, 1975.

Moses, J., Computer Science as the Science of Discrete Man-Made Systems, Knowledge: Creation, Diffusion, Utilization, Vol. 4, No. 2, pp. 219-226, December 1982, reprinted in The Study of Information: Interdisciplinary Messages, F. Machlup and U. Mansfield, eds., John Wiley and Sons, New York, NY, 1983.

Neisser, U., Cognition and Reality: Principles and Implications of Cognitive Psychology, W.H. Freeman and Co., San Francisco, CA, 1976.

Rabiner, L.R., Wilpon, J.G., Terrace, S.G., A Directory Listing Retrieval System Based on Connected Letter Recognition, Proc. of the IEEE Conference on Acoustics, Speech, and Signal Processing, San Diego, CA, pp. 3541-3544, March 1984.

Whitehill, S.B., Self Correcting Generalization, Proc. of the AAAI-80, pp. 240-242, 1980.

COMPUTERS VIEWING ARTISTS AT WORK

J.L. Kirsch, R.A. Kirsch (*), S. Ressler(**)

(*)The Sturvil Corporation
P.O. Box 157
Clarksburg, MD 20871
(**) National Bureau of Standards
Gaithersburg, MD 20899
U.S.A.

Abstract

Our title suggests an Artificial Intelligence approach to the use of computers in the fine arts. We consider computers to have capabilities beyond the utilitarian ones of aiding in art making. Rather, we will investigate the possibility of computers seeing, even understanding, significant form in art. This understanding cannot rise autonomously, but must be the product of careful tutelage by artists, critics, and historians. A powerful tutorial mechanism to use for computers to learn about art is the picture grammar, which allows large classes of compositional structures to be described to a computer by the scholar who has a deep understanding of the art works. In this paper, we illustrate how a machine can be taught the compositional structure of the paintings of the contemporary artist Richard Diebenkorn. With such grammatical instruction, the computer can analyze existing paintings, generate new ones of the same style, and provide a beginning to a computational theory of style.

Formalism in art and computers

Computers are devices for manipulating symbols in formal systems. This characterization is broad enough to include uses ranging from numerical mathematics to computer graphics, although for graphics in art making, the formal properties must be supplemented by the physical characterization of output devices. But such a stretching of the definition is not necessary if we consider the use of computers in the formal description of art works. Here, the notion of formalism as it is understood in computer science and in art criticism come into reasonably close correspondence.

NATO ASI Series, Vol. F45
Syntactic and Structural Pattern Recognition
Edited by G. Ferraté et al.
© Springer-Verlag Berlin Heidelberg 1988

The formal analysis of an art work deals with its hermetic visual properties, such as color, line, shape, materials and their arrangements, the so-called plastic elements. By contrast, there are extrinsic properties like feelings, stories, metaphor which are not so easily described to a computer. These are examples of what are often called the expressive properties of the art. Another common version of this distinction is the dichotomy between classical and romantic art. At this point, we can deal only with formal qualities and not extrinsic ones.

Actually, a bold body of criticism ranging from Roger Fry and Clive Bell to Clement Greenberg maintains that an aesthetic response to an art work depends solely on a purchase of its formal properties. Thus, describing the formal structure of art works to computers would seem a valid approach for making precise, to people and computers, the understanding of art.

Traditional methods of formal analysis have often been cumbersome. For example Loran[1] analyzed Cezanne's compositions with the use of schematic diagrams to account for space, planes, lines, volumes, and the other plastic elements. But a contemporary reader immediately realizes that such an analysis could be made more precise and complete with current computer graphic tools.

However still more powerful tools are available. They are drawn from the fields of image processing, pattern recognition, and computational linguistics, all parts of artificial intelligence. These tools arose [2] in 1964 when it was realized that the use of grammars for describing language could be generalized to describe images. Later, Stiny[3] further generalized these ideas, introducing the idea of a shape grammar. It is this form, slightly modified, that we have used in the study of painting.

Richard Diebenkorn's Ocean Park Paintings

Richard Diebenkorn is one of the most important and respected contemporary American painters. Between 1967 and 1986 he painted about 140 large oil paintings inspired by the look of the Ocean Park area of Santa Monica, California, where he has a studio. These painting appear, on first inspection, to be largely geometric and, hence, to lend themselves to the kind of formal analysis that we find most immediately possible. On further examination, one sees subtle and complex compositions built of the apparently informal and rich handling of layered colors, lines, and textures. Thus, a formal analysis restricted to the geometric qualities is, at best, a beginning of a complete grasp of the painting. It is this beginning that we have attempted.

A Diebenkorn Ocean Park Grammar

The grammar we have written [Kirsch 85] is a modified shape grammar. It contains 42 production rules, most of which offer options for how to subdivide regions of the painting. Some of these rules are recursive insofar as they produce subdivisions that, in turn, reinvoke the original rules. Such recursion enables the grammar to account for an infinite number of distinct compositions.

Although the grammar does not account for color, for example, it provides places where future rules for color may be added. This is in keeping with the practice, in computational linguistics, of providing "hooks" for the addition of features to an account of a corpus which is usually (in the case of language) arbitrarily large. In the case of Diebenkorn's painting, the complexity seems, at this early stage, arbitrarily large.

A shape grammar is a description of structure. For our grammar, it is a description of the linear compositional structure of this class of paintings. In any case, the grammar is equivocal with respect to two important possible uses. It may be used to analyze or to synthesize. The purpose of analysis is to reveal the structure of the paintings being described. Such an analysis can then be viewed by scholars to discover structure not necessarily evident in the finished work. The analysis is like the parse of a sentence. It conveys a significant part of the content of the object being analyzed. To see how such an analysis appears, Fig. 1 shows one of Diebenkorn's Ocean Park painting, number 111 of 1978. Fig . 2 shows the analysis assigned by the grammar to the painting. The rule numbers listed correspond to the grammar which assigns the analysis. Thus, for the grammar shown in the appendix we can see how some of the structural analysis is assigned to the painting by the grammar. We see several rules applied multiple times: rules 11,36,20, etc. These are some of the recursive rules. We also see how the grammar organizes the areas of the painting. Finally, we see how the highest level organization of the painting influences the lower levels. This is denoted by the notation /S that appears at the first level in OP/S, which denotes a kind of "suburban" landscape as opposed to the "rural" and "urban" alternatives provided by the grammar. In some of the final regions, like W/S we see the influence of this high level choice manifest in how the lower level regions remain, at this stage in the grammar. These properties are used, by later grammars, in choosing colors and other properties. These later properties represent more "surface" characteristics of the painting. The present grammar accounts for the "deep structure".

When the final analysis is produced by the grammar, it results in Fig. 3. We can compare this diagram with the original painting represented in Fig. 1, and see how the construction lines (the pentimenti) are both preserved by Diebenkorn, and generated by the grammar. This stage of the grammar only accounts for that part of the painting shown in Fig. 3. But it appears possible to "hang" further analyses on the superstructure assigned by this state of the grammar.

Testing the grammar

A grammar represents a theory, in this case, of the compositional structure of a class of paintings. There are a few ways to validate such a theory. The artist can be consulted to determine whether the grammatical analysis corresponds to his conscious plan of organization. There are perils in this approach, and of course, the uselessnes in the case of artists no longer living. Another form of validation consists of inspection by scholars with knowledge of the artist's oeuvre. They can compare their own intuitive analysis with that assigned by the grammar. Some day, it may even be possible for different grammars to be compared and evaluated, corrected and combined, just as computer programs are today. But the most powerful form of test available to us a present is the resynthesis test.

The grammar is a description of structure. Separate algorithms must be written to perform analysis with respect to the grammar. But algorithms can also be written which will synthesize pictures ab initio from the grammar. These pictures need not correspond at all to any extant paintings. In fact, since the grammar accounts for an infinite number of paintings, it is highly unlikely that a picture generated at random from the grammar will correspond to one that actually exists. But all such pictures have the structure that the grammar allows. We can exploit this property to test the grammar. By generating a random picture (making random choices when the grammar provides alternatives) we can inspect the final pictures for similarity with the artists oeuvre.

The picture of Fig. 4 is such a pseudo-Diebenkorn. It was (randomly) chosen to be a suburban landscape just as is that of Fig. 3. It thus has both a "busy" and "open" region. When Richard Diebenkorn saw this randomly generated picture he had "the immediate shock of recognition", notwithstanding the unusual history of its construction. The reader can with the grammar [Kirsch 85] in hand, perform such tests himself, testing the product with his own intuitive ability to recognize Diebenkorn's Ocean Park paintings.

Implementation

An additional testing mechanism is to implement the grammer. The rigours of producing a running system provide some interesting lessons. This particular implementation is concerned only with using the grammar to generate pseudo-Diebenkorns. One can control the system in a variety of ways. The rules are selected at random from among the valid ones at a given point in the production. Because many of the rules are recursive we had to weight the randomness towards the non-recursive rules so the program didn't always get trapped in recursion. Figure 5 explains the layout of the system and some of the capabilities.

The system can produce primarily two types of images. One is simply a single finished picture (fig 6) and the other is a series of images where each image represents the application of a single rule culminating in the finished picture (fig 7).

Fig.1 : Richard Diebenkorn, Ocean Park No. 111, 1978, Oil and Charcoal on Canvas, 336.2 x 336.7 cm , Courtesy Hirshhorn Museum and Sculpture Garden, Smithsonian Institution.

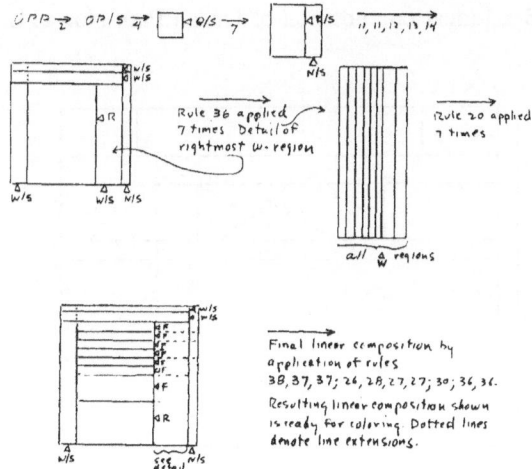

Fig.2 : Grammatical derivation of linear composition for Diebenkorn's Ocean Park No. 111. Rule numbers below arrows.

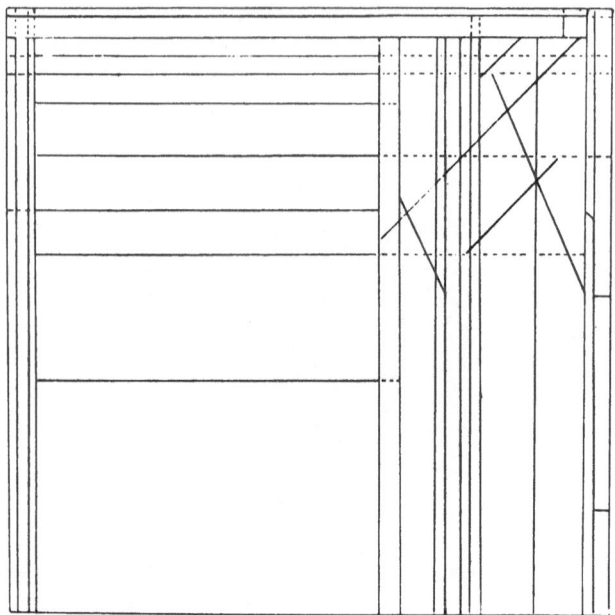

Fig.3 : Linear composition of Ocean Park No. 111.

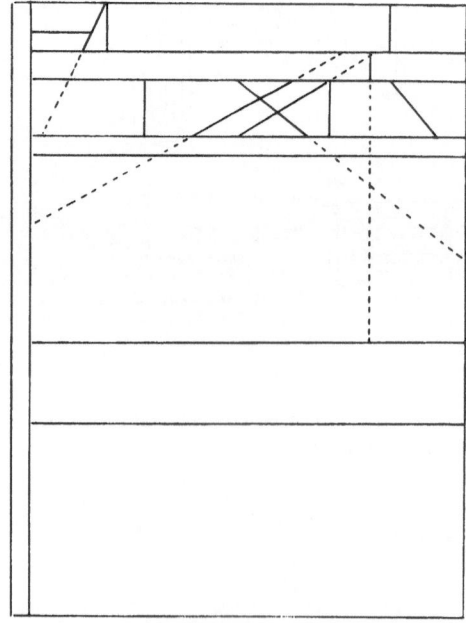

Fig.4 : A pseudo-Dibenkorn derived from the following sequence of rule applications : 2, 6, 17, 17, 11, 31, 31, 31, 30, 38, 37, 30, 31, 30, 30, 30, 32.

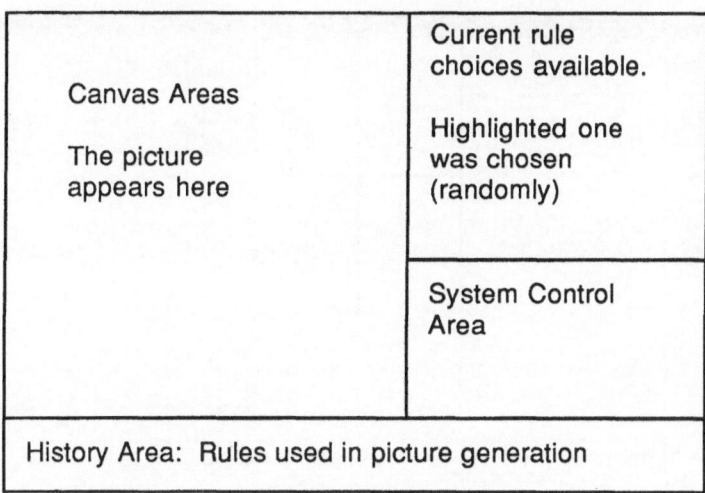

figure 5

After a picture has been generated the bottom row of the window displays a series of icons which represent the history of the rules used in the generation. More interestingly you may select the history rule and then see the geometry in the picture which corresponds to that rule. (It flashes) A extentions to the system would be to be able to modify the rule dynamically and then see the change in the picture as it propogates down the geometry. Such interactive methods may prove to be valuable in generating the grammar in the first place.

References
1.   Erle Loran, "Cezanne's Compositions", Berkeley, Ca, University of California Press, 1943.
2.   Russell A. Kirsch, "Computer Interpretation of English Text and Picture Patterns", IEEE Trans. Elect. Computers, EC13 (Aug 1964), p. 363.
3.   George Stiny, "Introduction to Shape and Shape Grammars", Environment and Planning B, 7:(1980) p. 343.
4.   Joan L. Kirsch, Russell A. Kirsch, "The Structure of Paintings: Formal Grammar and Design", Planning and Design 13:2(1986).

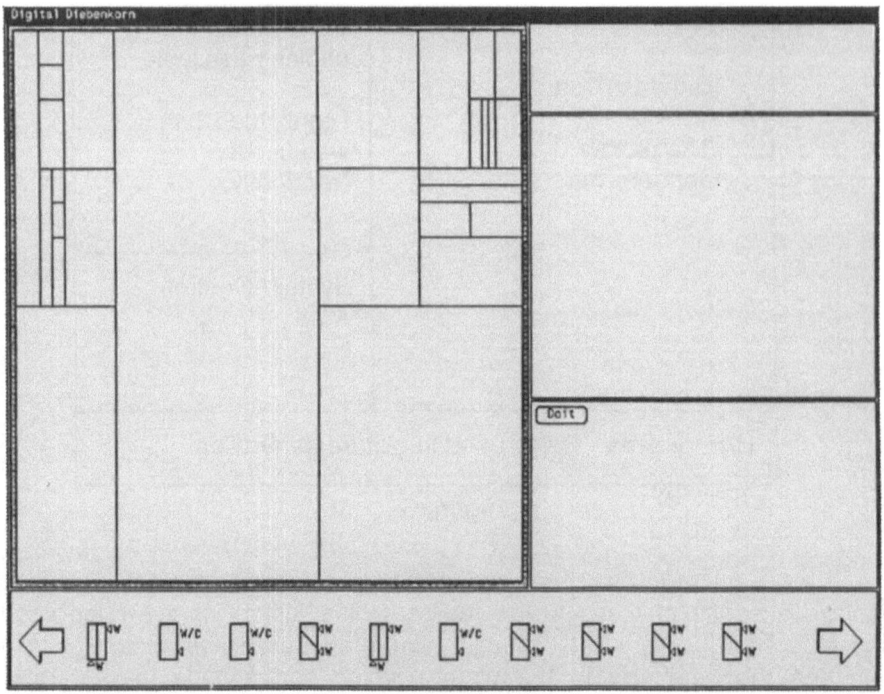

figure 6

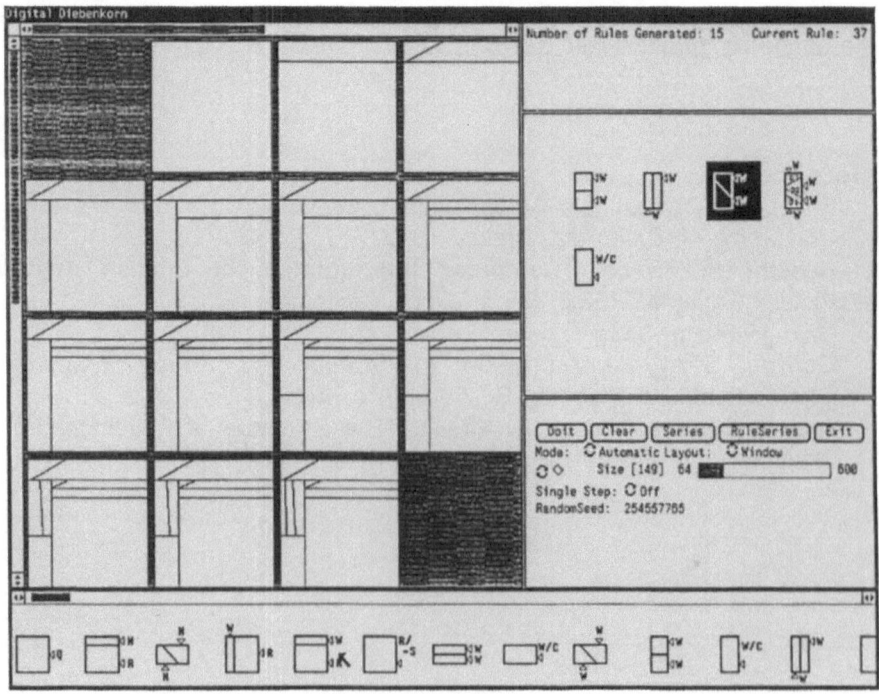

figure 7

Ocean Park grammar rules

Appendix

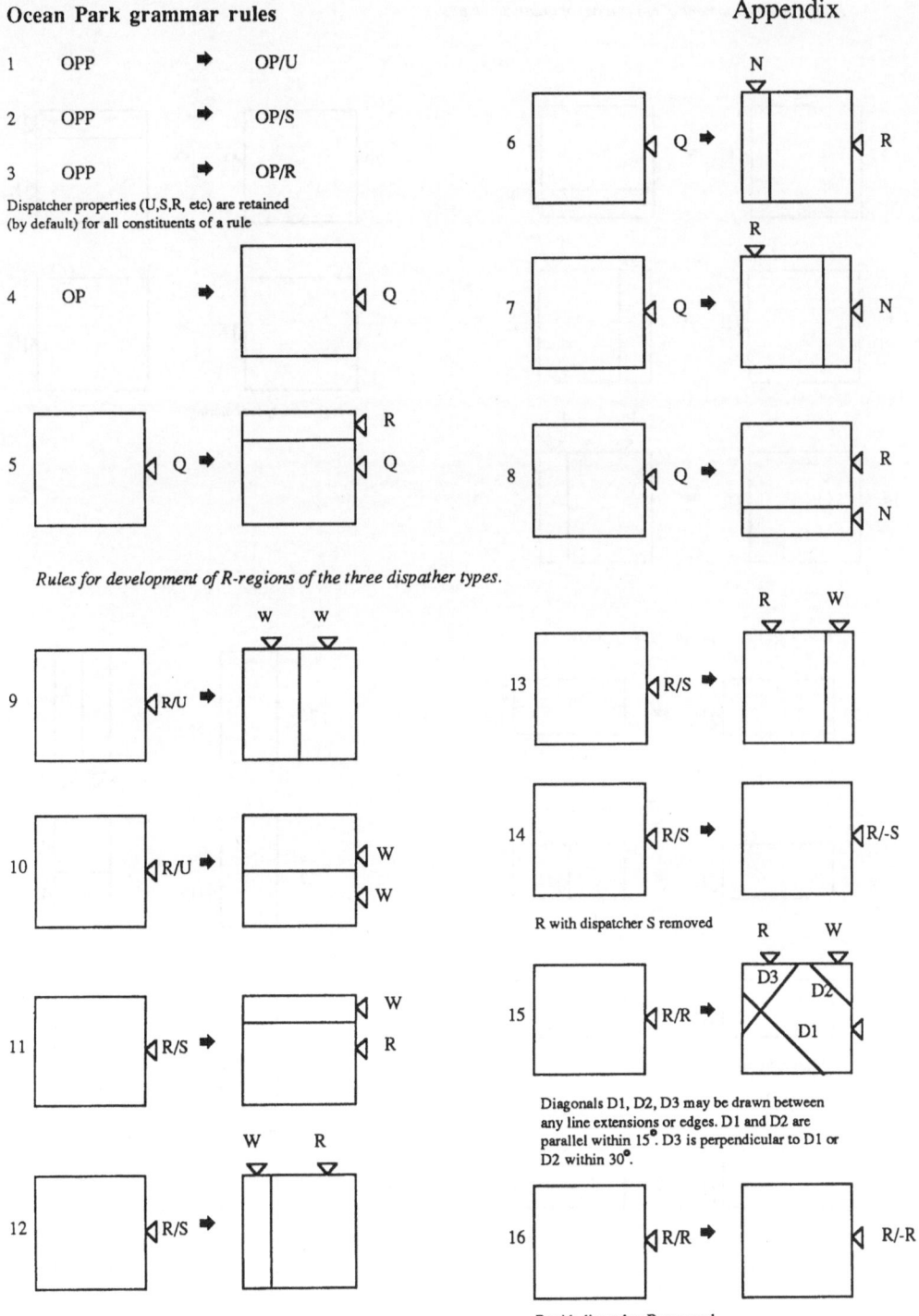

1   OPP   ➡   OP/U

2   OPP   ➡   OP/S

3   OPP   ➡   OP/R

Dispatcher properties (U,S,R, etc) are retained
(by default) for all constituents of a rule

4   OP   ➡

5   Q   ➡

Rules for development of R-regions of the three dispather types.

9   R/U ➡

10   R/U ➡

11   R/S ➡

12   R/S ➡

6   Q ➡

7   Q ➡

8   Q ➡

13   R/S ➡

14   R/S ➡   R/-S

R with dispatcher S removed

15   R/R ➡

Diagonals D1, D2, D3 may be drawn between
any line extensions or edges. D1 and D2 are
parallel within 15°. D3 is perpendicular to D1 or
D2 within 30°.

16   R/R ➡   R/-R

R with dispatcher R removed

300

Rules for development of R-regions of unlabeled type.

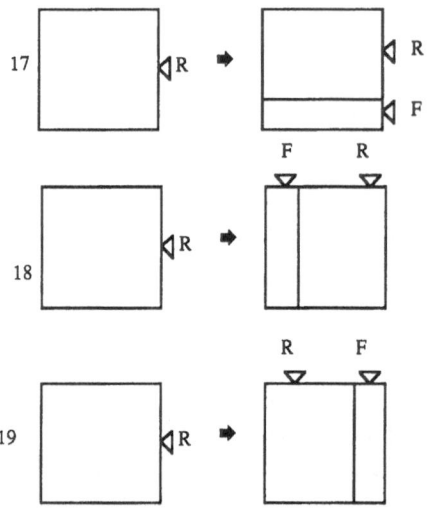

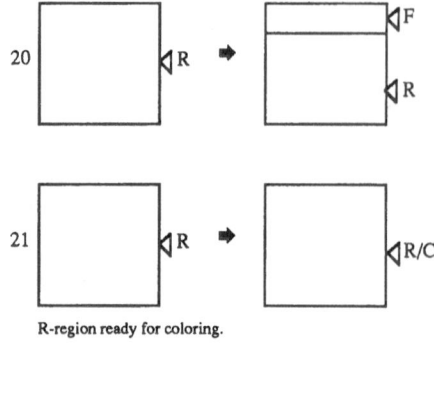

R-region ready for coloring.

Rules for development of N-regions.

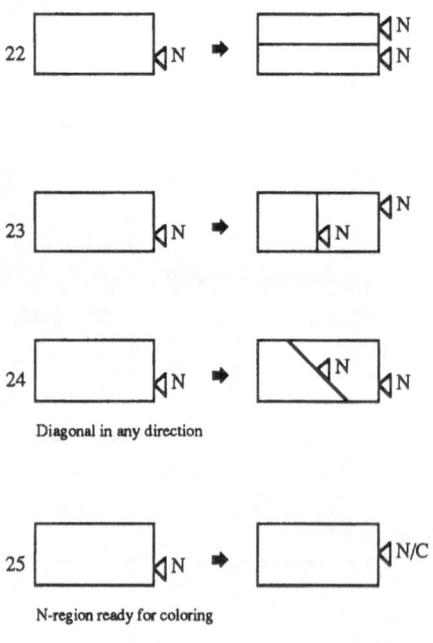

Diagonal in any direction

N-region ready for coloring

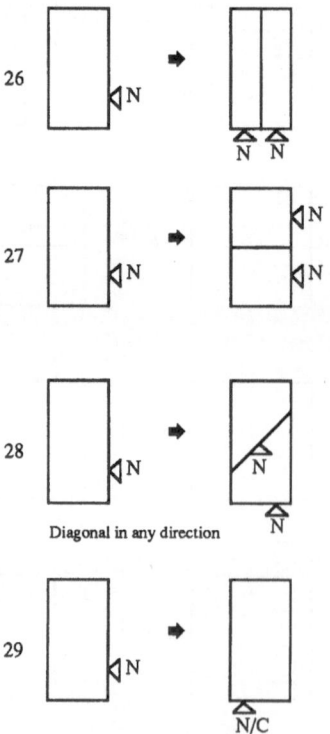

Diagonal in any direction

N-region ready for coloring

Rules for development of W-regions.

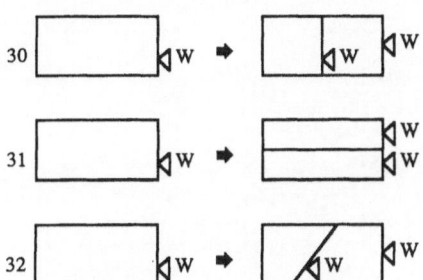

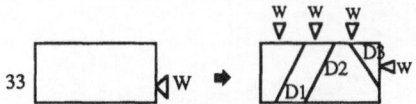

D1, D2, D3 may be drawn between any line extensions
or edges. D1 and D2 are parallel within 15°. D3 is
perpendicular to D1 or D2 within 30°.

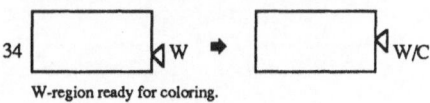

W-region ready for coloring.

Diagonal may be drawn between any line
extensions or edges.

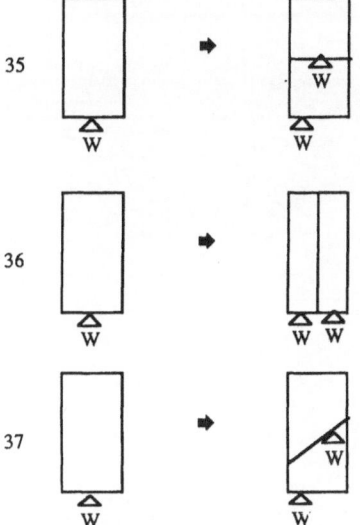

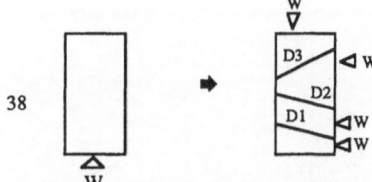

D1, D2, D3 may be drawn between any line extensions
or edges. D1 and D2 are parallel within 15°. D3 is
perpendicular to D1 or D2 within 30°.

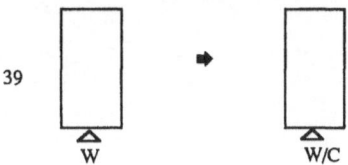

W-region ready for coloring

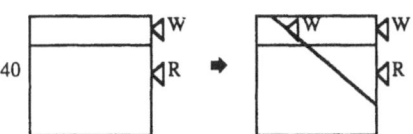

Rules for development of W and R-regions.

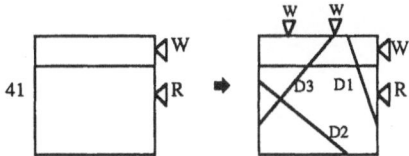

The diagonals may be drawn between any line extensions or
edges. D1 and D2 are parallel within 15°. D3 is perpendicular
to D1 or D2 within 30°. W is partitioned, but not R.

The diagonal may be drawn between any line
extensions or edges. Note that the W-region is
partitioned into two W-regions, whereas the
R-region is not partitioned and remains a single
rectangular region.

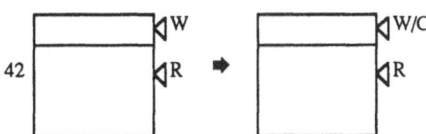

W is ready for coloring. R may be further developed.

FACE RECOGNITION FROM RANGE DATA
BY STRUCTURAL ANALYSIS

J.T. Lapresté, J.Y. Cartoux, M. Richetin

Electronics Lab., UA 830 of CNRS
University of Clermont II
B.P. 45
63170 Aubière
France

ABSTRACT

A primal approach to the analysis of a human face through its 3-D image is proposed. This analysis is based on local pattern concepts deduced from the intrinsic properties of a surface described by its local curvature characteristics. Some theoretical results about curvature are recalled and then used for the extraction of characteristic geometrical features from the face surface. Recognition can be achieved by using a pattern vector of distances calculated from these features. Experimental results are provided to illustrate the various steps of the approach.

1. INTRODUCTION

Though face identification checking is a natural action for human beings, the automation of the process to be performed by computer looks far more complex. Many authors have proposed monodimensional face characterization, mainly from profile curve [1,2], or bidimensional images [3]. This paper presents a primal approach to the 3-D analysis of face range data images.

The use of 3-D data has led to new image representations. Following differential geometry schemes, these representations [4,5] describe a surface by local pattern concepts obtained from local curvature information. The main advantage is, of course, that this data is intrinsic to the surface and thus does not rely upon a particular viewpoint. Thus unavoidable angular variations between various pictures are not damaging.

So, in this approach, the face surface is described by the principal curvature directions and values, whose computation is presented in section 3. In section 2, the need of data restoration and filtering is underlined.

The following sections give a sketch of the present state of the analysis and recognition algorithm. The basic idea of the procedure under development is to define in a learning phase and then to extract in the recognition one, a structural description of a face based on characteristic points of the face surface. These points must have strong properties so that they are stable under nearly rigid geometrical transformations of a face (it is not a solid !). It will be shown that such points have mainly been chosen , according to an anthropometric point of view, on extremal curvature lines. Moreover these points must be situated on the whole face surface in order to get reliable 3D information.

NATO ASI Series, Vol. F45
Syntactic and Structural Pattern Recognition
Edited by G. Ferraté et al.
© Springer-Verlag Berlin Heidelberg 1988

2. RESTORATION AND FILTERING

Two kinds of range data images have been used up to now. One comes from a Laser Range Finder developed at the C.N.R.C. [6] which gives very accurate 3D images, but with some "holes" due to pure reflexion phenomena on the face surface. The other kind has been obtained from photogrammetry.

Restoration and filtering are necessary because the curvature information used to describe the local structures of a surface must not be noisy, and because we want to keep a 5x5 mask to compute this information.

Restoration, i.e. filling gaps in the present case, is done by calculating a quadratic polynomial approximation in an area of holes, the size of this area being adjusted to the distribution of the holes. In fact, the image is completely analyzed with a 3x3 mask. At a given point which is not a hole, if at least one hole is found in the mask centered on that point, the analysis neighborhood is extended to a domain NxN. This extension is done until there is no holes on a 3x3 grid, centered on the given point, which has the size and the frontiers of the extended neighborhood. Then the quadratic approximation is done and each hole is filled with an approximated value. This procedure is fast and only requires one pass.

Filtering can be achieved with classical filters such as Gaussian or median ones. We also used the Structural Filter [7,8] developed in our laboratory that leads (besides possible noise elimination) to a good schematization of the principal structures of the image with almost no blurring (the almost is very small !). The main idea of this filtering is that the minimal curvature direction at a given point is a good hint of the orientation of some local structure. The filter replaces the point value by the means of the surrounding values taken in a square or oriented linear neighborhood. The choice is made according to the homogeneity of the field of minimal curvature directions around the point.

Recently we noted that another much more straightforward method leads to good results : scaling the image dynamics by a very small real factor, without any filtering seems to improve the accurateness of the curvature computation.

Examples of restoration and filtering are given in figure 1.

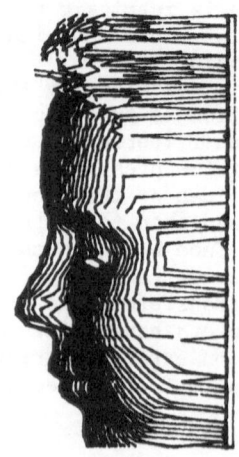 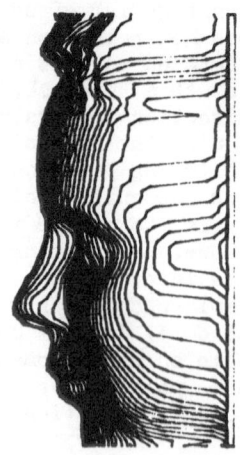

(a) face F1

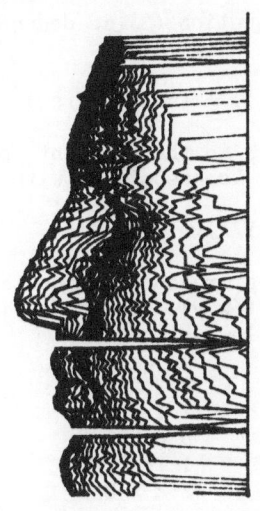 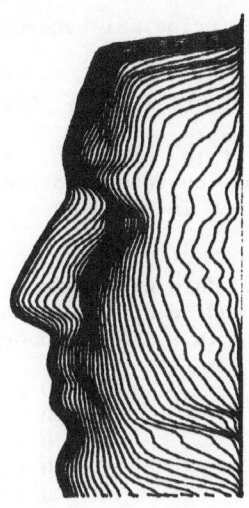

(b) face M1

Figure 1. Examples of restoration and filtering of 3D face images obtained (a) from a laser range finder, (b) by photogrammetry. The calibration of each kind of 3D-images is not the same.

3. COMPUTATION OF PRINCIPAL CURVATURES.

Let us recall some differential geometry results. Let S be a cartesian surface defined in an orthonormal reference system (i,j,k). Let N be a point of S. If S is a regular surface, there is one and only one tangent plane at N defined by the vectors $\delta N/\delta x$ and $\delta N/\delta y$, given by $\delta N/\delta x=(1,0,p)$ and $\delta N/\delta y=(1,q,0)$. Let (I,J,K) be a non orthonormal reference system centered at N, such that $I=\delta N/\delta x$, $J=\delta N/\delta y$, $K=IXJ=(-p,-q,1)$. Let $\delta^2 N/\delta x^2=(0,0,r)$, $\delta^2 N/\delta x\delta y=(0,0,s)$, $\delta^2 N/\delta y^2=(0,0,t)$, be the second derivative vectors. At each point of S, two fundamental quadratic functions are associated, which are :

$$P(X,Y) = (1+p^2).X^2 + 2pq.XY + (1+q^2).Y^2$$
$$Q(X,Y) = r.X^2 + 2s.XY + t.Y^2$$

The calculus of principal curvatures is achieved by optimization of the function $L(X,Y) = C.P(X,Y) + Q(X,Y)$. This leads to the resolution of the two following equations.

$$(C(1+p^2)+r).X + (Cpq+s).Y = 0$$
$$(Cpq+s).X + (C(1+q^2)+t).Y = 0 \tag{1}$$

System (1) has non trivial solutions if its determinant is null, and from that the principal curvature equation (2) is obtained [4,11] :

$$C^2.(1+p^2+q^2) + C.(r(1+q^2)+t(1+p^2)-2spq) + rt-s^2 = 0 \tag{2}$$

The principal curvature directions equation (3) is deduced from the system (1) :

$$(Y/X)^2 \cdot (tpq - s(1+q^2)) + (Y/X) \cdot (t(1+p^2) - r(1+q^2)) + (s(1+p^2) - rpq) = 0 \qquad (3)$$

In this equation, Y/X is the slope, in the tangent plane, of a principal curvature direction. The vector Vc associated with a principal curvature C is given by :

$$Vc = (1, Y/X, p+q \cdot Y/X) \qquad (4)$$

The data needed for the principal curvature computation is the first and second derivatives of the function $z=f(x,y)$. These derivatives are estimated in a 5x5 neighborhood with the following masks.

```
-2-1 0 1 2              2 2 2 2 2
-2-1 0 1 2              1 1 1 1 1
-2-1 0 1 2              0 0 0 0 0
-1-1 0 1 2             -1-1-1-1-1
-2-1 0 1 2             -2-2-2-2-2

    50.p                   50.q
```

```
2-1-2-1 2      -4-2 0 2 4        2 2 2 2 2
2-1-2-1 2      -2-1 0 1 2       -1-1-1-1-1
2-1-2-1 2       0 0 0 0 0       -2-2-2-2-2
2-1-2-1 2       2 1 0-1-2       -1-1-1-1-1
2-1-2-1 2       4 2 0-2-4        2 2 2 2 2

   70.r           100.s            70.t
```

These masks have been calculated from a polynomial approximation of a discrete function $z=f(x,y)$ in a 5x5 neighborhood centered at any point of the surface. This method allows to estimate the value of the derivatives even near the edges of the image since the neighborhood need not be symmetrical.

4. LABELLING THE SURFACE POINTS

The principal curvatures values and directions are a very rich set of information about the geometrical local structure of the surface. Knowing the sign of the two principal curvature values, it is possible to classify each point with a geometrical label [9]. This labelling gives partitions of the surface in regions of a given geometrical nature. The simplest is the distinction between convex and concave regions and will be strongly used here.

The nine fundamental surface types we have characterized, are the following, Cmax and Cmin being the maximal and the minimal curvatures :

```
if Cmax > 0 and
        (1°)  Cmin < 0     => Hyperbolic Convex Point,
        (2°)  Cmin = 0     => Parabolic Convex Point,
        (3°)  Cmin > 0     => Elliptic Convex Point,
        (4°)  Cmin = Cmax  => Umbilic Convex Point,
elsif (5°)  Cmax = 0 and Cmin = 0    => Plane Point,
```

```
      elsif Cmax < 0 and
            (6°)  Cmin > 0     =>  Hyperbolic Concave Point,
            (7°)  Cmin = 0     =>  Parabolic Concave Point,
            (8°)  Cmin < 0     =>  Elliptic Concave Point,
            (9°)  Cmin = Cmax  =>  Umbilic Concave Point.
```

Figure 2 shows this labelling for the concave domains of face Ml of figure 1.b. The types (5°),(6°),(7°),(8°),(9°) of the surface points are respectively represented by the symbols "-", "x"," ⊔ ","V", "." .

Figure 2. Labelling of the concave points of the 3D image of face Ml (see figure 1.b).

5. EXTRACTION OF CHARACTERISTIC CURVES

As previously defined, the convex/concave labelling allows to find face 'curves' denoting transitions between convex and concave areas. These curves can be simply obtained by a contour tracking algorithm.

In the parts of the image we are interested in, a scan of the lines is made and two stacks are used :

- a new points tracking stack in which new occurring contour points are chained to old curves, or new curves are created.

- a curve chaining stack in which linked curves are put (from the other stack), which are reputed complete.

Unsignificant curves, i.e. too short ones, are eliminated during the extraction procedure.

6. CONCAVE LINES OF THE FACE

Figure 2 shows the concave domains of the face. These domains, as pointed out by D.D. Hoffman and W.A. Richard [10], are significant in surface segmentation.

The aim of their theory is the recognition of an object according to a process copied on the human recognition process. The human vision system decomposes a shape into a hierarchy of parts and does it using Transversality Regularity. According to this principle, when any two arbitrarily shaped surfaces interpenetrate at random, they almost surely meet in a contour of concave discontinuity of their tangent plane. D.D. Hoffman and W.A. Richard thus propose to generalize this principle to smooth surfaces. They divide them into parts at loci of negative minima of each principal curvature along their associated family of lines of curvature. In the face case (Figure 2), three concave points areas would allow, using this theory, to extract characteristic lines : the lower border of the nose, the junction of the lips and the chin upper border. In these two domains, the search of points for which the maximal curvature is a local negative minima, would determine three lines we shall name 'nose lower line', 'mouth line' and 'chin line'.

According to this theory, it is not surprising that for anthropometric classification of faces some points have been chosen on these lines, precisely at their intersections with the face profile plane, which can be viewed as the plane of major global symmetry of a face. These points precise the borders of the different local surfaces which compose a face.

7. EXTRACTION OF CHARACTERISTIC POINTS

In our primal aprroach to face recognition, the goal is to extract on the face surface a set of characteristics points with which it will be possible to calculate some distances on this surface and then to compare and to discriminate two faces with the corresponding sets of distances.

The major characteristic points to be placed on the surface which have been presently chosen are :
 - the centers of the eyes,
 - points of high curvature on the profile of the face.

Many others could be also determined since the information given by a 3D image of a face is extremely rich. Some will be indicated in the following section.

7.1 Extraction of the profile plane and of the face profile

In a first attempt this plane had been defined as the midperpendicular plane of the segment joining the centers of the eyes. An eye center was defined as the mass center of the interior domain of a closed curve which is a concave-convex transition line situated in the eyes area of the face (see figure 3.a and 3.b). Unfortunately in some images, especially those obtained by photogrammetry which does require that the eyes must be closed, these eye curves do not exist.

Presently the profile plane is obtained from the principal inertia axes of a set of points. These points are all those included in a closed curve C which is a convex-concave transition line, and which define a domain mostly concave, possibly including the convex areas of the eyes, and surrounding the nose. Such a curve is given in figure 3 for different face images. Its extraction is quite easy because it is the longest closed convex-concave transition line which can be extracted from the curvature-labelled image of a face. Indeed we have supposed that such a closed curve C exists for any face. Figure 3.a corresponds with face Ml of figure 1.b. Figure 3a. and figure 3.b are related to the same face (Ml and Ml') but with different viewing conditions. It is the same for figures 3.c and 3.d (faces F2 and F2').

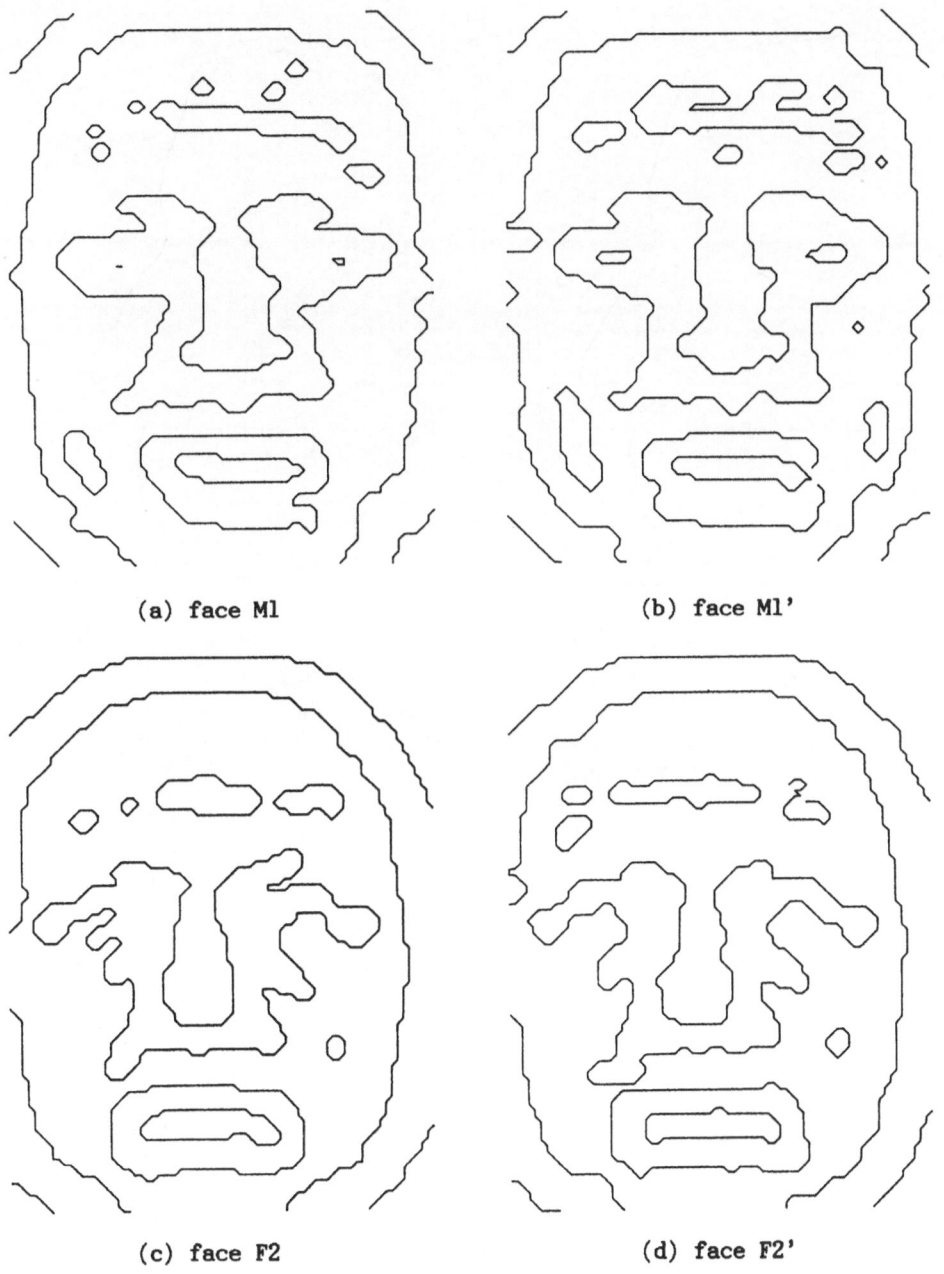

(a) face Ml

(b) face Ml'

(c) face F2

(d) face F2'

Figure 3. Convex/concave or concave/convex transition curves on faces.

Doing that we have also suppose that the profile plane is one of the principal inertia planes. This is true if the surface domain of the closed curve has a symmetry plane which is the profile plane. If not then the two planes will be slightly different.

Having the profile plane, then the face profile is the intersection of this plane with the face surface. Figure 4 gives some face profiles. Figure 4.a, 4.b, 4.c and 4.d are respectively related with faces Ml, Ml', F2 and F2'.

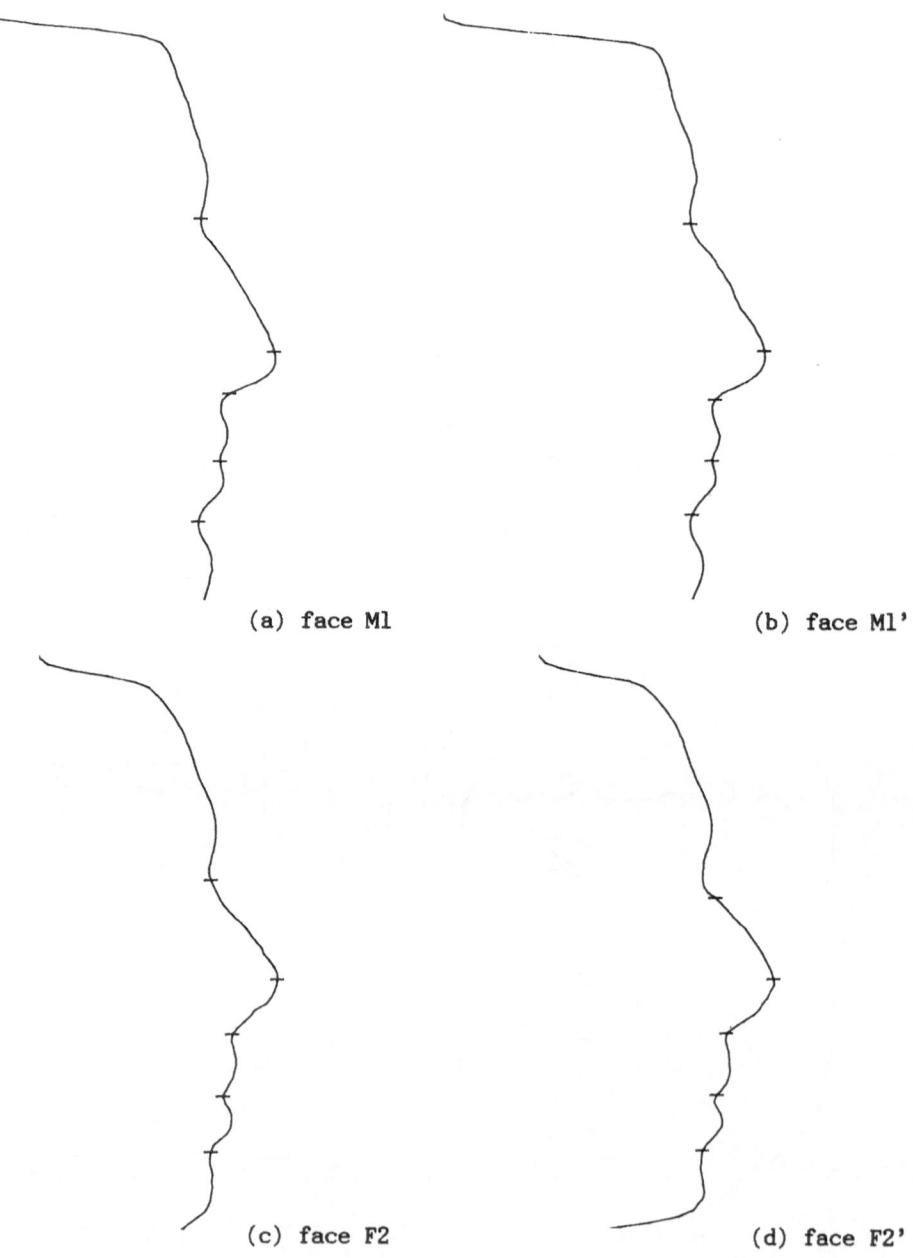

(a) face Ml

(b) face Ml'

(c) face F2

(d) face F2'

Figure 4. Face profiles.

7.2 Positioning the characteristic points

(i) the centers of the eyes : they are defined as indicated in § 7.1 if there exist two closed convex-concave transition lines inside the domain of closed curve C previously found.

(ii) the "nasion" : it is defined as the convex hyperbolic point on the upper part of the profile, for which the minimal curvature is minimal.

(iii) the "subnasal, lips junction, and infradental" (named "subnasal, lèvres and infradental" in figure 5) : these three points correspond with the minimums of the maximal curvature on the lower part of the profile. At these three points, the nature of the surface can be different: the subnasal is rather a hyperbolic point, while the two others are rather parabolic.

(iv) the "nose tip" (named "bout du nez" in figure 5) : it is the profile point situated between the nasion and the subnasal, and at the greatest distance of the line joining these two points.

(v) the "nose edge points" : these two points are on the closed curve C, at the bottom of the nose, on each side of the subnasal, and such as their distance to the nasion is maximal.

Figure 5 gives two face profiles with the corresponding Cmax and Cmin curves (they have been differently normalized), and the characteristic points determined on these profiles. Figures 5.a and 5.b are respectively associated with faces M1 and F2.

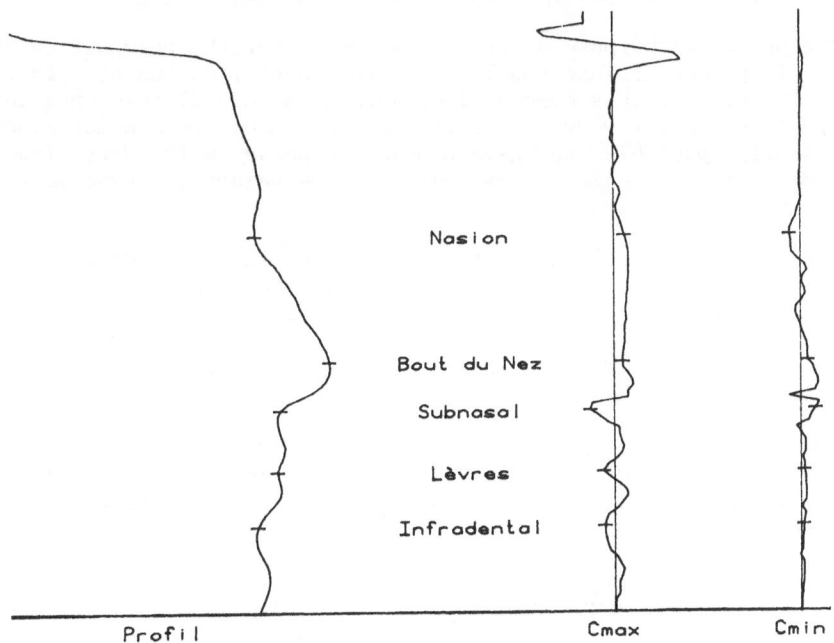

(a) face M1

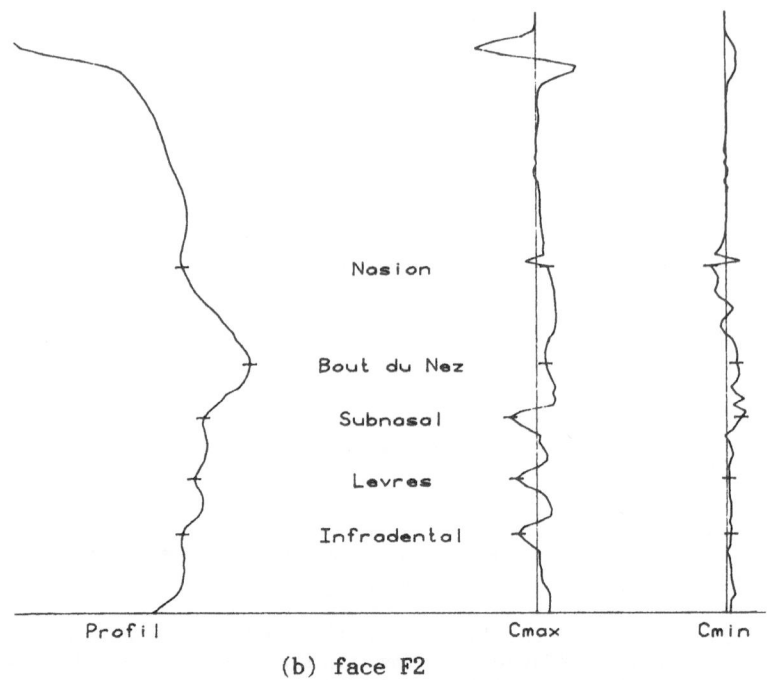

(b) face F2

Figure 5. Characteristic points on two profiles.

7.3 Comparing face images in view of identification

The practical purpose of this research is not the recognition of a face from a large set of face models, but the identification of a face person knowing the name of this person, i.e. knowing a face 3D reference image. The two problems are not very different but the second is simpler since it is theoretically possible to give a negative answer with very few indices measured on the face image, by comparing the values of these indices with recorded reference values.

Many indices can be extracted from a face image : distances between characteristic points, but also some others such as, for example, the principal inertia moment of the 3D subsurface defined by the closed curve C. At the moment only a distance vector V is constructed from the characteristic points defined in § 7.2. Nine distances are calculated giving a 3D geometry of a face.

Then two face images are compared by calculating the Euclidean distance D of the corresponding distance vectors V. This index is a global measure of the dissimilarity of two faces. Table 1 indicates the various distances D for faces Ml, Ml', Fl, Fl'.

| Ml/Ml' : D = 2.39 | Ml/F2 : D = 8.61 | Ml/F2' : D = 12.43 |
| F2/F2' : D = 4.85 | F2/Ml' : D = 6.64 | Ml'/F2' : D = 10.50 |

Table 1. Distances D for faces Ml, Ml', F2, F2'.

7.4 Comments on the present results and future extensions

Though the results of table 1 show that the lower distances D are obtained for the images of identical faces (M1,M1') and (F2,F2'), improvements can be brought in order to enhance them.

Comparing the profile of faces F2 and F2' (fig. 4.c and 4.d), it can be seen that the profile plane of face F2' seems to be wrong and gives a nasion which is shifted down. This is due to the asymmetry of the C-domain, i.e. the domain defined by the closed curve C (see fig. 3.d). In that case the principal inertia plane taken for the profile plane is not the symmetry plane of that face. This is why the distance D between F2 and F2' is nearly twice the distance between M1 and M1'. Thus it is necessary to extract from the C-domain, a symmetrical one. This can be done with an iterative procedure.

Moreover we noticed that the nose tip is unstable. This point can be replaced by another one, between the nasion and the subnasal, for which the Gaussian curvature is maximal.

Stronger indices can also be calculated. For example, when the profile curves are known, a distance between two profiles can be computed after a mean-square adjustment of the two profiles based on their characteristic points. A similar index for face surface can be defined in the same way.

8.CONCLUSION

This primal approach to the face characterization through 3-D images, is based on a structural analysis of the face surface described by local curvature features. The face representation extracted uses local patterns of different kinds, such as particular lines on the surface, plane of symmetry, profile curve, characteristic points. From the set of these later which are obtained through a procedural approach, at the present state of the study, a distance vector has been defined which already appears to be characteristic of a human face. However, it is necessary to examine a great deal of human faces before elaborating a robust identification strategy.

REFERENCES

[1] P.BAYLOU, E.H.BOUYAKHF, G.BOUSSEAU & A.MORA, "Analyse Automatique du Profil du Visage. Recherche du Meilleur Classifieur à Fin d'Identification.", Proc. 3rd AFCET Conf. on Pattern Recognition and Artificial Intelligence, Nancy(France), September 1981, 371-382.

[2] L.D.HARMON & W.F.HUNT, "Automatic Recognition of Human Face Profiles.", Computer Graphics and Image Processing, 6, 135-156 (1977).

[3] M.NAGAO, "Control Strategies in Pattern Analysis", Proc. of 6th ICPR, Munich, October 1982, 996-1006.

[4] M.BRADY, J.PONCE, A.YUILLE & H.ASADA, "Describing Surfaces", Proc. of 2nd Int. Symp. Rob. Res., H.Hanafusa & H.Inone Eds, MIT Press, Cambridge, 5-16.

[5] P.BESL & R.JAIN, "Intrinsic and Extrinsic Surfaces Characteristics", Proc. of CVPR, June 1985, San-Francisco, 226-233.

[6] M.RIOUX, "Laser Range Finder Based on Synchronized Scanners", APPLIED OPTICS, November 1984, Vol 23, N° 21, 3837-3844.

[7] M.RICHETIN, P.SAINT-MARC & J.T.LAPRESTE, " Describing Greylevel Textures through Curvature Primal Sketching", Proc. of ICASSP, Tokyo, April 1986.

[8] P.SAINT-MARC & M.RICHETIN, "Structural Filtering from Curvature Information", Proc. of 1985 Computer Vision and Pattern Recognition, Miami, June 1986.

[9] P.BESL & R.JAIN, "Range Image Understanding", Proc. of CVPR, San-Francisco, June 1985, 430-449.

[10] D.D.HOFFMAN & W.A.RICHARD, "Parts of Recognition", Cognition, 18, 1984, 65-96.

[11] M.E.MORTENSON, "Geometric Modeling", Wiley, New-York, 1985.

CRYPTOSYSTEMS FOR PICTURE LANGUAGES

R. Siromoney, K.G. Subramanian and Abisha Jeyanthi

Dept. of Mathematics
Madras Chistian College
Tambaram
Madras 600 059
India

0.   INTRODUCTION

There has been tremendous interest in the study and construction of hard to break public key cryptosystems, introduced
by Diffie and Hellman [3]. These depend on one-way functions
and provide secrecy in data transmission or storage. Different keys are used for the encryption and decryption processes,
the encryption key being made public in the form of a directory but the decryption key is kept secret. Thus, for a public key cryptosystem to be safe and effective, encryption and
decryption should be "easy" but cryptanalysis "hard" and this
is accomplished by hiding information in the trapdoor. Legal
recipient can decrypt easily with the help of the information
in the trapdoor but the eavesdropper finds it difficult to decode without this information.

Salomaa [8,9] has pointed out that results from formal language theory may be used successfully to construct safe and efficient public key cryptosystems [12]. The main purpose of this
paper is to indicate the usefulness of syntactic methods in cryp-
tography-specifically, in the construction of public key cryptosystems for picture languages i.e. pictures described syntactically.

A picture is treated as a composition of simple subpictures
whose structure is described by linguistic tools. Pictures
whose components are cells or unit lines from the cartesian
rectangular grid are suitable for a formal approach and are very
close to many situations in digital image processing. Sets of

NATO ASI Series, Vol. F45
Syntactic and Structural Pattern Recognition
Edited by G. Ferraté et al.
© Springer-Verlag Berlin Heidelberg 1988

pictures composed of cells are referred to as array languages
and there has been considerable interest in studying array grammars that generate or parse these [4,6,10]. The generative
grammatical array models fall into three basic types: Rosenfeld
[6], Siromoney [10] and Lindenmayer [10]. A survey of these
models is given in [10].

A very simple model for line pictures has been investigated
in [5]. A line picture is described by a word over the alphabet
$\{u,d,r,l\}$ where 'u' means go (and draw) one unit line up from
the current point and d,r,l are interpreted similarly. The traversal or drawing of a line picture is defined accordingly and sets
of such pictures are called 'chain code picture languages'.

We construct two types of public key cryptosystems (PKC)
for picture languages. In the first system, pictures to be encrypted are taken in the form of chain code picture languages
and hence described in terms of the four letter alphabet $\{l,r,u,d\}$.
The deterministic version of the L-array model introduced in [11]
is found useful since four tables viz. left, right, up and down
are used in the generation which will reflect the four letter
alphabet $\{l,r,u,d\}$. Thus in this PKC a chain code picture language is encrypted as a two dimensional array using the deterministic grammatical model. The chain code picture language can also
be encrypted as a string instead of as a two dimensional array.

In the second type of public key cryptosystem discussed in
this paper, the picture is considered as a digitized array (pixels)
and array morphisms and substitutions are used for decryption
and encryption. A public key cryptosystem based on L-systems is
constructed in [8,9] as an interesting application of theory of
formal languages. The public encryption key is a TOL system obtained from an underlying DTOL system which is required to be
unambiguous. But it is known [2] that this problem, viz. whether
a DTOL is unambiguous, is undecidable.

We first introduce a public key cryptosystem for strings in
which the unambiguity requirement is avoided. The encryption key
involves a TOL system based on a DOL system. The decryption based
on the trapdoor is simple and straightforward but cryptanalysis

is hard. Then, we extend this method by making use of array morphisms and array substitutions, to encrypt two dimensional arrays which represent digitized pictures.

1. PRELIMINARIES

In this section we review the basic concepts of crypto-systems [8], chain code pictures [5] and the definitions of array generating systems [11], needed here.

Messages stored in a medium or transmitted through an in-secure channel are referred to as plain texts. The process of transforming plaintexts and thereby locking its contents from being known to others is called encryption. The encrypted plain text is referred to as cryptotext. Decryption which is the inverse of encryption is the process of unlocking the crypto-text to get back the original plain text.

A cryptosystem specifies several keys. Each key K deter-mines an encryption function e_K and a decryption function d_K. A cryptotext c is obtained from a plain text w using e_K i.e. $e_K(w) = c$ and $d_K(c) = w$ so that $d_K(e_K(w)) = w$.

The act of analyzing the encrypted text with a view to obtain the contents of the plain text without the knowledge of the secret key is called cryptanalysis.

The idea of a public key cryptosystem is that the knowledge of e_K does not necessarily give away d_K, eventhough e_K is made public. The construction of a public key cryptosystem is based on the idea of a trapdoor. The information publicized is not enough to enable decryption of cryptotext without the knowledge of the secret information in the trapdoor. Thus the secret in-formation in the trapdoor enables the legal recipient of the message to decrypt the cryptotext easily but the cryptanalyst finds it difficult to decode the message.

We now informally describe the concept of chain code pictures. A drawn picture (dpic) is described by a word over the alphabet $\{l,r,u,d\}$, called the picture description, whose letters mean move (and draw) a unit line left (right, up, down respectively) from the current point. Languages of such pictures are called chain code picture languages.

EXAMPLE 1.1 Let w = rdlddrurdruulur. Then dpic(w) is given by Figure 1, where the start point is indicated by a circle and the end point by a square. We note that the figure is an approximation of the Hilbert curve.

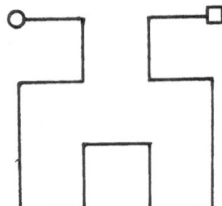

Figure 1 : Hilbert Curve

We next describe the array generating system [11] needed for the construction of public key cryptosystems.

The string DOL and DTOL systems are well-known [7].

DEFINITION 1.2 a) An array M with m rows and n columns over an alphabet T is of the form

$$M = \begin{matrix} a_{11} & \cdots & a_{1n} \\ . & \cdots & . \\ a_{m1} & \cdots & a_{mn} \end{matrix}, \quad a_{ij} \in T \text{ for } i=1,\ldots,m; j=1,\ldots,n$$

We denote by T^{++}, the set of all arrays over T.

b) A determinstic table OL array system (DTOLAS) with control is G = (T, P, C, M_o) where T is a finite alphabet; P = $\langle P_1, \ldots, P_k \rangle$. Each P_i is a finite set of rules of the form $a \rightarrow b_1 b_2 \ldots b_m$ where a, b_i (i=1,...,m) are in T. P_i is called a right (left, up or down) table. The right sides of rules within a table are of the same length. For each a in T, there is exactly one rule in a table i.e. P_i is deterministic. $C \subseteq P^*$ is called the control language. $M_o \in T^{++}$ is the axiom array.

For $M_1, M_2 \in T^{++}$, we write $M_1 \overset{\Rightarrow}{R} M_2$ if M_2 is obtained from M_1 by applying in parallel the rules in a right table R to all the symbols in the rightmost column of M_1. Similarly, we define $\overset{\Rightarrow}{L}$, $\overset{\Rightarrow}{U}$, $\overset{\Rightarrow}{D}$.

We write $M \overset{*}{\Rightarrow} M'$, for $M, M' \in T^{++}$ iff there exists a sequence of derivations $M = M_1 \underset{P_{i_1}}{\Rightarrow} M_2 \underset{P_{i_2}}{\Rightarrow} \cdots \underset{P_{i_n}}{\Rightarrow} M_{n+1} = M'$,

such that M_i $(i=1,\ldots,n+1) \in T^{++}$ and $P_{i_1} \ldots P_{i_n} \in C$ where each P_{i_j} $(j=1,\ldots,n)$ is in P.

If a table $P_i \in P$ is not deterministic, then we call G, a TOLAS with control.

We give an example of a DTOLAS.

EXAMPLE 1.2　Let $T = \{p, q\}$, $P = \{R, L, U, D\}$ where
$R = \{p \rightarrow ppp, q \rightarrow pqq\}$, $L = \{p \rightarrow pppp, q \rightarrow pqqp\}$,
$U = \{p \rightarrow qqqq, q \rightarrow qpqp\}$, $D = \{p \rightarrow qqq, q \rightarrow qpq\}$ and $M_o = \begin{matrix} pq \\ qp \end{matrix}$.
A sample derivation is as follows:

$$
M_o = \begin{matrix} p & q \\ q & p \end{matrix} \quad \overset{\Rightarrow}{R} \quad \begin{matrix} p & p & q & q \\ q & p & p & p \end{matrix} \quad \overset{\Rightarrow}{D} \quad \begin{matrix} p & p & q & q \\ q & q & q & q \\ p & q & q & q \\ q & q & q & q \end{matrix} \quad \overset{\Rightarrow}{L} \quad \begin{matrix} p & p & p & p & p & q & q \\ p & q & q & p & q & q & q \\ p & p & p & p & p & q & q \\ p & q & q & p & q & q & q \end{matrix}
$$

if RDL is in C.

2.　CONSTRUCTIONS OF PUBLIC KEY CRYPTOSYSTEMS (PKC)

(A) PKCs for plain texts of chain code pictures:

In the first type of public key cryptosystem introduced here, the plain text is a chain code picture or equivalently the picture description word over $\{1, r, u, d\}$ and the cryptotext obtained is an array. In constructing the PKC, we start with a deterministic TOLA system containing four tables, one for each letter in $\{1, r, u, d\}$ and convert that into a non-deterministic TOLAS containing two sets of four tables. This system is used for encryption in such a way that cryptanalysis amounts to membership problem for non-deterministic TOLAS which is NP-Complete. The conversion is done using a morphism, which is the trapdoor.

Let $G = (T, P, C, M_o)$ be a DTOLAS where the alphabet is T; P consists of a left (L), right (R), up (U), down (D) table; $C = P^*$; $M_o \in T^{++}$. The tables L,R,U,D define injective, uniform morphisms $h_i (i=1,2,3,4)$ from T^* to T^*. We require the DTOLAS to be unambiguous in the sense that no two words $w_1, w_2 \in C$ yield the same array from M_o i.e. if $M_o \xRightarrow{w_1} M$ and $M_o \xRightarrow{w_2} M$ then $w_1 = w_2$. Corresponding to these tables, we construct tables L', R', U', D' such that $a \rightarrow u$ is in L (R ,U ,or D) iff $a \rightarrow a_1 a_2 \ldots a_m$ is in L(R,U or D) and $u \in x^* a_1 x^* a_2 \ldots x^* a_m$ where x is a new symbol. $|u|$ is the same for all a in T and if $a \rightarrow u_1$, $b \rightarrow u_2$ are two rules in $L'(R', U'$ or $D')$ with $u_1 = b_1 \ldots b_n$, $u_2 = d_1 \ldots d_n$, then for each i, either b_i, $d_i \in T$ or $b_i = d_i = x$. We add a rule $x \rightarrow x \ldots x$, for x, so that the right side has the same length as the other rules in the table, since each table should be a complete set of rules for letters in $T \cup \{x\}$. Clearly, L', R', U', D' define morphisms h_1', h_2', h_3', h_4' on $(T \cup \{x\})^*$. A morphism g from V^* to $(T \cup \{x\})^{*1}$ is considered, such that $g(c) = a$ or x, for c in the alphabet V, which is of much greater cardinality than T and a in T. We assume $g^{-1}(a) \neq \emptyset$, for all a in T, g is the trapdoor information We form tables $L_i'' (R_i'', U_i'', D_i''$ respectively), $i = 1,2$ such that they define substitutions $t_{i1}, t_{i2}, t_{i3}, t_{i4}$ on V^* with $t_{ij}(d)$ as a finite, nonempty subset of $g^{-1}(h_j(g(d)))$, $j = 1,2,3,4$ for all d in V. For simplicity, we have taken two sets of four tables, $L_i'', R_i'', U_i'', D_i''$ $(i=1,2)$ but we can have n sets of tables, for some finite n.

We consider an array M_o' over T obtained from M_o by introducing a row or column of x's. We choose an array $M_o'' \in g^{-1}(M_o')$. The TOLAS $(V, L_i'', R_i'', U_i'', D_i'', M_o'')$, i=1,2 constitutes the public encryption key. The encryption is done, starting with M_o'' and applying the tables of the TOLAS. The sequence of application of the tables is decided by the plain text. In other words, if $x_1 x_2 \ldots x_m$ is the plain text, which is the picture description of a chain code picture, $x_i \in \{1, r, u, d\}$ for $i=1, \ldots, m$, then the ith table of the sequence of tables is L_1'' or L_2'' (R_1'' or R_2'', U_1'' or U_2'' or D_1'' or D_2'' respectively) according as x_i is l(r,u or d respectively). Encryption thus yields an array M over V.

The decryption of M begins with an application of g to M and then the resulting array is examined with L,R,U,D to obtain the sequence of tables that could be applied to M_o to give rise to g(M). The plain text is then recovered from the sequence, replacing L by l, R by r and so on.

We give algorithms for encryption and decryption. Algorithm for encryption:

Let sides 1,2,3 and 4 stand for left, right, up and down edges of an array respectively and side j is the length of side j. For i=1 to 4, a_i stands for the letter l,r,u,d and t_i stands for the table L,R,U,D respectively.

Input: A TOLAS $(V,t_i,t_i'$ (i=1,2,3,4),$M_o)$ where V is the alphabet, M_o is the axiom array, t_i, t_i' are two tables for each a_i(i=1,2, 3,4). A plain text $p = p_1 p_2 \cdots p_n$, $p_i \in \{a_1, a_2, a_3, a_4\}$.

Procedure: APPLY (M, t_j)

begib

    side $j = s_1 s_2 \cdots s_{|side\ j|}$
    for i=1 to side j do

        when $s_i \longrightarrow \alpha_1 \alpha_2 \cdots \alpha_m$ is a production in t_j
        if j=1 then replace in M, s_i by $\alpha_m \cdots \alpha_1$
        else if j=2 then replace in M, s_i by $\alpha_1 \cdots \alpha_m$
            else if j=3 then replace in M, s_i by $\begin{matrix} \alpha_m \\ \vdots \\ \alpha_1 \end{matrix}$

                else replace in M, s_i by $\begin{matrix} \alpha_1 \\ \vdots \\ \alpha_m \end{matrix}$

    return M

end;

begin

    Let $M = M_o$
    for i=1 to n do
        if $p_i = a_j$ then APPLY (M, t_j) or APPLY (M, t_j')

    return M_o

end;

If m is the maximum of right sides of rules in tables t_i, t_i', $i=1,2,3,4$, then the time taken for encryption is polynomial in m and n where n is the length of plain text.

If c is the maximum of the number of columns and rows in M_o then for each letter in plain text the space increases by atmost cm. Hence the encryption can be done in space polynomial in c and m.

Algorithm for decryption:

Let $RHS(t_i)$ denote the set of all words in the right hand side of table t_i of length l_i. Let $w_{ik}(l_i)$ denote the word of length l_i in the side i in the kth row (column) of M,i=1,2(i=3,4).

Input: A mapping $g:V \longrightarrow T \cup \{x\}$. A DTOLAS (T,t_1,t_2,t_3,t_4,M_o). The cryptotext C with m rows and n columns, say, $C=(c_{ij})$, $i=1,\ldots,m; j=1,\ldots,n$.

```
begin
    for i=1 to m do
      for j=1 to n do
        replace in C, cᵢⱼ by g(cᵢⱼ)
        remove all x in C
    return C
end;
Procedure: CHECK (M)
begin
    Let j=1
    for i=1 to 4 do
      k=1
      if k≤length of side i then
        if wᵢₖ(lᵢ)∈ RHS(tᵢ) then k=k+1
        else i=i+1
          PARSE (M,tᵢ)
        j=j+1
      repeat CHECK (M) until M=Mₒ
    return PⱼPⱼ₋₁···P₁
end;
```

Procedure: PARSE (M, t_i)

begin

    for k=1 to length of side i do

      replace in M, $w_{ik}(l_i)$ by a if $a \rightarrow w_{ik}(l_i)$ is in t_i

      $p_j = i$

    return M, p_j

end;

In this algorithm for the step PARSE (M, t_i), we make use of classical string matching techniques [1] which takes time linear in length of side $i \times l_i$. Time taken for decryption is polynomial in the maximum of l_i and the length of cryptotext. Space taken is linear in the size of the cryptotext. Cryptanalysis is equivalent to membership problem for TOL systems which is NP-Complete. We illustrate with an example.

EXAMPLE 2.1

Let $T = \{p, q\}$, $M_0 = \begin{matrix} p & q \\ q & p \end{matrix}$. The tables R,D,U,L are as in example1.2.

$R' = \{p \rightarrow xpxpp, \ q \rightarrow xpxqq\}$ $L' = \{p \rightarrow pppxp, \ q \rightarrow pqqxp\}$

$D' = \{p \rightarrow qqxq, \ q \rightarrow qpxq\}$ $U' = \{p \rightarrow xqqqq, \ q \rightarrow xqpqp\}$

    $g(a)=q$, $g(b)=g(c)=p$, $g(d)=g(e)=x$.

$R'' = \{b \rightarrow dbecc, \ c \rightarrow ecdbb, \ a \rightarrow dbeaa, \ d \rightarrow deddd, \ e \rightarrow deedd\}$

$D'' = \{a \rightarrow abde, \ b - aaea, \ c \rightarrow aaea, \ d \rightarrow dddd, \ e \rightarrow eeee\}$

$L'' = \{a \rightarrow caadb, \ b \rightarrow cbcdb, \ c \rightarrow bbbec, \ c \rightarrow bbcec, \ d \rightarrow deded,$
                                                   $e \rightarrow edede\}$

$U'' = \{a \rightarrow dacab, \ a \rightarrow eabac, \ b \rightarrow daaaa, \ c \rightarrow eaaaa, \ d \rightarrow edded$
                                                   $e \rightarrow deede\}$

$M_0' = \begin{matrix} p & x & q \\ q & x & p \end{matrix}$ and $M_0'' = \begin{matrix} b & e & a \\ a & d & c \end{matrix}$

If the plain text is rdld which is an initial part of the picture description word of the Hilbert curve in example 1.1., We have

$$\begin{matrix} b & e & a \\ a & d & c \end{matrix} \underset{R''}{\Longrightarrow} \begin{matrix} b & e & d & b & e & a & a \\ a & d & e & c & d & b & b \end{matrix} \underset{D''}{\Longrightarrow} \begin{matrix} b & e & d & b & e & a & a \\ a & d & e & a & d & a & a \\ b & d & e & a & d & a & a \\ d & d & e & e & d & e & e \\ a & d & e & a & d & a & a \end{matrix}$$

```
        c b c d b e d b e a a
        c a a d b d e a d a a
  ⟹     c b c d b d e a d a a
  L''   d e d e d d e e d e e
        c a a d b d e a d a a

        c b c d b e d b e a a
        c a a d b d e a d a a
        c b c d b d e a d a a
  ⟹     d e d e d d e e d e e
  D''   a a a a a d e a d a a      = M (the cryptotext)
        a b b d a d e b d b b
        e d d d e d e d d d d
        a a a d a d e a d a a
```

Applying g to M and removing x's we obtain the array

```
        p p p p p q q
        p q q p q q q
        p p p p q q q
        q q q q q q q
        q p p q p p p
        q q q q q q q
```

We note that this array could have been obtained only by applying a down table to the array

```
        p p p p p q q
        p q q p q q q
        p p p p q q q
        p q q p q q q
```

Likewise, this array could have been obtained only by applying a left table to the array

```
        p p q q
        q q q q
        p q q q
        q q q q
```

and so on. Thus the plain text recovered is rdld.

PKCs based on (D)TOL systems for plain texts of chain code pictures:

The plain text is again a chain code picture or the picture description word over l,r,u,d but the cryptotext is also a word.

Let T be an alphabet. Let w be a nonempty word over T. Corresponding to l,r,u,d injective morphisms h_1,h_2,h_3,h_4 from T^* to T^* are considered. We impose the unambiguity requirement that no two sequences of morphisms h_1,h_2,h_3,h_4 of the same length, should yield the same word from an axiom w, i.e. if

$$h_{i_n}(h_{i_{n-1}}(\ldots h_{i_1}(w)\ldots)) = h_{j_n}(h_{j_{n-1}}(\ldots h_{j_1}(w)\ldots)) \text{ where}$$

$h_{i_k}, h_{j_k} \in \{h_1,h_2,h_3,h_4\}$, then $h_{i_k} = h_{j_k}$, k=1,...,n. Let V be an alphabet of much greater cardinality than T, g, a morphism from V^* to T^* such that g(d) is in T or g(d) = λ and $g^{-1}(a) \neq \emptyset$, for all a in T. Let L,R,U,D be four subsititutions on V^*. For all d in V, L(d) is a finite, nonempty subset of $g^{-1}(h_1(g(d)))$. Similarly for R,U,D with h_1 replaced by h_2,h_3,h_4 respectively.

Let v be a word in $g^{-1}(w)$. The public encryption key is (V,L,R,U,D,v). The emcryption of a plaintext $x_1x_2\ldots x_n$, x_i in $\{1,r,u,d\}$ is done by choosing an arbitrary word from $f_n(\ldots f_2(f_1(v))\ldots)$ where each f_i is L,R,U or D according as x_i is l,r,u or d. Knowledge about the morphism g is the secret information hidden by the trapdoor. The decryption key consists of h_1,h_2,h_3,h_4,w and g.

We illustrete with an example.

EXAMPLE 2.2 Let T = $\{p,q\}$ and w = pq. h_1,h_2,h_3,h_4 and g are morphisms given by

$$h_1(p) = pq, \ h_1(q) = q \ ;$$
$$h_2(p) = p, \ \ h_2(q) = qp;$$
$$h_3(p) = pp, \ h_3(q) + qq;$$
$$h_4(p) = pq, \ h_4(q) = qp;$$
$$g(a) = g(c) = p, \ g(e) = q, \ g(b) = g(d) = \lambda$$

where V is the alphabet $\{a,b,c,d,e\}$.

Let L,R,U,D be the substitutions given by

L = {a⟶dae, b⟶b, c⟶abe, c⟶dabe, d⟶b, e⟶eb}
R = {a⟶cd, b⟶bd, c⟶a, d⟶d, e⟶ecd}
U = {a⟶ac, b⟶b, c⟶adc, c⟶daa, d⟶d, e⟶ede}
D = {a⟶ade, b⟶d, c⟶bae, d⟶d, e⟶dae}

Let v = cde. Suppose the plain text is the chain code picture in Figure 2.

Figure 2

The picture description word for the plain text is w = dru.
An encryption of w is as follows:

v = cde $\underset{D}{\Longrightarrow}$ baeddea

$\underset{R}{\Longrightarrow}$ bdcdecdddecdcd

$\underset{U}{\Longrightarrow}$ bdadcdededaadddedeadcdadcd = x,say.

This is an encryption of w.

The decryptoon of x is as follows:

We obtain g(x) = ppqqppqqpppp. On examining g(x) with h_1,h_2,h_3,h_4 we obtain a word pqpqpp so that h_3(pqpqpp) = g(x). Similarly, we obtain another word pqqp so that h_2(pqqp) = pqpqpp. Finally, we obtain the word pq so that h_4(pq) = pqqp. Thus the sequence $h_4h_2h_3$ yields g(x) from w = pq. Thus the plain text dru is recovered.

B) Public key cryptosystems based on DOL/TOL systems for plain texts of strings and arrays

We now discuss a public key cryptosystem where the public encryption key involves a TOL system [7] based on a DOL system with plain texts of strings of symbols. The advantage of the system is that the decryption based on the trapdoor is very atraightforward and much simpler than that of the public key cryptosystem in [8] but cryptanalysis is "hard". Also, we avoid the unambiguity requirement in [8]. In fact, without the trapdoor information, cryptanalysis is essentially an NP-complete problem, namely, the membership problem for TOL systems.

We now discuss the details of the construction of the system. Let $T = \{a_1,\ldots,a_n\}$ be an alphabet, $f:T^* \to T^*$ be an injective, λ-free morphism and u_i (i=1,...n) be n nonempty words in T^*. Let V be an alphabet of much greater cardinality than T amd $g:V^* \to T^*$ be a morphism such that for all d in V, g(d) is in T or $g(d) = \lambda$, and $g^{-1}(a) \neq \emptyset$ for all a in T. For i=1,...,m, for some $m \geqslant 1$, define substitutions t_i on V^* such that, for all d in V, $t_i(d)$ is a finite nonempty subset of $g^{-1}(f(g(d)))$.

Let x_i be any word in $g^{-1}(u_i)$, i=1,...,n respectively. $(V, t_1,\ldots,t_m, x_1,\ldots,x_n)$ is the public encryption key. The encryption of a plain text $w = p_1 \ldots p_r$, $p_j \subset T$, for j=1,...,r is done as follows: A word z is obtained from w by replacing each p_j by x_k, if p_j is a_k for $k \in \{1,\ldots,n\}$. The cryptotext is obtained by choosing an arbitrary word from $t_{j_k}(\ldots t_{j_2}(t_{j_1}(z)))\ldots)$, $k \geqslant 1$ where $j_1,\ldots,j_k \in \{1,\ldots,m\}$. Information concerning g is hidden by the trapdoor; decryption key consists of f, $u_1,,\ldots,u_n$, g. we note that if c is a cryptotext obtained, then $g(c) = f^k(w)$. The plain text can be recovered from g(c) using $u_1, \ldots, u_n$. The details regarding algorithms and the time taken for encryption and decryption and a suitable illustration are discussed in [13].

Having described the construction of cryptosystems based on DOL|TOL systems, we extend the system to array morphisms and array substitutions. This can be used to encrypt plain texts which are arrays of symbols representing digitized picture patterns.

Let T é $\{a_1, a_2,\ldots a_n\}$ be an alphabet, $f:T^{**} \to T^{**}$ ne injective, λ- free array morphism and let U_i, i=1,...n be nonempty arrays of same order in T^{**}. Let v be an alphabet of much greater cardinality than T and $g:V^{**} \to [T \cup \{x\}]^{**}$ be a morphism such that for d in V, g(d) is in T or g(d) = x and for all a in T. $g^{-1}(a) \neq \emptyset$. For i=1,...m, for some $m > 1$, define substitutions t_i on V^* such that for all d in V, $t_i(d)$ is a finite nonempty subset of $g^{-1}(f(g(d)))$.

Let X_i be an array in $g^{-1}(U_i)$, i=1,...n. $(V, t_1, t_2,\ldots,t_m, X_1,X_2,\ldots,X_n)$ is the public encryption key.

The encryption of a plain text

$$P = \begin{matrix} p_{11} & \cdots & p_{11} \\ \cdot & \cdots & \cdot \\ p_{11} & \cdots & p_{kl} \end{matrix} \; , \quad p_{ij} \in T, \; i=1,\ldots,k, \; j=1,\ldots,1$$

is done as follows: An array Q is obtained from P by replacing each p_{ij} by X_1 if p_{ij} is X_r, $i=1,\ldots,k$, $j=1,\ldots1$, $r=1,\ldots n$.

The cryptotext is chosen as an arbitrary array from

$$t_{j_r}(\ldots t_{j_2}(t_{j_1}(Q))\ldots), r \geqslant 1,$$

where

$$j_1, j_2, \ldots, j_r \in \{1, 2, \ldots, m\}.$$

g is the information hidden by the trapdoor and the decryption key consists of f, $U_1, U_2, \ldots U_n$ and g. We note that if C is the encrypted cryptotext $g(C)=f^k(P)$ and the plaintext is recovered from g(C) using T, f, U_1, $U_2 \ldots, U_n$. We gibe algorithms for encryption and decryption. In order to make the illustration simple we take T to be a binary alphabet $\{p,q\}$ and use only two tables. Algorithm for encryption of an array using system B:

Input: An array M over $\{p,q\}$ of order m x n. Two arrays M_p and M_q of equal order 1 x k over V. Tables t_1 and t_2 containing productions for letters in V. ||M|| is size of the array M.

```
begin
      for i=1 to m do
        for j=1 to n do
        replace the i-j th entry of M by M_p or M_q
          according as it is p or q
      return M
end;

begin
      Let x be a random number.
      Let s be a random sequence of x numbers where s_i=1 or 2
       i=1 to x
      for i=1 to x do
        if s_i=1 then APPLY RULES (M,t_1)  else APPLY RULES (M,t_2)
      return M.
end;
```

Procedure: APPLY RULES (M, t_i)

begin

    for i=1 to m do

      for j=1 to n do

        replace i-j th entry a of M by A where $a \rightarrow A$ is a

        production in table t_i.

    return M

end;

For the Procedure APPLY RULE (M, t_i) the time taken is $O(\|M\|)$, but $\|M\|$ increases in every step depending on $\|M_p\|$ and order of arrays in the tables t_i. Hence time and space taken for encryption are polynomial.

Algorithm for decryption:

Input: An array C of order m x n over V. A map $g: V \rightarrow \{p, q, x\}$. A deterministic injective morphism $f: \{p, q\} \rightarrow \{p, q\}^{**}$. Two arrays M_p and M_q.

begin

    for i=1 to m do

      for j=1 to n do

        replace the i-j th entry 'a' by g(a) in C

    return C

end;

Procedure: (PARSING, C)

begin

    remove all x from C

      find the unique array C' such that $f(C')=C$

      $C = C'$

        repeat (PARSING, C) until C is made up of only the

          arrays M_p and M_q.

          replace M_p by p and M_q by q.

    return C

end;

Clearly time taken for decryption is polynomial and space is linear in $\|C\|$.

We illustrate with an example.

Let $T = \{p,q\}$ and f be a deterministic array morphism

$f(p) = \dfrac{pq}{qp}$, $f(q) = \dfrac{pp}{qq}$ and let $M_1 = \dfrac{pq}{qq}$ and $M_2 = \dfrac{pp}{qp}$.

Define f′ by

$$f'(p) = \begin{matrix} p & \dot{x} & q \\ x & x & x \\ q & x & p \end{matrix}, \qquad f'(q) = \begin{matrix} p & x & p \\ x & x & x \\ q & x & q \end{matrix}$$

The trapdoor function g be given by $g(a) = g(b) = p$, $g(c) = q$, $g(d) = g(e) = x$. We then form array substitutions t_1, t_2 given by

$$t_1: a \longrightarrow \begin{matrix} a & d & c \\ d & e & e \\ c & d & b \end{matrix} \qquad d \longrightarrow \begin{matrix} d & e & d \\ e & e & e \\ e & d & e \end{matrix}$$

$$b \longmapsto \begin{matrix} b & e & c \\ e & e & e \\ c & e & a \end{matrix} \qquad e \longrightarrow \begin{matrix} e & e & d \\ e & e & e \\ d & e & e \end{matrix}$$

$$b \longrightarrow \begin{matrix} \iota & d & c \\ e & d & e \\ c & e & a \end{matrix} \qquad c \longrightarrow \begin{matrix} a & e & a \\ d & e & e \\ c & d & c \end{matrix}$$

$$t_2: a \longrightarrow \begin{matrix} b & e & c \\ e & e & e \\ c & e & a \end{matrix} \qquad c \longrightarrow \begin{matrix} a & e & b \\ d & e & e \\ c & d & c \end{matrix} \qquad e \longrightarrow \begin{matrix} e & d & e \\ d & d & d \\ e & e & d \end{matrix}$$

$$b \longmapsto \begin{matrix} b & d & c \\ e & d & e \\ c & d & a \end{matrix} \qquad d \longrightarrow \begin{matrix} d & d & d \\ e & e & e \\ d & e & d \end{matrix}$$

We then construct M_1' and M_2' from M_1 and M_2 using g.
Let $M_1' = \dfrac{a\ c}{c\ c}$, $M_2' = \dfrac{a\ b}{c\ b}$. The encryption is done using the tables t_1 and t_2.

Let the plain text be the array $M = \dfrac{pq}{qp}$. We first form an array replacing p by M_1' and q by M_2' , and obtain

$$\begin{matrix} a & c & a & b \\ c & c & c & b \\ a & b & a & c \\ c & b & c & c \end{matrix}$$

The tables t_1, t_2 are then applied a certain number of times.
An encryption can be done as follows:

```
                            b e c a e b b e c b d c
                            e e e d e e e e e e d e
                            c e a c d c c e a c d a
                            a e b a e b a e b b d c
          a c a b           d e e d e e d e e e d e
          c c c b   ⟹      c d c c d c c d c c d a
          a b a c    t₂     b e c b d c b e c a e b
          c b c c           e e e d e e e e d e e
                            c e a c d a c e a c d c
                            a e b b d c a e b a e b
                            d e e e d e d e e d e e
                            c d c c d a c d c c d c
```

We can thus continue and obtain the cryptotext using tables
t_1, t_2 a finite number of times.

The decryption is done by applying g to the cryptotext
and then examining the resulting array with f′ on hand. We can
thus obtain the axiom array from which encryption was done and
recover the plaintext array from it.

References

1. Aho, A.V and Corasick, M.J. (1975), Efficient string
   matching: An aid to bibliographic search,
   Comm. ACM, 18, 333-340.
2. Dassow, J. (1984), A note on DTOL systems,
   Bull. EATCS.,No.22, 11-14.
3. Diffie, W. and Hellman, M.E. (1976), New directions in
   cryptography, IEEE Trans. Inf. Theory IT-22, 644-654.
4. Fu, K.S. (1974). "Syntactic Methods in Pattern Recognition",
   Academic Press, New York.
5. Maurer, H.A., Rozenberg, G., and Welzl, E. (1982), Using
   string languages to describe picture languages,
   Inform. and Control, 54, 155-185.
6. Rosenfeld, A. (1979), "Picture Languages – Formal Models of
   Picture Recognition", Academic Press, London.
7. Rozenberg, G. and Salomaa, A. (1980), "The Mathematical
   Theory of L-systems", Academic Press, New York.
8. Salomaa, A. (1985), "Computation and Automata", Encyclopaedia
   of Mathematics and its applications, Vol. 25,
   Cambridge University Press.

9. Salomaa, A.(1985), Cryptography from Caesar to DES and RSA, Bull. EATCS, No.26, 101–120.
10. Siromoney, R. (1985), Array languages and Lindenmayer systems-a survey, in "The Book of L", Eds. G. Rozenberg and A. Salomaa, Springer-Verlag, Berlin, 413–426.
11. Siromoney, R. and Siromoney, G. (1977), Extended controlled table L-arrays, Inform. Control, 35, 119–138.
12. Siromoney, R. and Siromoney, G. (1986) A public key crypto-system that defies cryptanalysis, Bull. EATCS, No.28, 37–43.
13. Subramanian, K.G., Siromoney, R. and Abisha Jeyanthi (1986), A DOL/TOL public key cryptosystem, TR MATH 16/86, Madras Chriatian College, Tambaram.

VII.   HYBRID APPROACHES I

HYBRID APPROACHES

H. Bunke

Universität Bern
Institut für Informatik und
angewandte Mathematik
Länggassstrasse, 51
CH-3012 Bern
Switzerland

1. INTRODUCTION

The discipline of pattern recognition has a history of about thirty years now and a huge number of different techniques has been proposed. Most of these techniques fall into one of three major categories which are statistical (or decision theoretic), structural, and artificial intelligence based pattern recognition. Each of the different methods has its strength and its limitations. For overcoming these limitations, statistical, structural, and artificial intelligence based methods are mixed sometimes. This results in a hybrid approach.

This paper discusses various pattern recognition methods with a particular emphasis on the question how different methods are related with each other and how they can be combined into a hybrid approach. The organization of the paper follows the diagram shown in Fig. 1. Among the conventional approaches to pattern recognition, we will consider decision theoretic methods, syntactic methods, structural prototypes, and relaxation in section 2. In section 3, three important approaches to knowledge representation will be discussed, namely formal logic, production systems, and semantic nets. Further topics like search, control, and system organisation will be addressed in section 4.

NATO ASI Series, Vol. F45
Syntactic and Structural Pattern Recognition
Edited by G. Ferraté et al.
© Springer-Verlag Berlin Heidelberg 1988

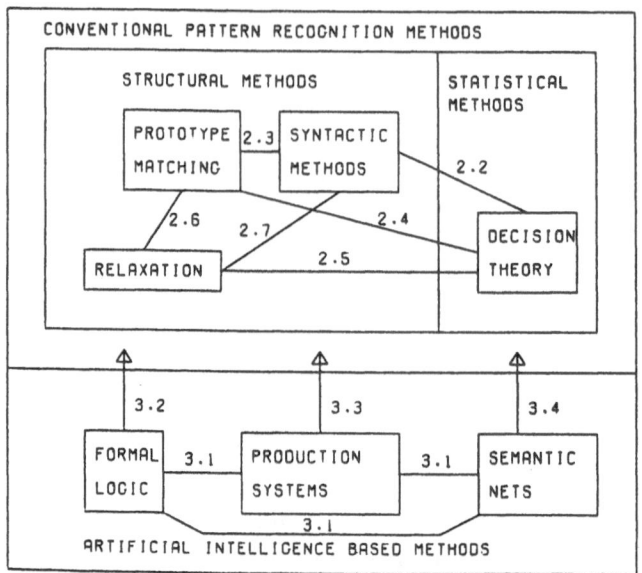

<u>Fig. 1</u>: A categorization of different approaches to pattern
recognition and relationships among them.

2. HYBRID APPROACHES BASED ON STATISTICAL AND STRUCTURAL METHODS

2.1. Preliminaries

Decision theoretic methods are applied primarily for the
purpose of classification. The task in pattern classification
is to assign an unknown input pattern to a class out of M
classes, where M⩾2. An individual pattern is represented by a
N-dimensional vector of features. Depending on the way the
classes are represented, we distinguish between nonparametric
and parametric statistical classification. The most important
subclasses of nonparametric and parametric statistical classification are nearest neighbor, or NN, classification, and
Bayes-classification, respectively. For more details on statistical classification see [FUKUNAGA 1972, DUDA/HART 1973,
DEVIJVER/KITTLER 1982].
Decision theoretic methods have a long tradition in pattern
recognition and are based on a well founded mathematical

theory. They have been proven useful in numerous applications. The algorithms are usually computationally inexpensive as compared to structural or artificial intelligence based methods. However, there are also several disadvantages and limitations of the statistical approach. Although a great deal of effort has been undertaken in deriving optimal algorithms with respect to classifying the extracted features, the features themselves are often choosen arbitrarily. Statistical methods provide only a class description of a pattern. They do not describe a pattern so as to allow its generation given its class, nor do they describe aspects of a pattern which make it ineligible for assignment to another class. Moreover, the various relations which may exist between the choosen features are completely neglected in the statistical approach.

The fundamental idea in <u>structural pattern recognition</u> is the explicit utilization of structural information in pattern representation. Structural information is used for modelling two important aspects, namely the hierarchical composition of a complex pattern based on simpler subpatterns, and the various relations which may exist between different subpatterns and/or characteristic features.

<u>Syntactic methods</u> are an important subclass of the structural approach. They are characterized by using formal grammars for pattern class representation. The terminals of the grammar correspond to primitive subpatterns which can directly be extracted from an input pattern by means of suitable preprocessing and segmentation methods. The set of grammar nonterminals corresponds to subpatterns of greater complexity which are successively built up from primitive elements. The process of building up complex (sub)patterns from simpler constituents is modelled by the grammar productions. Finally, the recognition process is based on a parser which analyzes an unknown input pattern according to the given grammar. Various types of grammars for a wide variety of applications have been proposed within this framework. For a detailled treatment, the reader is referred to [GONZALEZ/THOMASON 1978, FU 1982].

Syntactic methods are advantageous for many tasks since they allow, as recognition result, not only pattern classification but also the inference of a structural description of an unknown input pattern. Furthermore, these methods are based on the well founded theory of formal languages [HOPCROFT/ULLMAN 1979] which provides many useful results about power, limitations, and computational complexity of certain recognition procedures. On the other hand, the recognition of noisy patterns by means of syntactic methods has not yet been completely solved. Moreover, for dealing with complex patterns in real world applications, the approaches based purely on formal language theory are not always enough powerful.

Structural prototypes can be considered as a viable alternative to formal grammars for pattern class representation. The idea is to store, in an explicit way, a finite number of pattern prototypes. In contrast with statistical NN classification, patterns are not stored as N-dimensional feature vectors. Instead, structural representations based on strings, trees, or graphs are preferred. For the recognition of an unknown input pattern x it is necessary to match x with the prototypes in order to detect that prototype pattern which is most similar to x. Inevitably, unexact or error-correcting matching procedures are required for this purpose. A number of algorithms have been proposed in the literature. For an introduction see [FU 1982]. Further details on string matching can be found in [HALL/ DOWLING 1980, SANKOFF/KRUSKAL 1983]. For tree matching see [LU 1979, CHENG/LU 1985]. Two classical papers on graph matching are [TSAI/FU 1979, SHAPIRO/HARALICK 1981].

The use of structural prototypes is advantageous if pattern structure is important and if there are too few sample patterns available for deriving a grammar. A disadvantage is that these methods lack a well founded mathematical theory at present, and that their computational complexity - at least for graph matching - is high.

Relaxation is another class of structural methods. It is an iterative procedure which starts with an initial ambiguous labeling of primitive pattern components. Relaxation aims at deriving an unique interpretation of components, which is globally consistent under a given set of constraints. The relaxation procedures reported in the literature can be classified into discrete and continuous, or probabilistic, relaxation [WALTZ 1975, ROSENFELD/HUMMEL/ZUCKER 1976]. Many different schemes within the latter category have been proposed [PELEG 1980, FAUGERAS/BERTHOD 1981, HUMMEL/ZUCKER 1983].

Relaxation is suitable for reducing local ambiguities. It has been successfully applied to many pattern recognition problems. The method is appealing particularly for those app- lications where the available a priori knowledge is in the form of local constraints. On the other hand there are some problems with relaxation. From a theoretical point of view, there are several open questions, although interesting results have been reported recently [HARALICK 1983, HENDERSON 1984]. From an application oriented point of view, relaxation is li- mited since only local interpretations of pattern constituents are derived and no interpretation of a pattern as a whole can be achieved.

2.2. Syntactic – Statistical Methods

There are two major ways of combining statistical and syn- tactic pattern recognition methods, namely probabilistic gram- mars and attributed grammars. First, we will discuss probabi- listic grammars.

Probabilistic grammars can favourably be applied when dif- ferent pattern classes overlap, due to noise and distortions. The basic idea in probabilistic grammars is the attachment of a probability $p(r)$ to each production $r:X \rightarrow A_1...A_n$. Given an element $x \in L(G)$ (where $L(G)$ denotes the language generated by G, i.e. the set of terminal strings which can be derived), we can determine the probability $p(x/G)$ of x under grammar G by multiplying all probabilities of the productions which are used in the derivation $S \overset{*}{\rightarrow} x$ (where S denotes the grammar start symbol). For ambiguous strings, we have to sum over all possible derivations of x. If a particular pattern x is gene- rated by different grammars G_i, i=1,...,M, each representing a different pattern class, then we parse x according to each G_i, determining the probability $p(x/G_i)$, and decide for pattern

class j if $p(x/G_j)=\max\{p(x/G_i)/i=1,...,M\}$. If a priori probabilities $p(G_i)$ for pattern classes i = 1,..., M are available,
we determine the maximum of $p(x/G_i)p(G_i)$. What results is a
Bayes classification rule. So probabilistic grammars are a hybrid method based on techniques from statistical decision theory and syntactic pattern recognition.

Problems with probabilistic grammars are consistency, i.e.
the condition $\sum p(x/G)=1$ (where the sum is over all $x \in L(G)$)
and the assessment of the production probabilities $p(r)$. For
these topics, the reader is referred to [GONZALEZ/THOMASON
1978, FU 1982]. Recently Markov models have attracted much attention, primarily in the context of speech understanding
[LEVINSON et al. 1983]. As it is shown in [GONZALES/THOMASON
1978], a linear stochastic grammar can be considered as an
example of a Markov chain. Not only probabilistic string grammars, but also probabilistic tree grammars have been proposed
in the literature [Fu 1982]. The idea of error correcting parsing can be combined with probabilistic grammars. Here probabilities for elementary pattern deformations like substitution, insertion, and deletion are provided. The original grammar is augmented by error productions, the probabilities of
which are derived from the elementary pattern deformation pro-
bablities. Using this augmented grammar, a modified Earley
error-correcting parser can be constructed that determines the
most likely correction of a given distorted pattern [FU 1982].

Attributed grammars are the second major approach to combining syntactic and statistical pattern recognition methods.
A classical paper on this subject is [TSAI/FU 1980]. Further
applications can be found in [FU 1977, PAVLIDIS/ALI 1979, TANG
1979, BUNKE 1982]. In attributed grammars, each grammar symbol
A is augmented by a vector of attributes $\alpha(A) = (a_1(A), a_2(A)$
$,...,\alpha_N(A))$. Such a vector can be interpreted as a N-dimensional feature vector as used in statistical classification. Remark that the symbol A can denote a subpattern as well as a
relation. Considering a context free production $X \rightarrow A_1...A_n$, we
need an additional rule for expressing the relations between
the attributes of the left-hand side X and the right-hand side
$A_1...A_n$. Two cases can be distinguished. First, the attributes

of X can be dependent on those of $A_1...A_n$, i.e. $\alpha(X) = f(\alpha(A_1),...,\alpha(A_n))$. Secondly, the attributes of A_i can be dependent on X, i.e. $\alpha(A_i) = g_i(\alpha(X))$, $i=1,...n$. The first case (sythesized attributes) is most suitable for bottom-up analysis, while the second case (inherited attributes) is most suitable for top-down processing. In the extreme case, a (non-attributed) context-free grammar results when all attributes are deleted from an attributed grammar. In the other extreme case, a statistical classifier is obtained from an attributed grammar when each pattern is treated as a single entity and not decomposed into subpatterns. So attributed grammars can be considered as a very general approach containing as a special case pure syntactic and pure decision theoretic methods, respectively. A deeper discussion of this subject can be found in [FU 1983].

2.3 Syntactic Methods - Structural Prototypes

Pattern recognition by means of prototype matching has several aspects in common with syntactic methods. First, both methods emphasize structural properties of patterns, i.e. the composition of patterns from subpatterns including relations between subpatterns. Secondly, both approaches rely on the same data structures, namely strings, trees, and graphs. Finally, in both cases capabilities for error-correction are required in order to cope with noisy and distorted data.

Obviously, each finite set of prototype patterns $x_1...,x_n$ can be described by a grammar. In their simplest form, the grammar productions are $S_i \rightarrow x_i$, $i=1,...,n$. Conversely, if the language of a grammar is finite, it can directly be represented in such a way that each element is interpreted as a prototype pattern. Despite of this theoretical equivalence for finite sample sets, one method can be superior to the other for a specific application. Generally speaking, if the number of sample patterns is small, there is often no need for a grammar and a "direct" representation by means of prototypes is perhaps preferable. On the other hand, if a large number of sample patterns is involved, pattern recognition and pattern class representation is eventually more efficient if a grammar is used.

Formal grammars are more powerful than structural prototypes in the sense that they can generate an infinite number of elements. Furthermore, the derivation tree mirrors, as a result of parsing, the aggregation of subpatterns into a pattern along a potentially unbounded number of hierarchical levels. This is in contrast with structural prototypes, where the hierarchical composition of a pattern is limited to only two levels, namely the level of pattern primitives and the level where the pattern is considered as a whole. An exception to this limitation are substructures with priorities, like in PDL [FU 1982], or hierarchical graphs. The latter subject will further be discussed in section 3.4.

A hybrid approach combining structural prototypes and formal grammars has been proposed in [GERNERT 1981]. For the purpose of classification, an unknown pattern, which is represented by means of a graph, is compared with number of prototype graphs. The distance between a graph g and a prototype p is defined as the number of derivation steps of a grammar required in order to transform p into g. So inexact graph matching is accomplished by means of a transformation which is based on a grammar. A similar idea for string matching has been proposed in [TAI/FU 1982]. However, no practical experience demonstrating the utility of these hybrid approaches has been reported so far.

There are two other hybrid approaches based on graph grammars that combine techniques from syntactic and structural pattern recognition. In [BUNKE 1982], graph grammars for the recognition of schematic diagrams, including heavily distorted drawings, have been proposed. In contrast with the "classical" syntactic approaches where the recognition of an unknown pattern is based on parsing, the grammar is used as a tool that directly transforms an input pattern into the desired output descripton. The basic operation cycle of a graph grammar as proposed in [BUNKE 1982] consists of two steps, namely subgraph-isomorphism detection, and subgraph replacement. So this graph grammar approach can be considered as a typical example of a hybrid technique integrating graph matching with graph rewriting. Another approach is described in [KAUL 1987]. An error correcting precedence graph parser is introduced that can be used to compute the minimum error distance between two graphs. As a remarkable property, this procedure has a time complexity of only $O(n^3)$ where n gives the number of nodes of the input graph. However, the proposed approach is applicable only to a restricted class of graphs.

2.4 Structural Prototypes - Statistical Methods

There are many interesting relationships between pattern recognition based on structural prototype matching and decision theoretic methods. First, one notices that determining that prototype which is most similar to the input pattern is conceptually closely related with NN classification as discussed in section 2.1. From this observation it follows that the structural similarity measure should be a metric since then it is possible to employ clustering techniques for efficient prototype, i.e. sample pattern, organization [LU 1979, SHAPIRO/ HARALICK 1982, SHAPIRO 1986].

The inexact graph matching technique proposed in [TSAI/FU 1979] can be considered from a decision theoretic point of view. If probabilities for elementary graph deformations are provided, graph distance, or similarity, can be interpreted as deformation probability. Consequently, finding the most similar prototpye is equivalent to detecting the most probable deformation. Thus, prototype matching in the sense of [TSAI/FU 1980] can be interpreted as a special decision theoretic approach. Since deformation probabilites are often unknown in practical applications, deformation costs are used instead as an approximation [BUNKE/ALLERMANN 1983].

Another hybrid approach combining structural prototypes and decision theoretic methods is the use of stochastic prototypes. In what follows, we will restrict our discussion to prototype graphs. In contrast with [TSAI/FU 1979], where any node and any edge in a graph is present with a probability equal to 1 and where probabilities indicate only the likelihood of graph deformations, the nodes and edges themselves are stochastic in a stochastic graph. Intuitively, any node and edge is present only with a probaility $0 \leqslant p \leqslant 1$ in a stochastic graph. This concept has been proposed in [SHAPIRO/HARALICK 1982] for defining the "average" or the "median" graph of a cluster of similar prototypes. In another paper, probability distributions of node and edge attributes have been proposed, in addition to probabilities of node and edge occurrence [GROEN/SANDERSON/ SCHLAG 1985]. Thus object recognition is based on matching a non-stochastic graph representing the unknown object to each of the stochastic model graphs in order to find the most probable model. For another approach based on a similar idea see [WONG 1983].

Another very interesting relationship between NN-classification and matching of unknown patterns against structural prototypes is established in [GOLDFARB/CHAN 1984]. The authors consider a set of structural prototypes from different classes together with a pseudometric (i.e., a "distance" function not necessarily fulfilling the triangle inequality) and show how the prototypes can be mapped into a n-dimensional numerical vector representation in such a way that the distance between the prototypes is preserved. In a further step this representation is mapped into another vector representation of low dimensionality (dimensionality one in the example given in [GOLDFARB/CHAN 1984]). It is claimed that it should be possible by analysis of this low dimensional training set representation to select few characteristic sample patterns for each class, for example the element closest to the mean, or the elements corresponding to the piecewise linear boundaries. So by mapping structural prototypes into a vector representation, a reduction of the sample set size can be achieved. This is similar in its spirit to sample set condensation or sample set editing for numerical NN-classification as described in [DEVIJVER/ KITTLER 1982].

Another link between structural prototypes and statistical methods can be concluded from [KASIF/KITCHEN/ROSENFELD 1983] where subgraph isomorphism detection is accomplished by cluster detection in a parameter space.

2.5 Relaxation - Statistical Methods

In a number of papers, primarily addressed to the field of pictorial pattern recognition, the combination of statistical classification and relaxation has been proposed. The idea common to all these approaches is to make a classification of each pixel, using a statstical method, and to subsequently refine this initial classification by means of relaxation.

The initial classification can be based on a Bayes-classifier [KUBICHEK/QUINCY 1985], on a particular distance function [DAVIS/WANG/XIE 1983], on thresholding [BHANU/FAUGE-RAS 1982], or on other histogram based features [NAGIN/HANSON/RISEMAN 1982]. Notice that no contextual information is exploited in this first classification stage, i.e., the classification of a pixel is made without reference to the class of the neighboring pixels. Notice also that the statistical classification may yield ties or near ties between different classes. Relaxation following the initial classification is a suitable tool for making use of contextual knowledge and for breaking ties. In order to facilitate relaxation, it is required that the statistical classifier yields not only a classname w_j for each pixel i but a vector $[p_i(w_1),...,p_i(w_m)]$ where $p_i(w_j)$ is the probability that w_j is the correct class of pixel i. The principal idea for the exploitation of contextual constraints is to define compability coefficients in such a way that identical labels at neighboring pixels reinforce each other while different labels suppress each other. (This principle may be further refined if particular knowledge about the shape of the objects under consideration is available.) An extension to using also temporal context in image sequences has been reported in [DAVIS/WANG/XIE 1983].

An interesting relationship between probabilistic relaxation and Bayes-classification is derived in [HARALICK 1983]. It is shown that under some general conditional independence assumptions probabilistic relaxation can be interpreted as a process that computes conditional probabilities. More specifically, each iteration computes the conditional probability of each class for each object on the basis of a context which is the context of the previous iteration enlarged by one neighborhood width. Initially, only measurements made on an object itself are used as context. So relaxation iterations must only continue until either the conditional independence assumptions no longer hold or until the entire context is taken into account. Assigning finally the class with the highest probability to an object can thus be considered as a special Bayes desision rule.

2.6 Relaxation - Structural Prototypes

Relaxation is based on constraint exploitation for the purpose of reducing local ambiguities. In their most general form, constraints are given by expressions $R(x_1, \ldots, x_n)$, which indicate the likelihood that $x_1, \ldots, x_n$ is a correct interpretation of objects $O_1, \ldots, O_n$, if there is a relation R between $O_1, \ldots, O_n$. From this point of view, a structural prototype, e.g. a graph, can be considered as an aggregation of several unary and binary constraints. So matching an object with a prototype is nothing else but finding an interpretation of the components of the object which is maximally consistent with the constraints represented by the prototype.

On the other hand, relaxation can be used as a special technique for finding a match between a model and an unknown object. This idea has been applied in a system for aerial image understanding [FAUGERAS/ PRICE 1981]. The nodes in the model and the object graph correspond to lines and regions in an image. Node attributes represent features like color, texture, size, etc. Spatial relations between lines and regions, like adjacency, nearby, above, etc., are represented by graph edges. Initially, only a fixed number of best matching images nodes are assigned to each model node. For updating of probabilities, the relations between pairs of model nodes are systematically compared with those of corresponding image nodes. Good experimental results have been reported by the authors. Similar approaches to matching relational structures by means of relaxation have been proposed in [KITCHEN 1980, CHENG/HUANG 1981].

In conclusion, structural matching can be considered as a particular constraint satisfaction paradigm, while on the other hand constraint satisfaction, i.e. relaxation, can be used as a special technique for accomplishing structural matching.

2.7 Relaxation - Syntactic Methods

Relaxation follows the principle of least commitment, i.e. a maximum number of labels, or interpretations, for an object is considered initially. This idea can be adopted in syntactic pattern recognition in such a way that not a fixed symbol x_i, but a vector $(x_1, \ldots, x_m)$ of possible symbols is considered at each position in an input string. Eventually, a measure of confidence, or a probability is assigned to each possible symbol at each position. Assume an input string of lenth n with m alternative symbols at each position, representing an unknown pattern. The "brute force" approach to syntactic recognition is the application of a parser to all possible nm strings which can be formed thus and the determination of that sequence of n symbols which is compatible with the given grammar and has maximum probability among all compatible strings. If n and m are large, this becomes prohibitive. In order to reduce the effort required for parsing, discrete relaxation can be applied beforehand, eliminating successively the symbols in the input string which are incompatible with the given grammar. Using alternatively a probabilistic relaxation scheme, those strings among all possible nm strings which are compatible with the given grammar will be enhanced while all other strings will be suppressed.

This combination of probabilistic relaxation and syntactic pattern recognition was called "syntax-directed probabilistic relaxation" in a recent article [DON/FU 1985]. The authors proposed the following hybrid method. Given n string positions with m possible symbols at each position, where $p_i(x_j)$ denotes the probability that symbol x_j is correct at position i (j=1, ..;,m; i=1,...,n), probabilistic relaxation is applied first, in order to enhance the probability of those symbols which are consistent with the given grammar. After some iterations, all symbols x_j at position i with a low probability $p_i(x_j)$ are eliminated. From the remaining symbols all possible strings are formed and fed into an error-correcting parser. Using the results of this parser and the overall probability of an input string (before relaxation), a final decision, i.e. the most likely parse of the input pattern, is derived.

A direct solution to the problem of finding the string with maximum probability among the strings compatible with a given grammar, under the condition that the constraints are given by regular expressions, has been proposed in [BUNKE/GREBNER/ SAGERER 1984]. The solution is based on dynamic programming, and is very efficient with respect to time and space, and implementation effort, as well.

The application of grammatical constraints for reduction of ambiguities is limited to terminal symbols in [DON/FU 1985]. Another approach taking into regard also constraints between higher-level nonterminals is hierarchical relaxation [DAVIS/ HENDERSON 1981]. It is basically a bottom-up parsing procedure where, during the whole recognition process, reduction of symbols according to the rules of the given grammar and exploitation of contextual constraints for elimination of inconsistent symbols are intertwined.

3. HYBRID APPROACHES BASED ON ARTIFICIAL INTELLIGENCE METHODS

Artificial intelligence based methods are characterized by considering pattern classes as abstract concepts and individual patterns as instances thereof. Pattern classes are repre-

sented by explicity storing knowledge describing them. Applying this knowledge to the measurements made on an unknown
pattern, recognition is accomplished by drawing domain specific inferences, or logical conclusions. So artificial intelligence based pattern recognition emphasizes the issues of knowledge representation and control of problem solving. A very
important category of tasks where these ideas have been applied is diagnostic classification [SHORTLIFFE 1976, BARR/
FEIGENBAUM 1982, CHANDRASEKARAN 1986].

In section 3.1 we will briefly review the most important
techniques for knowledge representation and discuss how they
are related with each other. Relationship with conventional
pattern recognition methods will be studied in section
3.2-3.4.

3.1 Basic knowledge representation methods

Formal logic is a classical artificial intelligence approach to knowledge representation and inference. In what follows, we will consider only first order predicate calculus.
For other logics see [TURNER 1984]. The basic constituents of
first order predicate calculus are well-formed formulas which
are built up from predicates, functions, variables and constants according to particular syntactic and semantic rules.
The resolution priniple has become very important for drawing
logical conclusions from a set of axioms. This technique can
be applied to pattern recognition for inferring class labels
or structural interpretations from input patterns using a
priori knowledge that is expressed in terms of predicate calculus formulas.

Recently the logic programming language PROLOG has attracted much attention. PROLOG covers a major part of first order
predicate calculus (with almost no real restrictions with respect to practical applications) and has been shown useful for
the implementation of pattern recognition algorithms [DVORAK
1986]. For an introduction to PROLOG see [CLOCKSIN/MELLISH
1984]. General introductions to predicate calculus and its

applications are [CHANG/LEE 1973, WOS et al. 1984].

Predicate calculus can be understood as a high level knowledge representation language. As an advantage, PROLOG supports rapid prototyping and facilitates the formal verification of programs. Typically, emphasis in a PROLOG program is
on what the problem is rather on how the solution is algorith-
mically determined. Besides these advantages, there is a number of well known shortcomings and limitations in predicate
calculus and PROLOG. PROLOG is based on an exhaustive depth-
first search with backtracking. This is not elegant and results in slow execution. Another problem with predicate calculus in general is the property of monotony. This makes it difficult to cope with time varying problems, e.g. the analysis
of image sequences. Another well known problem is the so-
called frame problem. For more details, the reader is referred
to [NILSSON 1982].

Production systems are a well known tool for knowledge representation in expert systems and they have been used in many
pattern recognition applications. Basically, a production system consists of productions, or rules, of the form IF CONDI-
TION THEN CONCLUSION. Depending on the particular syntax of
CONDITION and CONCULSION, a great variety of different production system formalism results. Given an initial set of data stored in a database - the so-called short-term knowledge
base - conclusions can be derived by successive application of
rules. Control procedures include forward- and backward-chai-
ning as well as mixed-mode chaining of productions.

There are several advantages of production systems, for
example perspicuity and modularity. Shortcomings of production
systems include particulary the lack of expressive power and
sufficient control structures for certain applications. A general introduction to, and overview of, the field of production systems is given in [DAVIS/KING 1977, BARR/FEIGENBAUM
1981]. Two particular examples, namely the non-monotone forward chaining OPS5 and the backward chaining M.1 are described in [BROWNSTON et al. 1986, HARMON/KING 1985]. Further applications in pattern recognition are reported in [OHTA 1980,
NAGAO/MATSUYAMA 1980, NAZIF/LEVINE 1984, DUANE et al. 1985, Mc

KEOWN et al. 1985], among many others.

A semantic net can be considered as a graph. In contrast with "traditional" graphs, however, where the nodes and edges are atomic units, the nodes and edges in a semantic net (which are also called concepts and relations, respectively) are complex data structures constisting of a number of subunits, each. For example, a concept may have a name, a number of attributes, a number of conditions, a default value etc. With repect to edges, three standard relations are most common in semantic nets, namely "part", "specialization", and "instance". By means of the part-relation, an aggregation of a number of basic entities into a more complex object can be modelled. An important feature of the specialization and the instance relation is the inheritance property, which facilitates compact representation of knowledge. Other relations are problem dependent and may include relations of spatial, temporal, causal, etc. nature.

Inference procedures for semantic nets are mainly based on matching or on search (see also section 4). More details on semantic nets can be found in [BARR/FEIGENBAUM 1981]. Examples of applications in pattern recognition are [HANSON/RISEMAN 1978, TSOTSOS et al. 1980, BALLARD/BROWN/FELDMAN 1978, NIEMANN et al. 1985].

There are various relationships between predicate calculus, production systems, and semantic nets and these approaches can be integrated into a hybrid approach in a variety of ways. For a theoretical discussion of the relations between production systems and semantic nets, on the one hand side, and formal logic on the other hand side, the reader is referred to [NILSSON 1982, SCHUBERT 1976, HAYES 1979, WOS et al. 1984]. From a more practical point of view, the relationships become obvious from the fact that both semantic nets and production systems can easily be implemented in PROLOG. For more details see [BRATKO 1986].

There is also a close relationship between semantic nets and production systems. Given a set of production rules, a semantic net can be used as an aid for partitioning this set into smaller units. Usually, this results in a more efficient control procedure with respect to rule application [DUDA/ GASCHNIG/HART 1979]. On the other hand, given a semantic net, one can attach rules to the slots of a concept as procedural knowledge sources. An example is [NIEMANN et al. 1985].

Recently, many efforts in the artificial intelligence community have been devoted to the development of hybrid expert system shells and knowledge representation languages. They provide a combination of logic, productions, and semantic nets in one consistent framework. Examples of hybrid expert system shells are KEE [KUNZ et al. 1984], LOOPS [BOBROW/ STEFIK 1983], and BABYLON [DI PRIMIO/BREWKA 1985].

3.2 Formal logic - conventional pattern recognition

There is a close relationship between formal logic, at the one hand side, and formal grammars and structural prototypes,

```
a)    SUBMEDIAN:: = ARMPAIR, ARMPAIR.

      TELOCENTRIC:: = BOTTOM, ARMPAIR.

b)    submedian(S0,S,submedian(A1,A2)) :-
          armpair(S0,S1,A1),
          armpair(S1,S,A2).

      telocentric(S0,S,telocentric(B,A)) :-
          bottom(S0,S1,B),
          armpair(S1,S,A).

c)    if        armpair(S0) = [S1,A1]
         and    armpair(S1) = [S,A2]
      then      submedian(S0) = [S,submedian(A1,A2)].

      if        bottom(S0) = [S1,B]
         and    armpair(S1) = [S,A]
      then      telocentric(S0) = [S,telocentric(B,A)].
```

Fig. 2: a) Two productions of the chromosome grammar,
b) corresponding parts of the parser implemented in
PROLOG, c) corresponding parts of the parser implemented in M.1.

at the other hand side, insofar as parsing and structural matching can be implemented using logic programming. For an example look at Fig. 2a where two productions of the well known chromosome grammar (FU 1982) are given. In Fig. 2b, the corresponding PROLOG clauses (i.e. "statements") that are part of the parser are shown. There is a one-to-one correspondence between a grammar, as in Fig. 2a, and its parser in PROLOG. In other words, it is very easy to automatically generate a parser in PROLOG from a given context free grammar; see also [KOWALSKI 1979]. As it was shown in a recent thesis, also graph matching can be implemented in PROLOG in a very straightforward way [DVORAK 1986].

Another close relationship between formal logic and grammars may be concluded from the STRIPS-system [NILSSON 1982, chapter 7]. There are predicate calculus rules for manipulating other predicate calculus formulas. These rules are very similar to the rules in a grammar. In a sense, STRIPS-rules are more general since they may contain variables that can be matched with other variables or constants. However, these variables can be simulated, in many cases, by means of grammar attributes. Originally, the work on STRIPS was motivated by robot action planning. But the idea may be used in pattern recognition in a similar way for inferring conclusions about patterns.

Fuzzy logic can be considered as a link between predicate calculus and decision theory. For more details about this topic, including applications, the reader is referred to [ZIMMERMANN 1985].

3.3 Production systems - conventional pattern recognition

From a theoretical point of view, production systems are closely related with formal grammars. A rule in a production system may be interpreted as a grammar rule with the if-part (condition) corresponding to the left-hand and the then-part (conclusion) corresponding to the right-hand side of a grammar rule. Now consider a special type of a production system where the database is a linear string of facts. Each time a rule is applied, its if-part is removed from the database while its then-part is inserted at the place of the if-part. Assume furthermore, that the facts are divided into terminal and nonterminal facts. Initially, there is only one nonterminal fact in the database. The termination condition is that there are only terminal facts in the database. Control is by forward-chaining. Obviously, such a production system can simulate any formal grammar. Conversely, since type 0 grammars have the same computational power as Turning machines, any production system can certainly be simulated by a formal grammar. Because of this close relationship, all comments made in section 2.3 hold also for the relation between production systems and structural prototypes.

From a practical point of view, many production systems or expert-system shells can be applied like a programming language and are suitable for the implementation of particular pattern recognition algorithms. For example, look at Fig. 2c where the M.1 version of the grammar productions in Fig. 2a is shown. The rules in Fig. 2c are part of a parser for the chromosome grammar that has been written in M.1. One notices the one-to-one correspondence between Figs. 2a, 2b, and 2c, respectively. More examples, including PROLOG and M.1 implementations of string matching, graph matching, and discrete relaxation are given in [DVORAK 1986].

Many rule based expert systems are applied in diagnostic classification [CHANDRASEKARAN 1986] - a task that has traditionally been solved by means of statistical classification techniques. A closer look reveals that diagnostic classification using if-then rules can be interpreted as hierarchical classification, i.e. as a decision tree procedure where sequential measurements are made in order to successively rule out certain classes until only one class remains, or the input pattern is rejected. For more details about hierarchial classification see [MORET 1982].

Other links between production systems and syntactic pattern recognition can be concluded from [VERE 1977, TSATSOULIS/FU 1985]. Finally, it is to be noted that most production systems

include means for coping with uncertainty, i.e. uncertain data
and knowledge. Examples are certainty factors [SHORTLIFFE
1976], concepts from Dempster/Shafer's theory [ISHIZUKA et al.
1982], or Bayesian uncertainties [DUDA/GASCHNIG/HART 1979].
This can be considered as a link with statistical classification methods.

3.4 Semantic nets - conventional pattern recognition

Semantic nets, at the one hand side, and structural prototypes and grammars, at the other hand side, have much in common.
First, we will discuss relations between semantic nets and
structural prototypes. Any structural prototype represented by
a graph can be considered as a simple semantic net, without
part-, specialization-, and instance-relations. In this case,
inference is restricted to error correcting graph or subgraph
isomorphism detection. A hierarchical graph is a structural
prototype for modelling the aggregation of simpler constituents into more complex objects on several levels. Such a prototype can also be considered as a special type of a semantic
net, including part-relations but excluding specialization-
and instance-relations. Inference in a hierarchical graph needs means for inheriting problem dependent relations (for
example, spatial relations like above, right, etc.) up and
down the part-hierarchy, in addition to graph or subgraph
isomorphism detection. A semantic net in its full generality
includes, besides the features of a hierarchical graph, means
that can be used for making non-geometrical problem dependent
inferences. An example is [NIEMANN et al. 1985].

A grammar rule $X \rightarrow X_1 \ldots X_n$ has a direct analogy in a semantic net, namely a concept X with parts $X_1 \ldots, X_n$, and vice versa. Notice that a symbol X_i may not only represent an object
but also a relation between objects. In [HALL 1973] the equivalence between grammars and AND/OR graphs is discussed and it
is shown that parsing is equivalent to searching for a solution graph. An AND/OR-graph can be considered as a particular
type of a semantic net. So the derivation tree of a string can

be interpreted as a partial instantiation of the semantic net which corresponds to the grammar. Attributes, procedures for attribute calculation, application conditions, etc. are directly related with the corresponding components of a concept in a semantic net. An application example where a grammar operates on the slots (i.e., attributes) of a semantic net is [BONAMINI et al. 1982].

4. FURTHER DISCUSSION

The emphasis in section 3 was on knowledge representation. There are other issues from artificial intelligence which are very important in pattern recognition. One of them is search and control. The task of a pattern recognition system is the transformation of sensory input data into a pattern description, e.g. a class name. Such a transformation is accomplished in a series of processing steps with a number of intermediate results. In a complex system, those intermediate results are ambiguous and the overall sequence of processing steps cannot be uniquely determined beforehand. So a control procedure for optimal selection of processing steps is required. For achieving such an optimal selection, search methods from artificial intelligence can be applied [NILSSON 1982]. For an example, see [NIEMANN et al. 1985]. A discussion of the use of search strategies in pattern recognition can be found in [KANAL 1979]. In this paper it is shown from a very general point of view how parsing and classification can be understood as search paradigms.

Another important question is the overall organization of a pattern recognition system. The blackboard model according to [ERMAN et al. 1980] seems particularly useful for complex systems. It seems ideally suited for organizing large hybrid systems since the blackboard is the only shared global data structure and all the expert modules are completely independent from each other, from a conceptual and methodological point of view. Further examples of systems which are organized according to the blackboard model are [NAGAO/MATSUYAMA 1980,

LEVINE/SHAHEEN 1981].

The idea of using hybrid approaches in solving pattern recognition problems is not new [FU 1982]. The rapid growth of
pattern recognition and artificial intelligence and particularly many of the current research issues like complex knowledge based pattern recognition systems, knowledge based robotics assembly, or multi sensor systems as they are required,
for example, in autonomous vehicles will certainly stimulate
further research in this area.

LITERATURE

BALLARD, D.H. / BROWN, C.M. / FELDMAN, J.A.: An approach to
    knowledge directed image analysis, in [HANSON / RISEMAN
    1978], 664-670
BARR, A. / FEIGENBAUM, E.A. (Eds.): The handbook of artificial
    intelligence, Vol. 1, Pitman Books, London, 1981
BARR, A. / FEIGENBAUM, E.A. (Eds.): The handbook of artificial
    intelligence, Vol. 2, Pitman Books, London, 1982
BHANU, B. / FAUGERAS, O,D.: Segmentation of images having uni-
    modal distributions, IEEE Trans. PAMI-4, 1982, 408-419
BOBROW, D.G. / STEFIK, M.: The LOOPS manual, Xerox Corp., Palo
    Alto, Ca., 1983
BONAMINI, R. / DE MORI, R. / LETTERA, A. / SANDRETTO, E.: An
    electrocardiographic signal understanding system, in
    [KITTLER/FU/PAU 1982], 443-464
BRATKO, I.: Prolog programming for artificial intelligence,
    Addison Wesley, Reading, Ma., 1986
BROWNSTON, L. / FARELL, R. / KANT, E. / MARTIN, N.: Program-
    ming expert-system in ORS5, Addison Wesley Publ. Co.,
    Reading, Ma., 1986
BUBROW, D.G. / STEFIK, M.: The LOOPS manual, Xerox Corp., Palo
    Alto, Ca., 1983
BUNKE, H.: Attributed programmed graph grammars and their app-
    lication to schematic diagram interpretation, IEEE Trans.
    PAMI-4, 574-582, 1982
BUNKE, H. / ALLERMANN, G.: Inexact graph matching for struc-
    tural pattern recognition, Pattern Recognition Letters 1,
    1983, 245-253
BUNKE, H. / GREBNER, K. / SAGERER, G.: Syntactic analysis of
    noisy input strings with an application to the analysis
    of heart-volume curves, Proc. 7th ICPR, Montreal, 1984,
    1145-1147
CHANG, C. / LEE, R.C. : Symbolic logic and mechanical theorem
    proving, Academic Press, New York, 1973
CHANDRASEKARAN, B.: From numbers to symbols to knowledge
    structures: Pattern recognition and artificial intelligen-
    ce perspectives on the classification task, in GELSEMA,
    E.S. / KANAL, L.N. (Eds.): Pattern recognition practice
    II, Elsevier Science Publ. B.V., 1986, 547-559

CHENG, J.K. / HUANG, T.S.: Image recognition by matching relational structures, IEEE Proc. PRIP, Dallas, 1981, 542-547

CHENG, Y.C. / LU, S.Y.: Waveform correlation by tree matching, IEEE Trans. PAMI-7, 1985, 199-305

CLOCKSIN, W.F. / MELLISH, C.S.: Programming in Prolog, Springer-Verlag, 1984

DAVIS, L.S. / HENDERSON, T.C.: Hierarchical constraint processes for shape analysis, IEEE Trans. PAMI-3, 1981, 265-277

DAVIS, L.S. / WANG, C.Y. / XIE, H.C.: An experiment in multi-spectral, multitemporal crop classification using relaxation techniques, Comp. Vision, Graphics, and Im. Proc. 23, 1983, 227-235

DAVIS, R. / KING, J.: An overview of production systems, in ELOCK, E.W. / MICHIE, D. (Eds.): Machine Intelligence 8, Ellis Horwood, Chichester, 1977, 300-332

DEVIJVER, P. / KITTLER, J.: Pattern reconition: A statistical approach, Prentice Hall Int., 1982

DI PRIMIO, F. / BREWKA, G.: Babylon, kernel system of an integrated environment for expert system development and operation, Proc. 5th Int. Workshop on Exp. Systems and their Applications, Avignon, 1985, 573-583

DON, H.S. / FU, K.S.: A syntactic method for image segmentation and object recognition, Pattern Recognition 18, 1985, 73-87

DON, H.S. / FU, K.S.: A parallel algorithm for stochastic image segmentation, IEEE Trans. PAMI-8, 1986, 594-603

DUDA, R.D. / GASCHNIG, J. / HART, P.: Model design in the propector consultant system for mineral exploration, in MICHIE, D. (Ed.): Expert systems in the micro-electric age, Edinbourgh Univ. Press, 1979, 153-167

DUDA, R.O. / HART, P.E.: Pattern classification and scene analysis, Jon Wiley & Sons, 1973

DUANE, G.S. / VENABLE, S.F. / RICHTER, D.J. / WIEDEMANN, A.M.: A production system for scene analysis and semantically guided segmentation, SPIE Vol. 548 Applications of Art. Intell. II, 1985, 35-45

DVORAK, J.: Artificial intelligence programming with rule based systems, MS-Thesis, Dept. of Computer Science, University of Berne, Switzerland, 1986 (in German)

ERMAN, L.D. / HAYES-ROTH, F. / LESSER, V.R. / REDDY, R.: The HEARSAY-II speech-understanding system, Comp. Surveys 12, 1980, 213-253

FAUGERAS, O. / BERTHOD, M.: Improving consistency and reducing ambiguities in stochastic labeling: An optimization approach, IEEE Trans. PAMI-3, 1981, 412-424

FAUGERAS, O.D. / PRICE, K.E.: Semantic description of aerial images using stochastic labeling, IEEE Trans. PAMI-3, 1981, 633-642

FU, K.S.: Syntactic pattern recognition, applications, Springer Verlag, 1977

FU, K.S.: Syntactic pattern recognition and applications, Prentice Hall, 1982

FU, K.S.: Hybrid Appoaches to Pattern Recognition in [KITTLER / FU / PAU 1982], 139-155

FU, K.S.: A step towards unification of syntactic and statistical pattern recognition, IEEE Trans. PAMI-5, 1983, 200-205

FUKUNAGA, K.: Introduction to statistical pattern recognition, Academic Press 1972

GERNERT, D.: Distance or similarity measures which respect the internal structure of the objects, Methods of Operations Research 43, 1981, 329-335

GOLDFARB, L. / CHAN, T.Y.T.: On a new unified approach to pattern recognition, Proc. 7th ICPR, Montreal, 1984, 705-708

GONZALEZ, R.C. / THOMASON, M.G.: Syntactic pattern recognition, Addison-Wesley, 1978

GROEN, F.C.A. / SANDERSON, A.C. / SCHLAG, J.F.: Symbol recognition in electrical diagrams using probabilistic graph matching, Pattern Recognition Letters 3, 1985, 343-350

HALL, P.A.N.: Equivalence between AND/OR graphs and context-free grammars, CACM 16, 1973, 444-445

HALL, P.A.V. / DOWLING, G.R.: Approximate string matching, Comp. Surveys 12, 1980, 381-402

HANSON, A.R. / RISEMAN, E.M.: Visions; a computer system for interpreting scenes, in [HANSON / RISEMAN 1978a], 303-333

HANSON, A.R. / RISEMAN, E.M. (Eds.): Computer vision systems, Academic Press, New York, 1978(a).

HARALICK, R.M.: An interpretation for probabilistic relaxation, Comp. Vision, Graphics, and Image Processing 22, 1983, 388-395

HARALICK, R.M.: Decision making in context, IEEE Trans. PAMI-5, 1983, 417-428

HARMON, P. / KING, D.: Expert-systems-artificial intelligence in business, Jon Willy, New Youk etc., 1985

HAYES, P.H.: The logic of frames. In Metzing, D. (Ed.): Frame conceptions and text understanding, de Gruyter, Berlin, 1979, 46-61

HENDERSON, T.C.: A note on discrete relaxation, Comp. Vision, Graphics, and Im. Proc. 28, 1984, 384-388

HOPCROFT, J.E. / ULLMAN, J.D.: Introduction to automata theory, languages and comptation, Addison Wesley, 1979

HUMMEL, R. / ZUCKER, S.: On the foundations of relaxation labeling processes, IEEE Trans. PAMI-5, 1983, 267-287

ISHIZUKA, M. / FU, K.S. / YAO, T.P.: SPERIL: an expert system for damage assessment of existing structures, Proc. 6th ICPR 1982, Munich, 932-937

KANAL, L.N.: Problem-solving models and search strategies for pattern recognition, IEEE Trans. PAMI-1, 1979, 193-201

KASIF, S. / KITCHEN, L. / ROSENFELD, A.: A Hough transform technique for subgraph isomorphism, Pattern Recognition Letters 2, 1983, 83-88

KAUL, M.: Similarity of graphs - a grammar-driven divide-and conquer approach, in this volume.

KITCHEN, L.: Relaxation applied to matching quantitative relational structures, IEEE Trans. SMC-10, 1980, 96-101

KITTLER. J. / FU, K.S. / PAU, L.F. (Eds.): Pattern recognition theory and applications, D. Reidel Publ. Co., Dodrecht etc., 1982

KOWALSKI, R.: Logic for problem solving, North-Holland, 1979

KUBICHEK, R.F. / QUINCY, E.A.: Idenfification of seismic stratigraphic traps using statistical pattern recognition, Pattern Recognition 18, 1985, 440-458

KUNZ, J.C. / KEHLER, T.P. / WILLIAMS, M.D.: Applications development using a hybrid ai development system, AI Magazine 5, 1984, 41-54

LEVINE, M.D. / SHAHEEN, S.I.: A modular computer vision system for picture segmentation and interpretation, IEEE Trans. PAMI-3, 1981, 540-556

LEVINSON, S.E. / RABINER, L.R. / SONDHI, M.M.: An introduction to the application of the theory of probabilistic functions of a Markov process to automatic speech recognition, Bell System Techn. Journal 62, 1983, 1035-1074

LU, S.Y.: A tree-to-tree distance and its application to cluster analysis, IEEE Trans. PAMI-1, 1979, 219-224

McKEOWN, D.M. / HARVEY, W.A. / McDERMOTT, J.: Rule-based interpretation of aerial imagery, IEEE Trans. PAMI-7, 1985, 570-585

MORET, B.M.E.: Decision trees and diagrams, Comp. Surveys 14, 1982, 593-623

NAGAO, M. / MATSUYAMA, T.: A structural analysis of complex aerial photographs, Plenum Press, New York, 1980

NAGIN, P.A. / HANSON, A.R. / RISEMAN, E.M.: Studies in global and local histogram guided relaxation algorithms, IEEE Trans. PAMI-4, 1982, 263-277

NAZIF, A.M. / LEVINE, M.D.: Low level image segmentation; an expert system, IEEE Trans. PAMI-6, 1984, 555-577

NIEMANN, H. / BUNKE, H. / HOFMANN, I. / SAGERER, G. / WOLF, F./ FEISTEL, H.: A knowledge based system for analysis of gated blood pool studies, IEEE Trans. PAMI-7, 1985, 246-259

NILSSON, N.J.: Principles of artificial intelligence, Springer Verlag, 1982

OHTA, Y.: A region oriented image-analysis system by computer, Ph. D. diss., Dept. of Inform. Sciences, Kyoto Univ., Japan, 1980

PAVLIDIS, T. / ALI, F.: A hierarchical shape analyzer, IEEE Trans. PAMI-1, 1979, 2-9

PELEG, S.: A new probalistic relaxation scheme, IEEE Trans. PAMI-2, 1980, 362-369

ROSENFELD, A. / HUMMEL, R.A. / ZUCKER, S.W.: Scene labelling by relaxation operations, IEEE Trans. SMC-6, 420-443

SANKOFF, D. / KRUSKAL, J.B.: Time warps, string edits, and macromolecules: The theory and practice of sequence comparisons, Addison Wesley, Reading, Ma., 1983

SCHUBERT, L.R.: Extending the expressive power of semantic networks, Art. Intell. 7, 1976, 163-193

SHAPIRO, L.G. / HARALICK, R.M.: Structural descriptions and inexact matching, IEEE Trans. PAMI-3, 1981, 501-519

SHAPIRO, L.G. / HARALICK, R.M.: Organization of Relational Models for Scene Analysis, IEEE Trans. PAMI-4, 1982, 595-602

SHAPIRO, L.G.: The use of numerical relational distance and symbolic differences for organizing models and for matching, in ROSENFELD, A. (Ed.): Techniques for 3-D machine perception, North-Holland, 1986, 255-270

SHORTLIFFE, E.A.: Computer-based medical consultations: MYCIN, American Elsevier, New York, 1976

TAI, J.W. / FU, K.S.: Semantic syntax-directed translation for pictorial pattern recognition, Proc. 6th ICPR, Munich, 1982, 169-171

TANG, G.Y.: A syntactic-semantic approach to image understanding and creation, IEEE Trans. PAMI-1, 1979, 135-144

TSAI, W.H. / FU, K.S.: Error-correcting isomorphisms of attributed relational graphs for pattern analysis, IEEE Trans. SMC-9, 1979, 757-768

TSAI, W.H. / FU, K.S.: Attributed grammar-a tool for combining syntactic and statistical approaches to pattern recognition, IEEE Trans. SMC-10, 1980, 873-885

TSOTSOS, J.K. / MYLOPOULOS, J. / COVVEY, H.D. / ZUCKER, S.W.: A framework for visual motion understanding, IEEE Trans. PAMI-2, 1980, 563-573

TSATSOULIS, C. / FU, K.S.: Modelling rule-based systems by stochastic programmed production systems, Inf.Sci. 36, 1985, 207-230

TURNER, R.: Logics for artificial intelligence, Ellis Horwood Ltd., Chichester, 1984

VERE, S.A.: Relational production systems, Art. Intell. 8, 1977, 47-68

WALTZ, D.: Understanding line drawings of scenes with shadows, in WINSTON, P.H. (Ed.): The psychology of computer vision, Mc Graw Hill, 1975, 19-91

WONG, A.K.C. / YOU, M.: Entropy and distance measures of random graphs, IEEE Comp. Soc. Conf. on PRIP, 1983, 371-376

WOS, L. / OVERBEEK, R. / LUSK, E. / BOYLE, J.: Automated reasoning introduction and application, Prentice Hall, Englewood Cliffs, 1984

ZIMMERMANN, H.J.: Fuzzy set theory - and its applications, Kluwer-Nijhoff Publishing, Bosten etc., 1985

AN AI-STRUCTURAL APPROACH TO EDGE DETECTION

J.-y. Zhou and L.-d. Wu

Dept. of Computer Science
Fudan University
Shanghai
P.R. China

1. Introduction

Edge detection is an important issue in computer vision and many works have been done in this area[1,2,3]. A new approach to edge detection is presented in this paper. It is based on the intuition of edge: edge should be at the place where the gray level of picture has great change and which has a long but narrow shape, i.e. it uses both gray level and structural information in picture. The key contribution of this paper is the finding of a special data structure of tree to represent both gray level and structural information in picture conveniently. The approach is reliable and of general purpose. An overview of the approach is given in §2. The details about the various aspects of the approach are presented in §3-6. Five examples including both artificial ones and natural ones are given in §7. The paper is concluded with the discussion about the advantages and potential improvements in §8.

2. Overview of the Approach

The approach consists of four steps(see Fig. 1):
Step 1: It transforms input image $P(i,j)$, $1 \leqslant i \leqslant I$, $1 \leqslant j \leqslant J$ into another so-called edge image $D(i,j)$, $1 \leqslant i \leqslant I$, $1 \leqslant j \leqslant J$, where $D(i,j) > 0$ and represents the magnitude of the change of gray level of input picture P at pixel (i,j), i.e. the greater

NATO ASI Series, Vol. F45
Syntactic and Structural Pattern Recognition
Edited by G. Ferraté et al.
© Springer-Verlag Berlin Heidelberg 1988

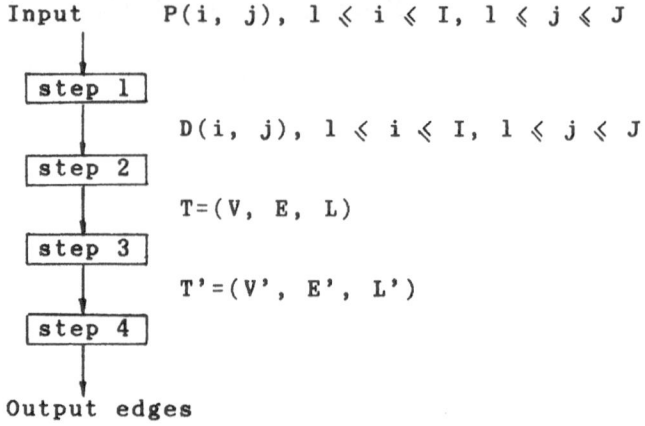

Fig. 1

the D(i,j) is, the more likely the pixel (i,j) is on edge. This is a well known process in edge detection and can be done by a lot of operators such as Robert's. What we use is a difference of two integrals of P(i,j) over two neighbours of different sizes. The detail will be given in $3.

Step 2: It transforms edge image D(i,j), $1 \leqslant i \leqslant I$, $1 \leqslant j \leqslant J$ into a labeled tree T=(V, E, L). The tree is as follows:

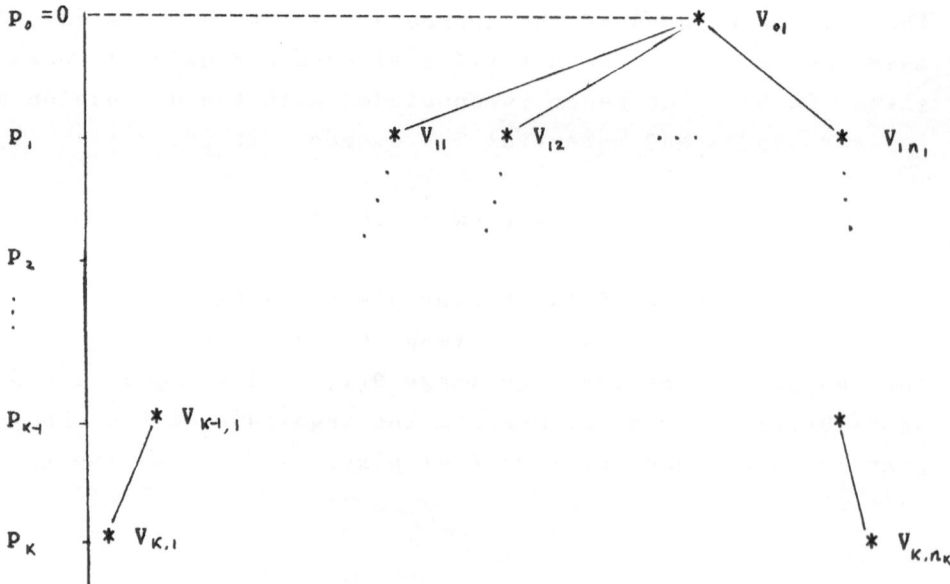

Fig. 2

i.e. the area of A(k,l) or $V_{k\ell}$ and

e= M/a (called extension of A(k,l) or $V_{k\ell}$)

where

$$M= \sum_{(i,j) \in A_{k\ell}} [(i-\bar{i})^2 + (j-\bar{j})^2] = \sum (i^2 + j^2) - [(\sum i)^2 + (\sum j)^2]/a$$

and

$$i = \sum_{(i,j) \in A_{k\ell}} i /a, \qquad j = \sum_{(i,j) \in A_{k\ell}} j /a$$

It will be shown in section 4 that e characterizes the extension of A(i,j), i.e. the greater the e of A(k,l) is, the longer and more narrow the A(k,l) is. An efficient one-pass algorithm to implement the transformation is given in $5.

Now it should be clear that a node V having large k and e values corresponds to a domain A(k,l) which is very likely an edge. Since it has large k value it means that there are great change of gray levels in this area and since it has large e value and it means that the domain is long and narrow, like an edge. Therefore the labeled tree T introduced here is a very useful global representation for edge detection. It combines both the semantic information (how large is the change of gray levels ?) and the structural information (Is it long and narrow, just like an edge ?).

Step 3 and 4: It cuts off the tree T to get a subtree T' such that only those leaves having large k and e values are left. Now let the pixels in the domains corresponding to the leaves of T' have gray value 1 and the others zero, we get the edge output. The detail will be given in section 6.

3. Edge Image D

There are a lot of operators to get D from input P such as Robert's. But they are sensitive to noise and so not reliable. What we use is the follows:

$$D(i,j) = \frac{1}{\beta} \sum_{(k,\ell) \in B(i,j)} \sum P(k,l) - \frac{1}{\beta'} \sum_{(k,\ell) \in B'(i,j)} \sum P(k,l)$$

where
$$0 = p_0 < p_1 < \ldots < p_K < \max_{i,j} D(i,j)$$
are given set of parameters and for each p , let
$$A(k) = \{(i,j): D(i,j) \geqslant p_K\}$$
$$= \bigcup_l A(k,l), \qquad 1 \leqslant k \leqslant K$$
where $A(k,l)$, $1 \leqslant l < L_K$, is disjointed 8 -- connected components of $A(k)$ and let each $A(k,l)$ corresponds to a node V in tree T.

Since $A(0) = \{ (i,j): 1 \leqslant i \leqslant I, 1 \leqslant j \leqslant J \}$ and $A(0) \supsetneq A(1)$ ... $\supsetneq A(K)$, for each $A(k,l)$ there is an unique $A(k-1, m)$ such that
$$A(k,l) \subseteq A(k-1, m), \quad 1 \leqslant l \leqslant Lk, 1 \leqslant k \leqslant K$$

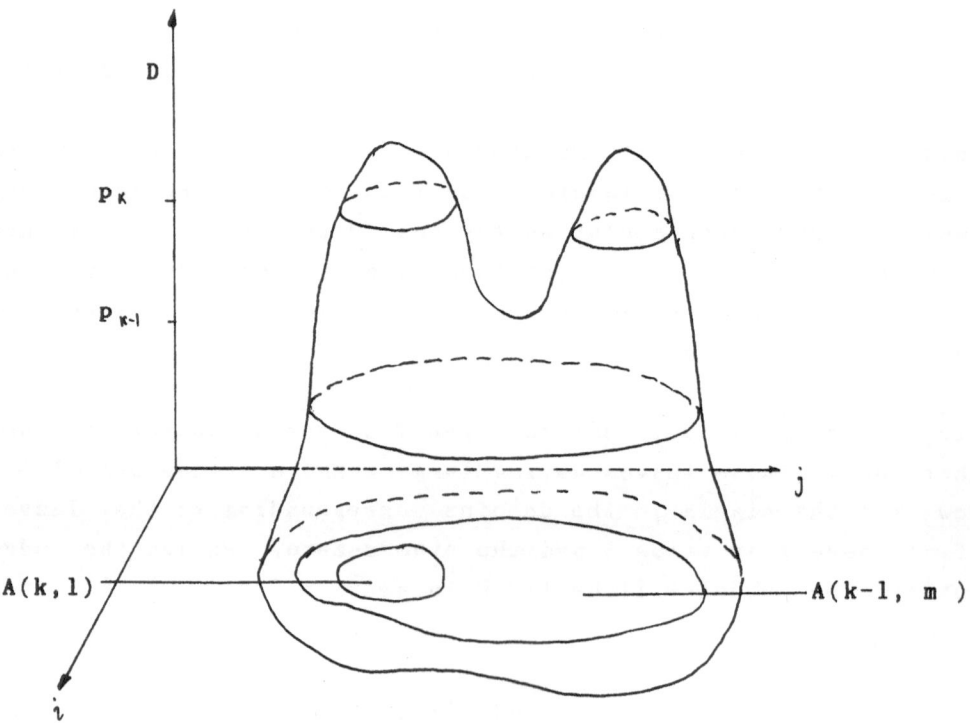

Fig. 3

In this case we will call $V_{k,l}$ the son of $V_{k-1,m}$ and $V_{k-1,m}$ the father of $V_{k,l}$, and there is an edge connecting $V_{k-1,m}$ and $V_{k,l}$.

Besides, for each node V define a label $l=(a, e)$ where
    a= number of pixels in $A(k,l)$

where B(i,j) and B'(i,j) are the disk-like neighbours of (i,j):

$$B(i,j) = \{(k,l): |k-i|^2 + |l-j|^2 \leqslant R^2\}$$

$$B'(i,j) = \{(k,l): (|k-i|^2 + |l-j|^2 \leqslant r^2) \cap (|P(k,l)-P(i,j)| < \Delta)\}$$

with r < R, and B and B' are the area of B(i,j) and B'(i,j). is a given small positive number. The introduction of integrals is

to suppress the noise. The continuous analogue are

$$D(x,y) = \frac{1}{B} \iint\limits_{B(x,y)} p(u,v) \; d\,d - \frac{1}{B'} \iint\limits_{B'(x,y)} p(u,v) \; d\,d$$

and

$$B = \{(u,v): (u-x)^2 + (v-y)^2 \leqslant R^2\}$$

$$B' = \{(u,v): ((u-x)^2 + (v-y)^2 < r^2) \cap (|p(u,v)-p(x,y)| < \Delta)\}$$

with r < R.

They are sensitive to various kinds of edges and can be seen from the responses to the typical edges:

(i) step edge and its response

$$p(x,y) = \begin{cases} H1, & a*x +b*y +c \geqslant 0, \\ \\ H2, & a*x +b*y +c < 0, \end{cases} \qquad H1 \neq H2$$

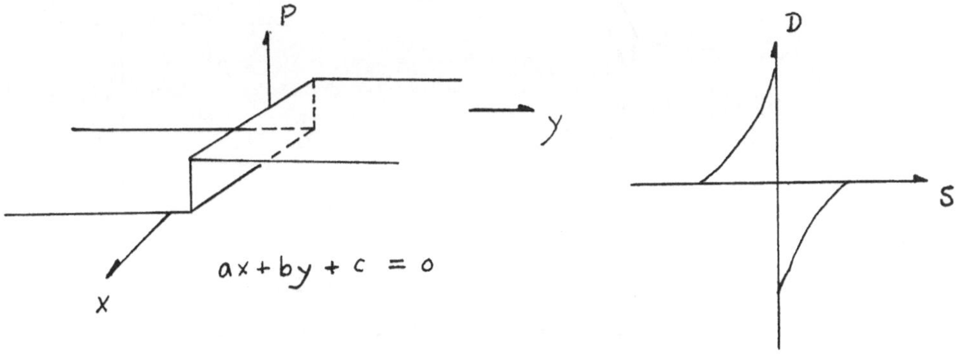

Fig. 4

where the meaning of s is as follows:

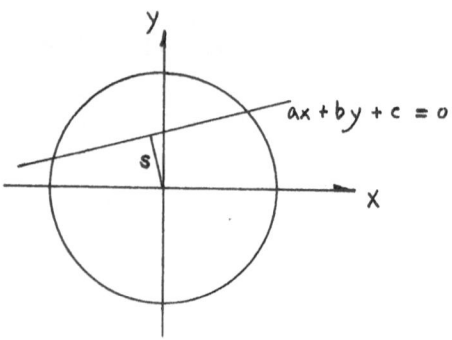

Fig. 5

(ii)   Roof edge and its response:

$$p(x,y)= \begin{cases} a*x + b*y + (a*x+c)*b/b, & a*x+b*y+c > 0 \\ a*x + b*y + (a*x+c)*b/b, & a*x+b*y+c < 0 \end{cases} \quad b = b$$

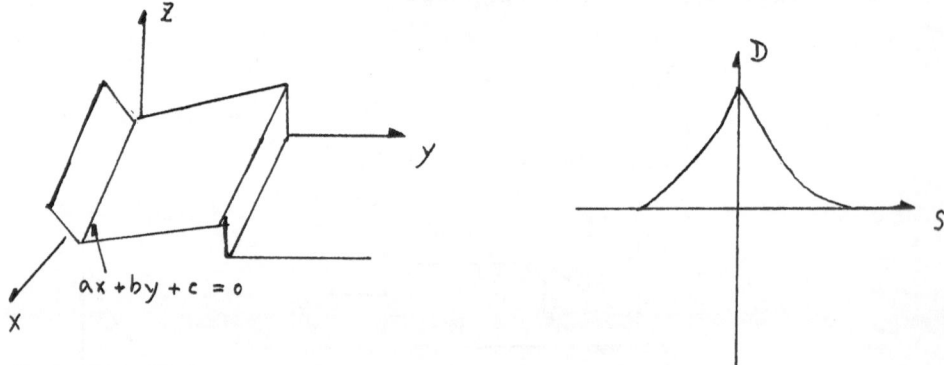

Fig. 6

where the s has the same meaning as above.

In general D(i,j) has the following properties

(i) It is sensitive to edges

(ii) It is not sensitive to the direction of edges

(iii) It is not very sensitive to the noise due to the integral processes.

4. The Extension e

It can be shown that label e has many properties that characterizthe extension of domains. Some typical cases are as follows:

Property 1: Let D be a rectangle with length of L, width of W and L > W, and D' be another rectangle with the same length L but a smaller width W' < W then e(D) < e(D')

Proof: let L=2*l+1, W=2*w+1, then according to the definition, we have

$$e(D) = \frac{(2l+1)*2*w(w+1)(2w+1)/6+(2l+1)*2*2*l(l+1)(2l+1)/6}{(2l+1)^2 *(2w+1)^2}$$

$$= \frac{l(l+1)+w(w+1)}{3*(2l+1)(2w+1)}$$

fix l and take the derivative respect to w, we have

$$e'_w(D)= \frac{1}{3}\left\{ \frac{2w^2+2w+1 - 2l^2-2l}{(2l+1)(2w+1)} \right\} < 0$$

Since l > w (or l > w +1), $2l^2+2l > 2(w+1)^2+2(w+1) > 2w^2 +2w+1$ so $e'_w(D) < 0$ and the property is proved. Besides, we can prove:

Property 2: Among the domains of the same area the digital straight line segment has the maximum e value and the circular disk has the minimum e value.

5. The Tree T= (V,E,L)

First of all it should be mentioned that the parameter set

$$0 = p_0 < p_1 < \ldots < p_K < \max_{i,j} D(i,j)$$

is not very important and so it is not a problem of using this approach. For example, we can simply let

$$p_0 = 0, \quad p_1 = 31, \quad p_2 = 63, \ldots , \quad p_K = 255$$

where D(i,j) has an 8-bits representation as usual.

Now let us turn to the algorithm to implement the

transformation from D to T. The algorithm is adapted from [1]
with minor change and presented as follows:

Input: $D(i,j)$, $1 \leqslant i \leqslant I$, $1 \leqslant j \leqslant J$; $p_0 < p_1 < \ldots < p_K$

Output: T

step 1: Let r be the root of T and

   $e(r) = 0$ ($e(\cdot)=$ the extension of $\cdot$)

   $a(r) = 0$ ($a(\cdot)=$ the area of $\cdot$)

   $f(r) = 0$ ($f(\cdot)=$ the length of path from r to $\cdot$)

   $i = 1$, $j = 1$

step 2: Let $q = (i,j)$ and q_1, q_2, q_3, q_4 be the left-above
neighbours of q as shown in Fig. 7
and b be the one having maximum D value among them, i.e.

$$D(b) = \max \{ D(q_1), D(q_2), D(q_3), D(q_4) \}$$

Compute 1 and 1' such that:

$$p_\ell \leqslant D(a) < p_{\ell+1}, \qquad p_{\ell}' \leqslant D(b) < p_{\ell}' +1$$

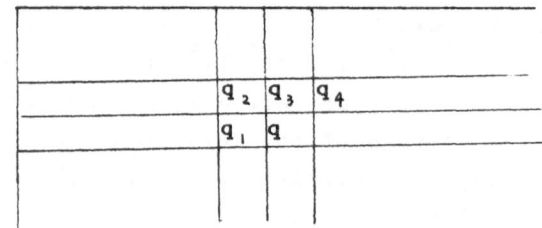

Fig. 7

and denote:

   $f(a)=1$, $f(b)=1'$, $n=1-1'$

step 3: if $n > 0$ then generate n new nodes in T: [b1],
[b2],..., [bn], and let

   [b] --> [b1] --> [b2] -->...--> [bn] be a branch on the tree
where b is the node (domain) containing pixel b and let

   $f(b_i) = f(b) + i$, $1 \leqslant i \leqslant n$

   if $n \leqslant 0$ then find the node c from the ancestral nodes of
[b] such that $f(c) = f(q)$ and let

   [q] = [c]

step 4: if $f(q) < f(q_3)$ then goto step 6

If $f(q_4) < f(q_3)$ then goto step 6

If $f(q_1) < f(q_3)$ and $f(q_2) < f(q_3)$ then goto step 6

step 5: Let [c] be the ancestral node of [q] satisfying $f(c) < f(q)$.

If there is a node [d] in the ancestral nodes of [q_1] and [q_2] such that $f(c) = f(d)$ then let

[c] = [d]

step 6: Compute

$a([q]) = a([q]) + 1$

$S([q]) = S([q]) + i^2 + j^2$

$mx([q]) = mx([q]) + i$

$my([q]) = my([q]) + j$

Let [d_v] are the all ancestral nodes of [q], for all [d_v] do

$a([d_v]) = a([d]) + 1$

$S([d_v]) = S([d]) + i^2 + j^2$

$mx([d_v]) = mx([d]) + i$

$my([d_v]) = my([d]) + j$

step 7: If $i = I + 1$ then goto step 8 else

If $j = J + 1$ then begin $i = i + 1$, $j = 0$ end else

begin $j = j + 1$, goto step 2 end

step 8: For each node [n] of T compute

$$e([n]) = \frac{S([n_v]) - [\ mx\ ([n_v])^2 + my\ ([n_v])^2\ ]}{a([n_v])^2}$$

and stop.

6. Cutting -- off the Tree T

According to the intuition of edge what needs to be reserved is those nodes which have large D value (i.e. those nodes which are far from the root of tree T) and large e value and cut off the others. Besides, we would like to cut off those nodes which have very small a value, since it is very likely that they are false due to noise.

Because we prefer to reserve those nodes having large D value, we travel the tree T in depth-first manner and it leads the following algorithm.

Input: Tree T and a threshold
Output: The subtree T'

Travel the tree T in depth-first manner, et u be the present node and v be the farther of u, do
   If a(u) < $\mathcal{E}$ or e(u) < e(v) then
      begin cut-off node u from tree T,
        If u has brother u' then let u' be present node
           else let v be present node (u=v) and find new v
     end

7. Examples

In the following there are five sets of pictures. The first two are artificial and the next three are natural. The last one is a microscopic picture of DNA and the purpose here is to find the line, not the edge.

The pictures with (a), (b) and (c) are the original one, the one only using the gray level information, i.e. using D with a thresholding, and the final one, i.e. using both gray level information and structural information by through whole the steps in our approach.

From these examples it is easy to see that the approach we suggested is of general- purpose and works quite well. It can remove much noise by using the structural information. It shold be mentioned that the use of structural information in our approach does not cost a lot since the construction of tree T is of linear computational complexity and after that the number of nodes is much less than the number of pixels, about two orders of magnitude lower in our examples.

8. Discussion

Intuitionally an edge is a domain in picture where the local gray level are changing rapidly and which shape is long but narrow. It is clear that these two aspects both have been taken accounted of in our approach. And so the approach is quite attractive conceptionally.

The approach is also fast, since the additional computation cost due to the consideration of structural aspect is low as explained in the last section.

The approach is of general-perpose, since it works well in finding both edge and line in both artificial and natural pictures as shown in the last section.

The approach has much room for improvement since it consists of several parts and each part can be adapted and improved seperately. For instance, different edge operators, labels and cutoff strategies might be used to improve the quality or to reduce the computaton cost. Besides, an additional step might be added to do some post-processing such as thinning etc., and get nicer looking result. In other words the approach proposed here actually is a frame of a class of approaches to edge and line detection.

(a)

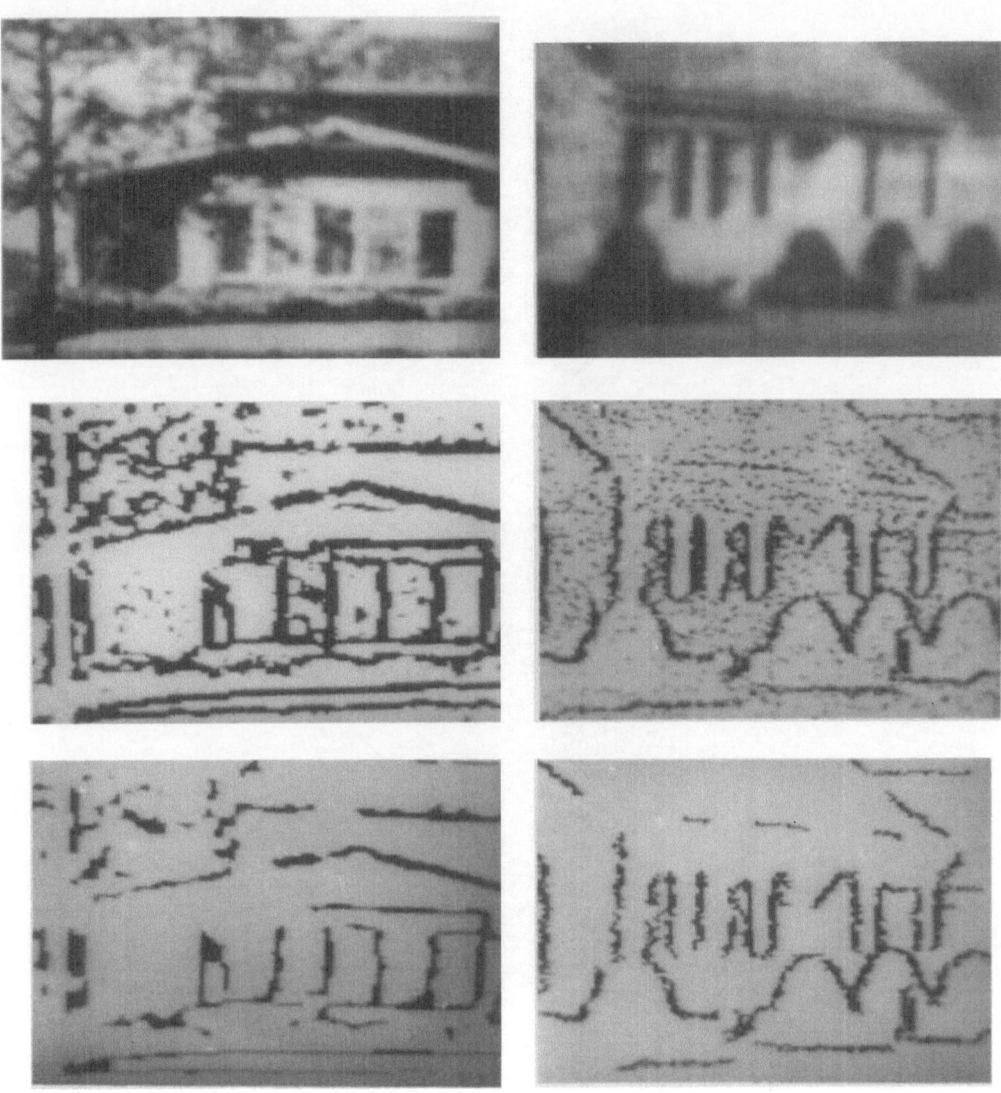

(b)

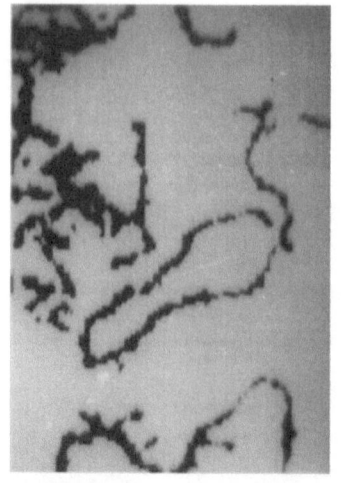

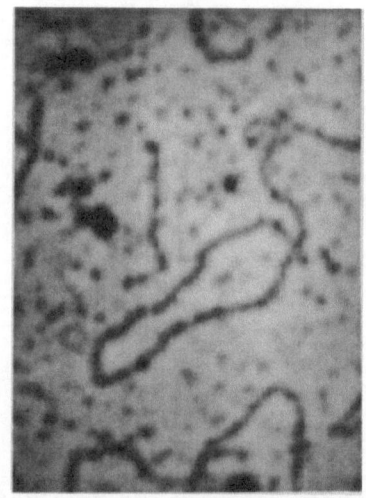

(c)

Reference

[1] Pavlidis T. (1977), "Structural Pattern Recognition", Springer-Verleg.

[2] Rosenfeld A, Kak A (1982), "Digital Picture Processing", 2nd Eds, Acad. Press.

[3] Ballard D, Brown C (1982), "Computer Vision", Printice-Hall.

[4] Levialdi S. (1980), "Finding the edge", in Simon J. C., Haralick R. M. (eds.), "Digital Image Processing", D. Reidel Publishing Company.

[5] Wei S. (1984), "H-model of Image and Its Application to Cell Image Analysis", Proc. 4th National Conference on Pattern Recognition, Hefei, Anhui, P. R. China.

BUILDING HIERARCHIES - AN ALGORITHMIC APPROACH

C. Sielaff

Universitat Hamburg
Fachbereich Informatik
Bodenstedtstr. 16
2000 Hamburg 50
Germany

Introduction

Image analysis is a problem which requires powerful and flexible systems to handle the large amount of various data and complex procedures which are needed for recognizing objects in images.

Relational structures are an appropriate formal tool for the union of these two requirements [1],[2]. An essential task of image analysis is the generation of mappings between relational prototype- and image-descriptions to identify objects in a scene. These mappings are provided by R-morphisms [2]. Due to the large extent of a symbolic description of complex prototypes and images the matching process between model and image may become relatively slow. To speed up this process and to clarify the complex descriptions we use the concept of hierarchical synthesis introduced by Barrow and Milner[3].

We extend their concept by introducing significant substructures for each part of the hierarchical model. Thus the descriptions of similar (sub-) objects with identical significant substructures may be condensed into a single node of the hierarchy. To further reduce redundancy, a sub-object may contribute to several higher-level objects. The different descriptions condensed in a single node will have different sub-parts in the hierarchy. Therefore the hierarchically inferior nodes of a node have to be grouped in sets of nodes, each set describing one of the sub-structures in the hierarchically superior node. This way the introduction of the significant sub-structure of a node leads directly to the use of AND/OR processes [4] in the hierarchy and in the recognition process. Each description in a node of the hierarchy is connected to its inferior nodes by a set of

NATO ASI Series, Vol. F45
Syntactic and Structural Pattern Recognition
Edited by G. Ferraté et al.
© Springer-Verlag Berlin Heidelberg 1988

arcs. Thus a set of arcs will be denoted as AND-arc because only the recognition of all sub-parts will lead to a perfect recognition of a model object.

The constructed *modelgraph* is an extension of the tree-like part-of hierarchy to an AND/OR-graph combined with AND/OR-processes which provides the mappings between prototype and images and the communication of the nodes of the hierarchical graph [5],[6].

The second essential task of image analysis is the generation of appropriate models for the re-cognition-process. The formal structure of a hierarchy provides only an algorithmic frame for an image interpretation process. The contents of each node strongly influences the performance and efficiency of a system. Due to the part-of relation and the requirement of minimal redundancy the hierarchical structure is 'naturally' determined. For the contents of the nodes there are many different possibilities independent of the formal structure.

We introduce an algorithmic process which constructs a hierarchical description – a modelgraph – given a *flat* (nonhierachical) relational description of an object. In each step of this process, we select one of the nodes of the actual hierarchy and call a subprocess which provides a set of different decompositions of the selected sub-object. Each decomposition leads to another extension of the actual hierarchy.

The main problem is to find an appropriate weight-function which guides the selection among the constructed hierarchies and the decomposition-process. We estimated the costs of a match between the actual hierarchical model and an unknown image and chose this estimate as our weight-function for the selection. The decomposition-process is realized by a set of different subprocesses each providing low-cost decompositions in the sense of different terms of our weight-function. Due to these choices the constructed modelgraph permits fast and efficient mappings of the described object. Other tasks may be supported by other weight-functions and appropriate decomposition-processes.

The entire system consists of three parts : a low-level vision-process which provides a flat relatio-nal description of a prototype or an image, a hierarchy-builder which constructs a modelgraph given a flat description and a set of primitives and a matcher which provides the mappings between a constructed modelgraph and a relationally described image.

The Modelgraph

A natural decomposition of an object is provided by the part-of hierarchy. To avoid the narrow frame of hierarchical levels [7],[8] and treelike structures [9], which often lead to unnecessary redundant and repetitive descriptions, we define the structure of decomposition as a graph. We attain that a single substructure may contribute to several greater parts and even to parts of different objects.

Each node of the *modelgraph* contains a relational description of a part of a model object which is composed of the relational descriptions of subparts provided by hierarchically inferior nodes. The part-of relations between hierarchically inferior and superior nodes are provided by R-monomorphisms –injective R-morphisms. If the subparts are 'glued' these connections are realized by identification of relations of different subparts. If not, the connections are provided by relations between parts of different subparts. Therefore it is sufficient to restrict the repetition of inferior descriptions to those parts which support the connecting relations and thereby are important for the recognition process. Thus it is possible to mark parts of the relational structures as being *significant* for the recognition process. Only these *significant substructures* will be passed to hierarchically superior nodes of the modelgraph. Furthermore, we use the significant substructures to reduce redundancy by condensing the descriptions of similar (sub-)objects with the same significant substructure into a single node of our hierarchy.

To formally introduce the significant substructures, we define

Definition 1 *Let R a relational description and $R_1, ..., R_i$ hierarchically inferior parts with R-monomorphisms $M_j : R_j \longrightarrow R$. If*

$$\bigcup_{j=1}^{i} M_j(R_j) = R$$

then we denote R as an agglutination *of the R_j.*

Let $r \in R$ a relation tupel of R. r is denoted as agglutination tupel *of R if and only if*

i) There are at least two relation tupels $r_j \in R_j$ and $r_k \in R_k$, $j \neq k$, such that

$$M_j(r_j) = M_k(r_k) = r \ or$$

ii) There are relation tupels $\tilde{r}, r' \in R, r_j \in R_j$ and $r_k \in R_k$ such that

$$M_j(r_j) = r, M_k(r_k) = r'$$

and $\tilde{r}$ describes an interrelationship between r and r'.

Definition 2 *Let R a relational description of a model part and $r \in R$ a relation tupel of R. r will be denoted as a* significant tupel *of R, if there exist R-monomorphisms $M, M', M'', ...$*

$$M : R \longrightarrow R', M' : R' \longrightarrow R'', ...$$

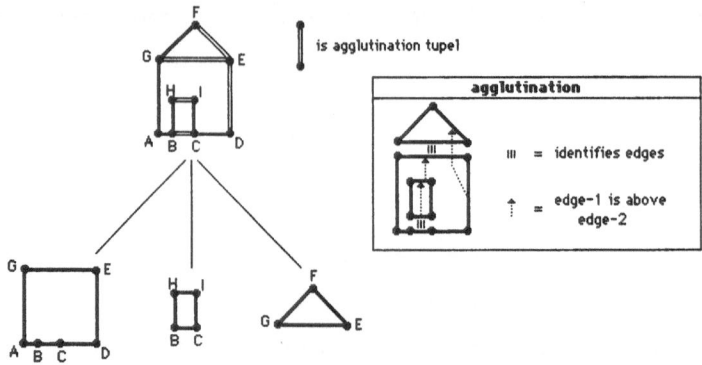

Figure 1— Example of agglutination tupels

such that the image of r under the R-monomorphisms $M^{(i)}$ is an agglutination tupel of a hierarchically superior part (see fig. 2).

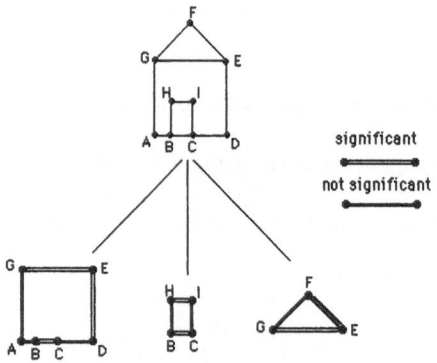

Figure 2— Example of significant tupels

Now we can define the modelgraph :

Definition 3 *A* modelgraph *is a weighted hierarchical AND/OR-graph*
$(N, E, \lambda, \leq, AND, p)$ *with special properties i), .., iii).*

N *is a final set of nodes.*

E *is a final set of directed arcs.*

λ *is a left total function assigning a joining arc $e_{i,j}$ to a pair of nodes n_i, n_j, if $n_i \leq n_j$ in the partial ordering defined by the relation $\leq$.*

AND *is a mapping $N \times E^{n_i} \longrightarrow E \wedge E \wedge ... \wedge E$, assigning an AND-arc $e_{j,i} \wedge ... \wedge e_{k,i}$ to a set* *of arcs $e_{j,i}...e_{k,i}$ with a common node n_i.*

p is a function which assigns a weight to an AND-arc of the graph.

i) each node consists of similar relational descriptions of one or several model parts with a common significant substructure.

ii) Two nodes n_i, n_j are denoted as $n_i \leq n_j$ if there exists an R-monomorphism $M : n_i \longrightarrow n_j$ which maps the significant substructure of n_i into the relational description of n_j.

iii) The set of nodes $n_j, ..., n_k$ connected with a node n_i by a single AND-arc $e_{j,i} \wedge ... \wedge e_{k,i}$ is the minimal set which provides a complete description of the node n_i.

Building Hierarchies

The formal structure of a hierarchy provides only an algorithmic frame for an image interpretation process. The contents of each node and the dependencies of the nodes strongly influence the performance and efficiency of a system. The hierarchical structure of the modelgraph is 'naturally' determined by the part-of relations provided by R-monomorphisms, by the agglutination of inferior substructures and the resulting significant substructures and by the requirement of minimal redundancy. The contents of the nodes of the modelgraph are independent of the formal structures and are only determined by the tasks of the image interpretation processes.

There are two possibilities to model an object:

- The modeling is done by a user of the system or

- the modeling is done by a prototypical image of the object.

Both techniques provide a flat –a nonhierarchical– symbolic description of the object. To construct a hierarchy from this flat relational structure there are again two possibilities:

- The hierarchy is constructed by the user or

- the hierarchy is constructed by an algorithmic process.

The construction of hierarchies independent of a system user by an algorithmic process has some advantages in comparison with interactivly building a hierarchy:

- It allows to adapt the hierarchy to various tasks and different algorithms of an image interpretation process.

- It is faster than an interactive process. Often the time-consuming modeling is the bottleneck in developing and testing a system.

- It is not confused when confronted with a prototypical real-world image as a model. Due to imperfectly preprocessed images the symbolic description of a prototype may be too obscure and confusing for being interactively modeled.

If a flat model consists of k relation tupels, the number of possible hierarchies is of $O(2^{(2^k)})$. The aim of a hierarchy building process is not only to construct one (random) hierarchy, the aim is to find an appropriate or "best" hierarchy among the possibilities. Due to the large number of possible hierarchies, it is not possible to construct all hierarchies first and then to select the best one. It is necessary to design an iterative process which expands the flat model step by step and which selects in each step the actual best hierarchy for further expansion.

Following this approach, the process of constructing a hierarchy may be regarded as a simple treesearch [10]:

1▷ *The simplest hierarchy given by the flat model and the set of primitives is the root of the tree. It is the first node in a set OPEN. There is another set CLOSED which is initially empty.*

2▷ *Select the best weighted node of OPEN and move it to CLOSED.*

3▷ *Expand the hierarchy contained in the selected tree node as follows :*

    3a▷ *Select one non-primitive and yet undecomposed node of the actual hierarchy if possible.*

    3b▷ *Construct a set of different decompositions of the selected subobject.*

    3c▷ *For each decomposition, extend the actual hierarchy by the new sub objects and add the new hierarchy as a new node of the tree to OPEN (see fig. 3).*

4▷ *If OPEN is empty, select the best weighted node of CLOSED as the desired hierarchy and terminate, else goto 2▷.*

The main problem is to find an appropriate weight function which guides the treesearch and the decomposition process. As an example let us consider the problem of constructing a hierarchy which provides an efficient and fast interpretation process.

According to the partition of the construction process we divide the construction or selection of

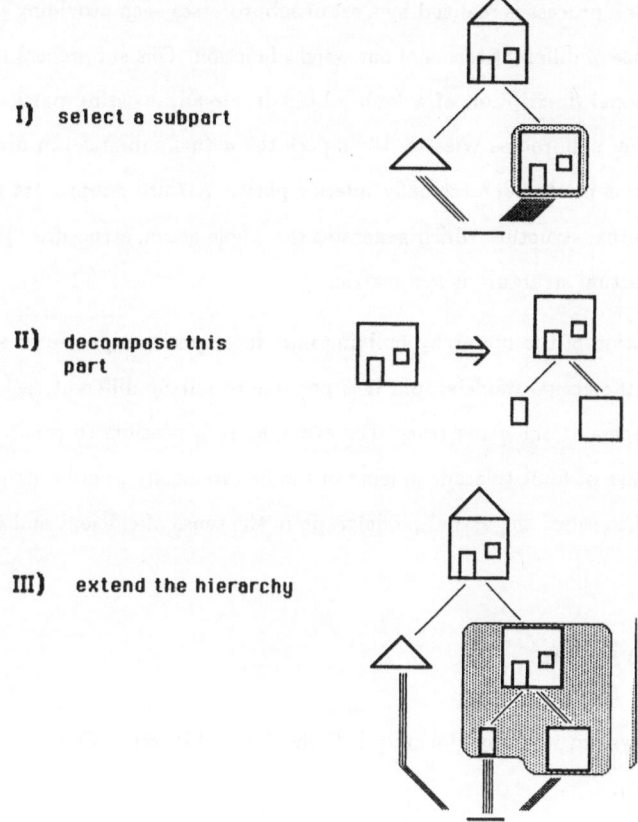

I) select a subpart

II) decompose this part

III) extend the hierarchy

Figure 3— Example of an extension of a hierarchy

the weight function into two parts. The treesearch which selects the presently best hierarchy is guided by an estimate of the costs arising from a mapping between a selected hierarchy and an image which contains the actual object. The estimate is provided by a combination of several requirements:

Considering two different modelgraphs of the same object we require the better weight for the hierarchy to be one

- • - which consists of fewer relations and therefore has less redundancy;

- • - which marks fewer relation tupels as agglutination tupels;

- • - which causes lower costs glueing inferior parts providing superior model parts of the hierarchy;

- • - which consists of fewer nodes.

The decomposition process is realized by a set of subprocesses each providing low-cost decompositions in the sense of different terms of our weight function. One subprocess tries to decompose the actual relational description of a (sub-)object in already existing parts of the actual modelgraph. Another subprocess tries to decompose the actual submodel in disjunct subparts to avoid redundancies in the hierarchically inferior parts. A third subprocess may try to find a subpart of the actual structure which generates the whole actual structure. This is possible, for example, if the actual structure is symmetric.

Due to the partition of the hierarchy building into decomposition processes and the treesearch like selection of the actual modelgraph, it is possible to pursue different tasks of the hierarchy construction process at the same time. For example, it is possible to construct the decompositions in the sense of fault-tolerant glueing of the hierarchically inferior structures to superior model parts and to select the actual modelgraph in the sense of efficient and fast interpretation of images.

First Experiences

A prototypical system was implemented in COMMON LISP on a VAX 11/780. It consists of four parts (fig. 4):

- The first module provides pre-processing and edge-detection.

- The second module provides the flat model of the actual (prototypical) image.

- The third module builds an optimal hierarchy of the flat model.

- The last module matches the actual flat model or the actual hierarchy against a (new) image provided by the first module.

For this first implementation we restricted the symbolic description and the images to simple line-drawings. The single primitive is an edge given by the coordinates of its endpoints. The relations between these primitives are restricted to the angle between two edges and, if two polygones are not connected by an edge, to the relative position of two edges, each belonging to one connected part of the object.

The decomposition of a model or modelpart is provided by two subprocesses. The first tests whether one or more already existing parts of the actual hierarchy may contribute to the de-

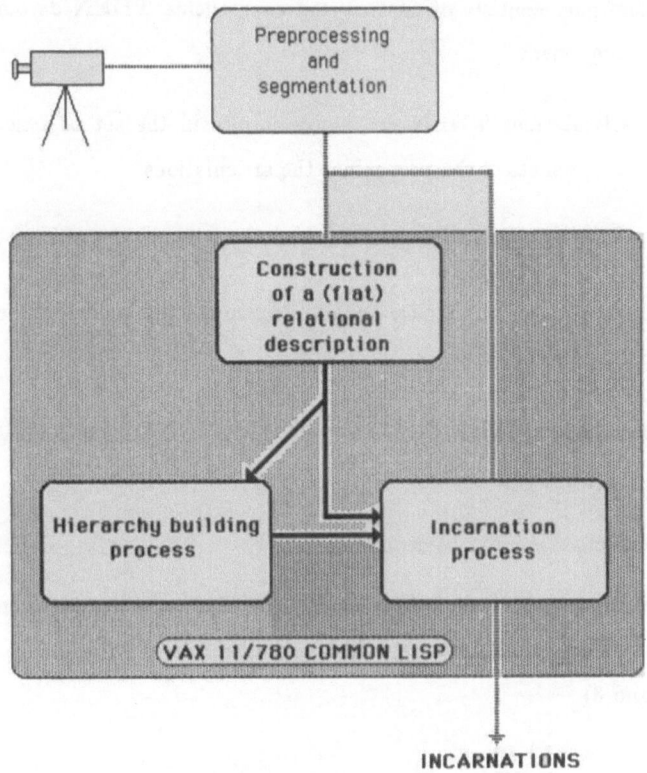

Figure 4— The architecture of our system

composition of the actual part. If there are complete covers of the actual part by other parts only these decompositions will be returned. If there are only partial covers, all non-redundant combinations of such parts together with the connected components of the actual part minus the partial cover will be provided. This subprocess works according to the "meta-rule":

Trie to decompose the actual part (partially) in already existing parts of the actual hierarchy to minimize the number of different nodes in the resulting modelgraph.

The second subprocess works according to another "meta-rule":

Decompose the actual part so that each part has a complex structure and strong intra-connections and the inter-connections between different parts are only weak.

According to the form of the actual part, one of the following rules will be applied (see fig. 5 also):

a) **IF** the actual part consists of unconnected components **THEN** decompose it into its connected component.

b) **IF** it has articulations **THEN** decompose it into in the set of articulations and the connected components of the part minus the articulations.

c) **IF** it consists of a number of regions **THEN** decompose it into several different regions.

d) **IF** it consists of exact one region **THEN** decompose the border of the region into two disjunct polygones.

e) **IF** it is star-shaped **THEN** decompose the star into two disjunct parts of similar same size

f) **ELSE** decompose it into two disjunct parts

After the decomposition process each decomposition is inspected whether it contains isomorphic parts. If there are isomorphic parts they will be condensed in one single part. For an example see figures 6 2) and 3).

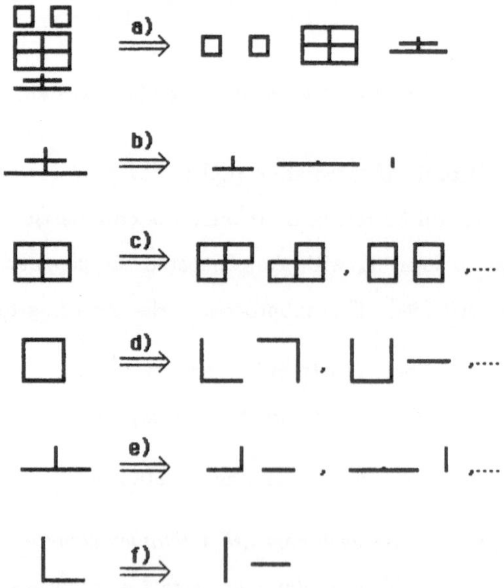

Figure 5— Example for the decomposition rules

The choice of an appropriate weight funktion is a more complex subject.

We want to construct hierarchies which are efficient and fast in the interpretation process. So the first step in finding an appropriate weight function is to estimate the costs of interpreting an image using a hierarchical model. We estimate only the needed time, because the needed space is of the same size in important parts of the process and negligible in other parts.

The segmentation of the image causes the first costs but these costs are unimportant because they are constant for all images.

Next we have to estimate the costs for each nonprimitive. node and we have to add the costs of each nonprimitive node of the hierarchy.

The costs of each node consists of four parts :

- First the costs of constructing associations between the tupels of the actual part and the tupels provided by inferior parts

  $= t_1 a$, where a denotes the number of associations.

- Second the costs of the compatibility tests

  $= t_2 \frac{1}{2} a(a-1)$.

- Third the costs of finding the cliques in the compatibility graph which means constructing injective morphisms between modell and image

  $= t_3(\frac{1}{2} a(a-1), d)$ where $t_3(x, y)$ denotes the time needed by the clique search in a graph with size x and dense y.

  – Now we have identified the inferior parts wich may lead to an incarnation of the actual part –

- Fourth and last, the costs of verifying the relations, which describe the connections between different inferior parts

  $= t_4 r_i S$ where r_i denotes the number of relations which describe interrelationships between inferior parts and S denotes the number of different sets of incarnations of inferior parts which may provide an incarnation of the actual part.

Now we can identify the parameters of our searched weight function:

- One parameter should be the number of nonprimitive nodes.

- Another parameter should be the number of associations. It is needed to compute the costs of associations, of compatibility tests and of finding the morphisms. A better choice is the square of the number of associations regarding the costs of compatibility tests and the costs of constructing the morphisms.

- Finally the number of tests for interrelationships shoud be a parameter of our weigth function.

So then we have to add over all nonprimitive nodes, we have to estimate the number of associations and we have to estimate the costs of agglutination.

The number of associations is provided by the number of tupels in the actual node of the hierarchy, by the number of tupels in the inferior nodes and the number of incarnations of each inferior node.

The number of interrelationship tests is provided by the number of agglutination tupels of the actual part and the number of different sets of inferior incarnations which lead to an incarnation of the actual part. This last number is equal to the number of incarnations of the actual node.

Thus, we count the nonprimitve nodes, the number of tupels in each node and we compute the costs of agglutination of each node. At last we estimate the number of incarnations of a part of a hierarchy by a simple heuristic :

Let r_p and r_n the number of primitive and nonprimitive tupels in the actual part. If we consider two different parts with the same number r_n but r_p of the first part is smaller then r_p of the second, we assume that in general the number of incarnations of the first part will be smaller than the number of incarnations of the second. The same is true if the two parts have the same r_p and r_n of the first is smaller than r_n of the second.

In short, if one part is more complex than a second then the number of incarnations of the second part will be smaller.

Let r_p' and r_n' the number of primitive and nonprimitive tupels in the flat model. Let n and n' the number of primitives in the image and the number of incarnations of the set of primitives of the flat model in the image. The numbers n and n' will be the parameters of our function which will estimate the number of incarnations. They are estimated from typical images, from the experiences of the system user or established by some experiments.

Now we have some constraints :

- If r_p is 1 and r_n is 0 , we will have n incarnations.

- If r_p equals r_p' and r_n is 0 , we will have n' incarnations.

- If r_p equals r_p' and r_n equals r_n' we will have a single incarnation of our model.

- Furthermore we require that the number of incarnations decreases if r_p or r_n increases according to a function as $f(x) = \frac{a}{(b+x)^c}$.

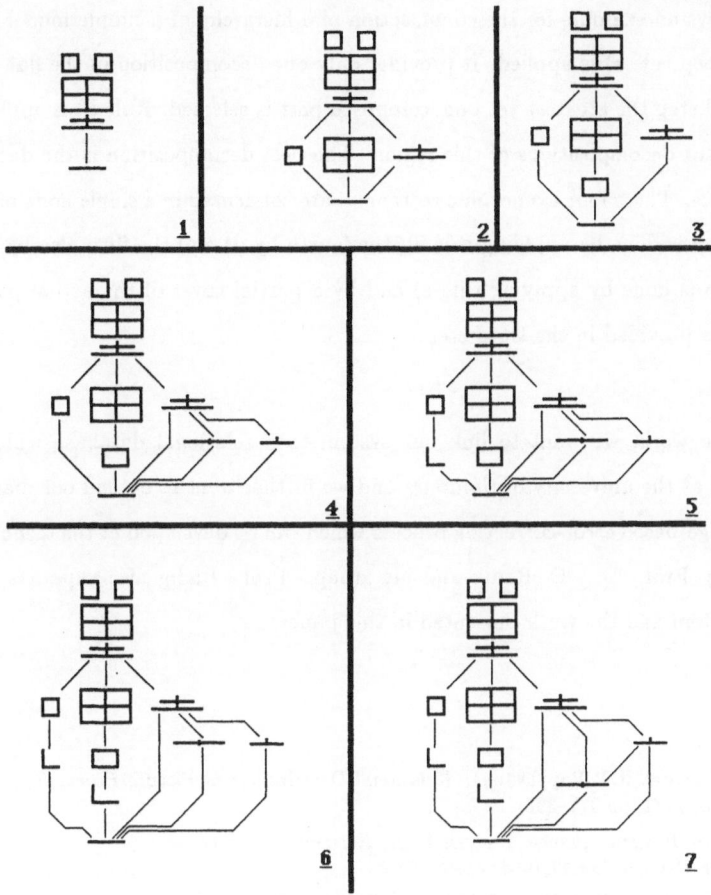

Figure 6— Example for a construction process

With the notation introduced and from our constraints follows :

$$I(r_p, r_n) = n r_p^{-\frac{\log n - \log n'}{\log r_p^l}} (r_n + 1)^{-\frac{\log n'}{\log (r_n^l + 1)}}$$

where $I(r_p, r_n)$ denotes the number of incarnations of a part with r_p primitive and r_n nonprimitive tupels. Summarized, our weight function has the following form :

$$w(\mathcal{H}) = \sum_{i=1}^{k} \left(r_{p_i} \sum_{j=l_1}^{l_i} r_{p_j} I(r_{p_j}, r_{n_j}) \right)^2 + r_{n_i} I(r_{p_i}, r_{n_i})$$

where k denotes the number of nonprimitiv nodes of the hierarchy $\mathcal{H}$, and $l_1 ... l_i$ denote the precedessors of the actual node.

Figure 6 shows an example for the construction of a hierarchy of a simple model object :

In the first step rule a) is applied. It provides only one decomposition of the flat model.

In the second step the greatest yet undecomposed part is selected. Rule b) is applied and provides 47 different decompositions of this region. The best decomposition is the decomposition in four rectangles. These four isomorphic rectangles are condensed in a single node of the hierarchy The third step is done by applying rule c), the fourth by d) and the fifth also by d).

The last step is done by applying rule e) and by a partial cover of the actual part by the two collinear lines provided in the third step.

In our future work, we want to link our system to a relational database which is partially implemented at the university of Hamburg and we further want to extend our matching process by a knowledge-based error-correcting process which will be developed at the technical university Müenchen by Prof. Dr. B. Radig and his group. Prof. Radig also supports the relational database system and the work presented in this paper.

References

[1] H.G. Barrow, R.J. Popplestone: *Relational Descriptions in Picture Processing*, Machine Intelligence 6, 1971, pp.377–396

[2] B. Radig:*Image Sequence Analysis Using Relational Structures*,
Pattern Recognition 17,1984, pp.161–167

[3] H.G. Barrow, A.P. Ambler, R.M. Burstall: *Some Techniques for Recognising Structures in Pictures*, in: Frontiers of Pattern Recognition, S. Watanabe (ed), Academic Press New York, San Francisco, London, 1972, pp. 1–29

[4] J.S. Conery, D.F. Kibler: *AND Parallelism in Logic Programming*, Proc. IJCAI-8, Karlsruhe, 1983, pp. 539–543

[5] C. Sielaff: *Hierarchische Dekomposition und Sythese von Objekten*,
in: GWAI-85, Informatik Fachberichte 118, H. Stoyan(ed), Springer-Verlag, Berlin Heidelberg New York Tokyo, 1986, pp. 348–355

[6] C. Sielaff: *Hierarchical Decomposition And Synthesis of Relational Descriptions –The Modelgraph*, Proc. ICPR-8, Paris, 1986, pp. 1207–1209

[7] L.S. Davis, T.C. Henderson: *Hierarchical Constraint Processes for Shape Analysis*, IEEE PAMI-3, 1981, pp. 265–277

[8] A.R. Hanson, E.M. Riseman: *VISIONS: A Computer System for Interpreting Scenes*, in: Computer Vision Systems, A.R. Hanson, E.M. Riseman (eds), Academic Press New York, San Francisco, London, 1978, pp. 303–334

[9] A.P. Ambler et al.: *A Versatile System for Computer Controlled Assembly*, Artificial Intelligence 6, 1975, pp. 129–156

[10] N. Nilsson: *Principles of Artificial Intelligence*, Tioaga Publ., Palo Alto, 1980

VIII. HYBRID APPROACHES II

COMBINING LOGIC BASED AND SYNTACTIC TECHNIQUES: A POWERFUL APPROACH

A. Sanfeliu

Instituto de Cibernetica
Diagonal, 647
08028 Barcelona
Spain

1 Abstract

Artificial intelligence techniques, mainly logic based techniques, and syntactic pattern recognition techniques are different in nature, but both can be used to solve similar type of problems, usually of classification, verification or recognition. Each one of these techniques brings special features which are suitable for solving concrete aspects, for example AI techniques are very useful for interpretation processes, for example on illness diagnostic, where there exist uncertain data. On the other side, pattern recognition techniques (mainly structural and syntactic ones) are very suitable to be applied on problems which need fast process time or where the models are described by structures. Although there exist differences between both techniques, they could share knowledge representation and some of the AI methodology can be seen as a special case of formal grammars. These features make possible to integrate both techniques in an efficient way. In this work some types of integration will be presented. First, classical AI techniques will be presented as a special case of formal grammars. Then the problem of inference on grammars will be shown as a combination of AI and syntactic techniques, and the integration of inference and classification techniques will be described for 3D object recognition. Finally, a systems called PIRS which incorporates part of the integration techniques will be described.

NATO ASI Series, Vol. F45
Syntactic and Structural Pattern Recognition
Edited by G. Ferraté et al.
© Springer-Verlag Berlin Heidelberg 1988

2 Introduction

One of the big problems in the development of a hybrid system applied to interpretation problems, is the integration of different kind of techniques with several types of representations. At the present time very few systems show such integration and the reason is that many of the known techniques use different types of knowledge representation. On classification, verification or interpretation problems, two very well known group of techniques are usually applied: the AI techniques (mainly logic based) (Nilsson, 1980; Barr and Feigenbaum, 1982) and the pattern recognition techniques (Fu, 1982; Pavlidis, 1977; Fukunaga, 1972). Both group of techniques can solve similar problems, but they are more suitable for specific problems.

For example, the AI techniques are appropriate for solving problems where there is uncertain data or interpretation processes are needed, for example on illness diagnostic (Shortliffe, 1976). For problems where the domain is known and the features or structures are well specified, these methods are very slow and usually the results are not as good as with the pattern recognition techniques. However, they are very useful when the data is not well defined and reasoning with uncertain data is needed.

On the other side, there are the pattern recognition techniques which are very suitable for problems where the domain, the features and structures are very well specified (many of the non complex industrial problems). In these cases, pattern recognition techniques are very fast and reliable since the classifier is built to fit the requirements. The problem arise when there is a need to interpret or deduce part of the information, then pattern recognition techniques are not adequate to solve these problems.

In this paper we present a way of integrating both group of techniques to be used in an efficient way, and how to structure the information in a unique data representation. In order to get this data representation, we will show how some classical AI techniques can be formulate as formal grammars and how to use AI techniques to make inference through formal grammars. This last technique is one of the fundamental elements to integrate both techniques in an efficient way. With respect to the utility of the formulation of AI techniques as formal grammars, all the theory of formal grammars and automatas can be applied to AI techniques, which gives an interesting way of their study. Moreover, this formulation show how both group of techniques can be combined.

In order to see how both groups of techniques are integrated, a method for joint together classification and inference is presented, and is applied to the recognition of 3D objects. Finally, a system which uses these methods, called PIRS, is presented.

3 Classical AI techniques as formal grammars

Some of the classical methods on artificial intelligence field could be formalize as formal grammars, and some of them, will be described in this section. The main purposes are to formalize the techniques in a way which will be suitable for their integration, and to use the results of the classical theory of formal grammars . In this section, we will show how the predicate calculus and the production systems can be described as formal grammars. The case of the semantic networks will not be presented here, although they can also be described as formal grammars.

3.1 Predicate calculus

As was shown by Sickel (Sickel, 1977), all proofs of some theorems in the predicate calculus can be seen as the language generated by a context free string grammar. This representation allows to transform a predicate calculus characterization of a problem, into a regular algebra characterization of the solutions. In order to get such a grammar the set of variable disjoint clauses are described by a clause interconnectivity graph (CIG) which incorporates the clause literals as nodes of the graph. The edges are the symmetric relation between pairs of nodes iff the literals associated with the nodes have opposite signs and unifiable atoms. The substitution is a mapping : $Edges \rightarrow Directed\ Substitutions$ such that $Subst(<A, B>) = \theta_{s1s2}$ where θ is the most general unifier, and a clause is a mapping : $Nodes \rightarrow Powerset(Nodes)$. Finally, $Residual\ literals$ is a mapping : $Nodes \rightarrow Powerset(Nodes)$ where $Residual\_literals(B) = Clause(B) - \{B\}$ (Sickel, 1977). Fig.1 shows an example of a CIG for clauses $\{A(x)B(f(x)), \overline{A}(y)\ \overline{C}(g(y)), D(b), \overline{A}(f(z))C(g(u))\ \overline{D}(w), \overline{B}(f(f(a)))\ \overline{E}(g(e))F(a), E(v), \overline{F}(a)\}$.

The idea of the method is to transform the input (a CIG) into a grammar which generates exactly the complete set of proofs of the theorem. For example in the ground case, the

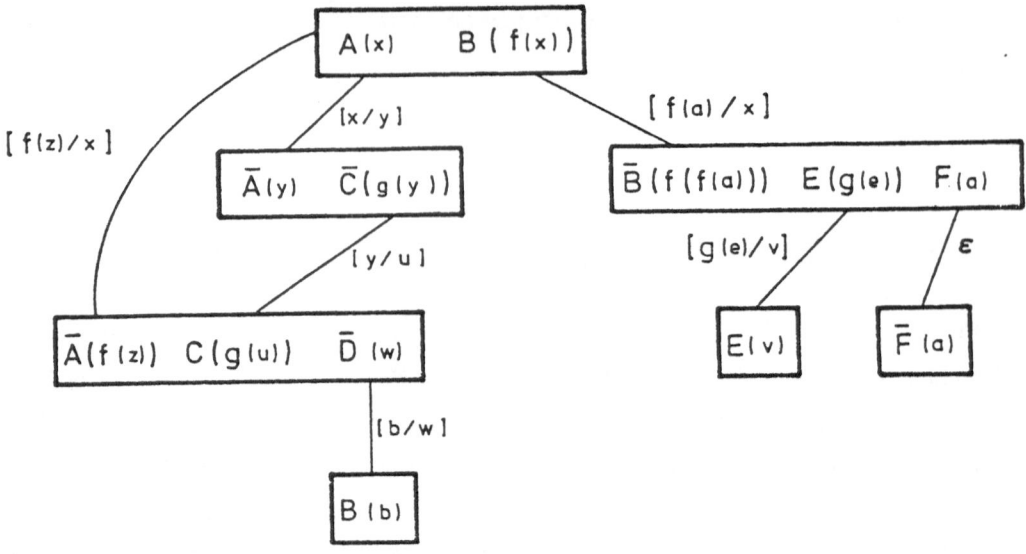

Fig.1 : An example of a CIG

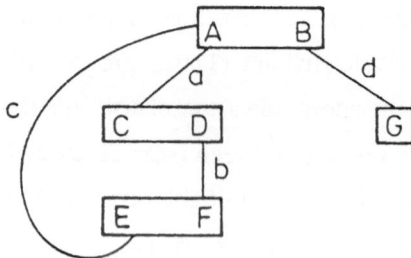

Fig.2 : A ground case of a CIG

construction of the grammar is the following one (Sickel, 1977).

Let a ground case CIG : C = ($Nodes$, $Edges$, $Subst$, $Clause$). The context-free grammar which generates all the proofs of this set of clauses is constructed as follows: G = ($Nodes \cup \{S\}$, $Edges$, P, S)

where $S \notin Nodes \cup Edges$. And P is

1. $S \to L1 \cdots Lk$ for each starting clause $\{L1 \cdots Lk\}$.

2. $B \to eC1 \cdots Ck$ for each edge $e = < B, C >$ and $Residual\_literals(C)$ = $\{C1 \cdots Ck\}$.

3. $B \to mL1 \cdots Lk$ for each merge loop m and where $L1 \cdots Lk$ are the undeleted residuals of m.

For example, for the ground case CIG shown in Fig.2 the grammar is :

$$G = (\{S, A, B, C, D, E, F, G\}, \{a, b, c, d\}, P, S) \text{ where } P \text{ is :}$$

$$
\begin{array}{llll}
S & \to & AB & \qquad C \to a'B \\
A & \to & abc(a\ merge\ loop) & \quad D \to bE \\
A & \to & aD & \qquad E \to cB \\
A & \to & c'F & \qquad F \to b'C \\
B & \to & d & \qquad G \to d'A
\end{array}
$$

and the language is $L(G) = \{abcdd, c'b'a'dd, abcd\}$, where an edge denoted by "a" means $< x, y >$ and "a'" means $< y, x >$.

A derivation tree for $abcd$ is shown in Fig.3, where the string $abcd$ is one of the proofs. Other information concerning this method can be seen in (Sickel, 1977).

3.2 Production systems (PS)

Following the definition given by Nilsson (Nilsson, 1980) for production systems, some PS can be seen as special cases of formal grammars, where the different reached states are nothing but the language of the grammar. The idea of the method is to transform the production rules of PS into production rules of a grammar with some minor changes,

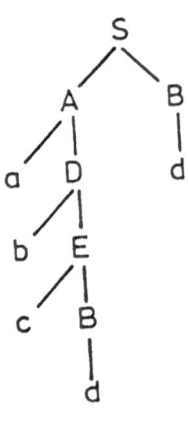
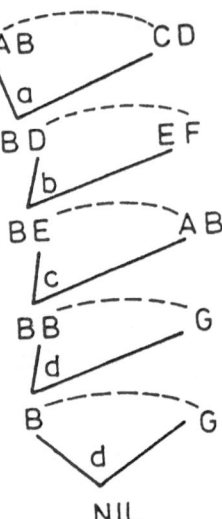

Fig.3 : A derivation tree for the sample *abcd*

and to incorporate the initial states as S derivation rules. However, it will not always be feasible to incorporate the initial states as S, for example, when there are infinite number of initial states, although there some cases where they could be introduced. For example, when an initial state can be reached from another initial state and this is in the S derivations, then there is not need for considering the state on the S derivations. There are some modifications with respect to the general procedure of making a derivation, basically on the way of matching the left member β ($\beta \rightarrow \alpha$). Before any operation of matching is performed, the terminals of the left member have to be generalized to include the universal quantifier ($\forall x$), so

$$HO(A) \text{ will be transformed into } HO(x) \; \forall x$$

Then the matching can be done without taking into account the order of the nonterminals and terminals. On the other hand, once the rightmost member of the production rule has been applied, then at least one of the ground instances must be considered. So, for example, if $HO(x)$ is on the right member, and its ground instances are $HO(A)$ and $HO(B)$, then one of them is included in the derivation as a substitution of $HO(x)$. The method to make the transformation is shown below.

Let a Production System with the following rules of derivation :

$$Rule : Name(x, y, ...)$$
$$Preconditions : \quad P_1(X, ...)P_2(X, ...)...$$
$$Action : \qquad\qquad A_1(X, ...)A_2(X, ...)$$

or

$$Rule : Name(x, y, ...)$$
$$Preconditions \; and \; deletions : \quad P_1(x, ...)...$$
$$Action : \qquad\qquad\qquad\qquad A_1(x, ...)...$$

then the grammar which will generate the same derivations of the PS is constructed as follows :

Let $G = (V_N, V_T, P, S)$ the grammar. The rules of P are obtained by :

Method :

1. Include as many derivations of S as initial states are needed (W)

$$S \to EI \cdot W \quad \text{where } W \ \varepsilon \ (V_T \cup V_N) \ and \ EI\varepsilon \ V_N$$

2. For each production rule from PS add :

$$EI < prec_j > \quad \to \quad EI < action_j >$$

$$EI < prec_j > \quad \to \quad < action_j >$$

where

$$< prec_j >= ...x_{ij}[a_i, b_i]...$$

$$< action_j >= ...y_{ij}(C_i)...$$

$j = 1, m$, $x_{ij} \ \varepsilon V_T$, $[a_i, b_i]$ is an interval, $y_{ij}\varepsilon V_T$, and CF_k is a certainty coefficient

The language generated by this grammar is

$$L(G) = \{x \mid S \overset{G}{\Rightarrow} x \ with \ \beta \to \alpha \ and \ |\beta| \leq |\alpha|\}$$

Another possible way of step 2 is :

$$EI < prec_j > \quad \to \quad EI < action_j > \qquad (add)$$
$$< prec_j > \qquad (delete)$$

$$EI < prec_j > \quad \to \quad < action_j > \qquad (add)$$
$$< prec_j > \qquad (delete)$$

and in this case the generated language is :

$$L(G) = \{x \mid S \overset{G}{\Rightarrow} x \ with \ \beta \to \alpha\}$$

The following example shows the transformation of a PS into a grammar.

Example : Modeling robot actions (Nilsson, 1980)

Production rules :

(1) $PI(x)$:

$$P\&D \quad : OT(x), CL(x), HA$$
$$A \qquad : HO(x)$$

(2) $PD(x)$:

$$P\&D \quad : HO(x)$$
$$A \qquad : OT(x), CL(x), HA$$

(3) $ST(x,y)$:

$$P\&D \quad : HO(x), CL(y)$$
$$A \qquad : HA, ON(x,y), CL(x)$$

(4) $UST(x,y)$:

$$P\&D \quad : HA, CL(x), ON(x,y)$$
$$A \qquad : HO(x), CL(y)$$

where

PI :	*pickup*
PD :	*putdown*
ST :	*stack*
UST :	*unstack*
CL :	*clear*
HA :	*handempty*
HO :	*holding*
OT :	*ontable*
ON :	*on*

The grammar is the following one :

$$G = (V_N, V_T, P, S)$$

where

$$V_N = \{EI, S, ON(x,y), CL(x), HO(x), OT(x)\}$$
$$V_T = \{ON(z,v), CL(z), HA, HO(z), OT(z)\}$$

with

$$z, v = A, B, C$$

and

$P:$

(1)	S	$\rightarrow EI.CL(A).CL(B).CL(C).OT(A).OT(B).OT(C).HA$
(2)	$EI.OT(x).CL(x).HA$	$\rightarrow EI.HO(x)$
(3)	$EI.OT(x).CL(x).HA$	$\rightarrow HO(x)$
(4)	$EI.HO(x)$	$\rightarrow EI.OT(x).CL(x).HA$
(5)	$EI.HO(x)$	$\rightarrow OT(x).CL(x).HA$
(6)	$EI.HO(x).CL(y)$	$\rightarrow EI.HA.ON(x,y).CL(x)$
(7)	$EI.HO(x).CL(y)$	$\rightarrow HA.ON(x,y).CL(x)$
(8)	$EI.HA.CL(x).ON(x,y)$	$\rightarrow EI.HO(x).CL(y)$
(9)	$EI.HA.CL(x).ON(x,y)$	$\rightarrow HO(x).CL(y)$

An example where a robot has to be moved from the initial state to the final state is shown below:

The initial state : $CL(B).ON(C, A).CL(C).OT(A).OT(B).HA$

The final state : $HA.ON(A, B).CL(A).ON(B, C).OT(C)$

The derivation to reach the final state from the initial state is :

$$S \rightarrow EI.CL(A).CL(B).CL(C).OT(A).OT(B).OT(C).HA \rightarrow ...$$

$$\rightarrow EI.CL(B).ON(C, A).CL(C).OT(A).OT(B).HA \rightarrow$$

applying the production rules (8),(4),(2),(6),(2),(7) the following result is obtained

$$\rightarrow HA.ON(A, B).CL(A).ON(B, C).OT(C)$$

4 Inference through structures described by formal grammars

In syntactic and structural pattern recognition the basic representations of structures are strings, trees or graph grammars, which have several interesting advantages (Fu, 1984). The main tool for getting information of the patterns represented by grammars, in pattern recognition, are the grammar parsers, which are used to know if a string belongs to a specific language. However, this tool is not enough in order to get information concerning the structures, the substructures or the family (generated by the grammar) properties. For example, at the present time, the symmetries of a pattern generated by a grammar only can be examined over the specific sample, but not over all the members of the family, since the language can be infinite. A similar case happens when, for example, we want to know if there exist parallel lines in the structures described by the grammar which can only be obtained by means of the geometric properties of the structure. The substructures or in general, the subgraphs belonging to a graph generated by a grammar are also another open question, since the way of getting the subgraphs is by means of reasoning in the structure, but not in the grammar. Finally a very important question is how to obtain the general properties of the grammar rules for grammatical inference.

All these comments lead to a new kind of tool needed to work with patterns described by grammars, which we call the "Inference through a structure described by formal grammars (ITSDFG)". The ITSDFG needs to include induction and deduction procedures through a structure described by a string, tree or graph grammar, by means of the proper grammar rules. The main issue of this tool is to deduce or to induce facts (like there exist symmetric substructures in the family patterns), or there exist general rules in a grammar which are common to a set of samples (typical procedure in grammatical inference). Besides the interest of this technique on pattern recognition, the ITSDFG is one of the basic elements for integrating artificial and syntactic pattern recognition techniques.

4.1 General methodology of ITSDFG

The technique needed for doing inference on grammars is directly dependent on the type of the family grammar, so we are going to describe the general methodology of these techniques without going in detail. A deeper description of the different methods can be seen in (Sanfeliu, 1988). The reason why there is a need of different techniques for doing inference come from the fact that grammars which generates strings, trees and graphs are based in different implicit rules for constructing a pattern. For example, the string of Fig.4 is generated by the grammar of Fig.4 using the following assumptions : the generated string has to be read from left to right, and the construction of the pattern (an object for example) proceeds by connecting the head of the previous primitive to the tail of the next one. This is essential if symmetric substructures have to be discovered or if we try to proof that two substructures of the string are parallel.

The same reasoning can be applied for trees or graphs were some conventions are taking into account to reproduce the pattern described by the respective grammars. The tree case is shown in Fig.5 and two versions for graphs are shown in Fig.6 (nonhierarchical case) and in Fig.7 (hierarchical case). In these cases, the way of making inference depends on the type of structure and the special embedding conditions. On the other hand, the production rules have partial information of the structure contained in the right member which consist of primitives and attributes that are the elements that incorporate the links and the properties of the primitives. The last point shows that the production rules have only part of the structure explicit information, by construction, which produce, as we will see later on, that the order of the inference mechanism have to be included to make inferences.

For example, if we want to infer that face f_6 is parallel to chair CH_1 of Fig.7 , we need to reach the production rule $BA_1 \rightarrow \cdots$ to be able to compare both elements, since their representative vectors need to be recomputed with respect to the new coordinates of the reference frames (each element - e.g., bank, chair, leg, $\cdots$ - has its own reference frame). In this case, we realize that we only have part the information in each one of the production rules, although some of them have a concrete definition (chair, leg).

The aforementioned comments lead to see that in order to make inferences, we need several levels of reasoning and description. Fig.8 shows the general methodology which is presented in this work. The first level is used to translate the query , where

$G = (\{S, A, B, D, H, J, E, F\}, \{a, b, c, d\}, P, S)$

P :

$S \rightarrow AA$

$A \rightarrow PB$

$B \rightarrow FBE$

$B \rightarrow HDJ$

$D \rightarrow FDE$

$D \rightarrow d$

$F \rightarrow d$

$E \rightarrow b$

$H \rightarrow a$

$J \rightarrow a$

$P \rightarrow c$

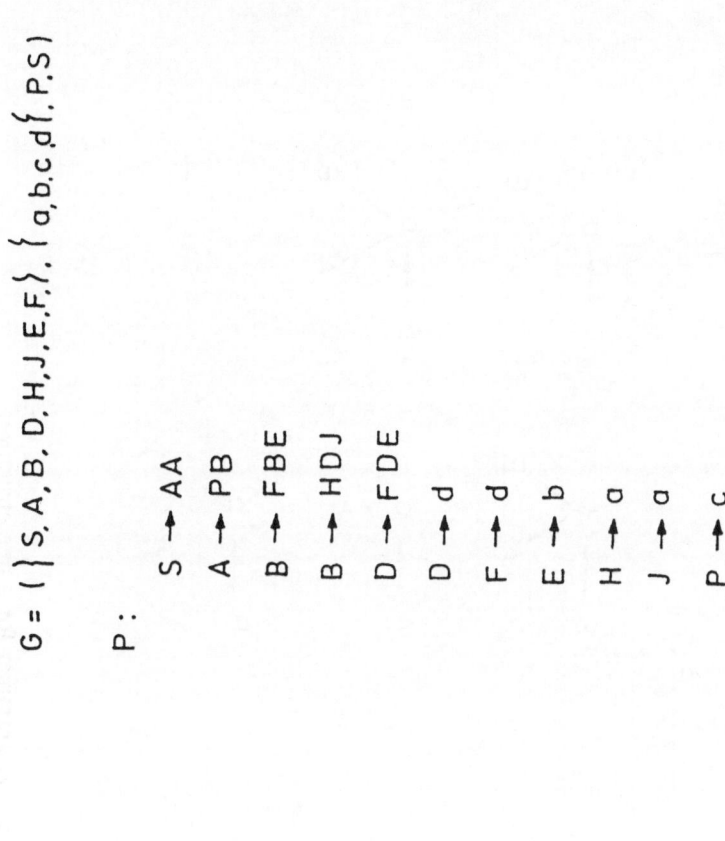

Fig.4 : A grammar for a chromosome pattern

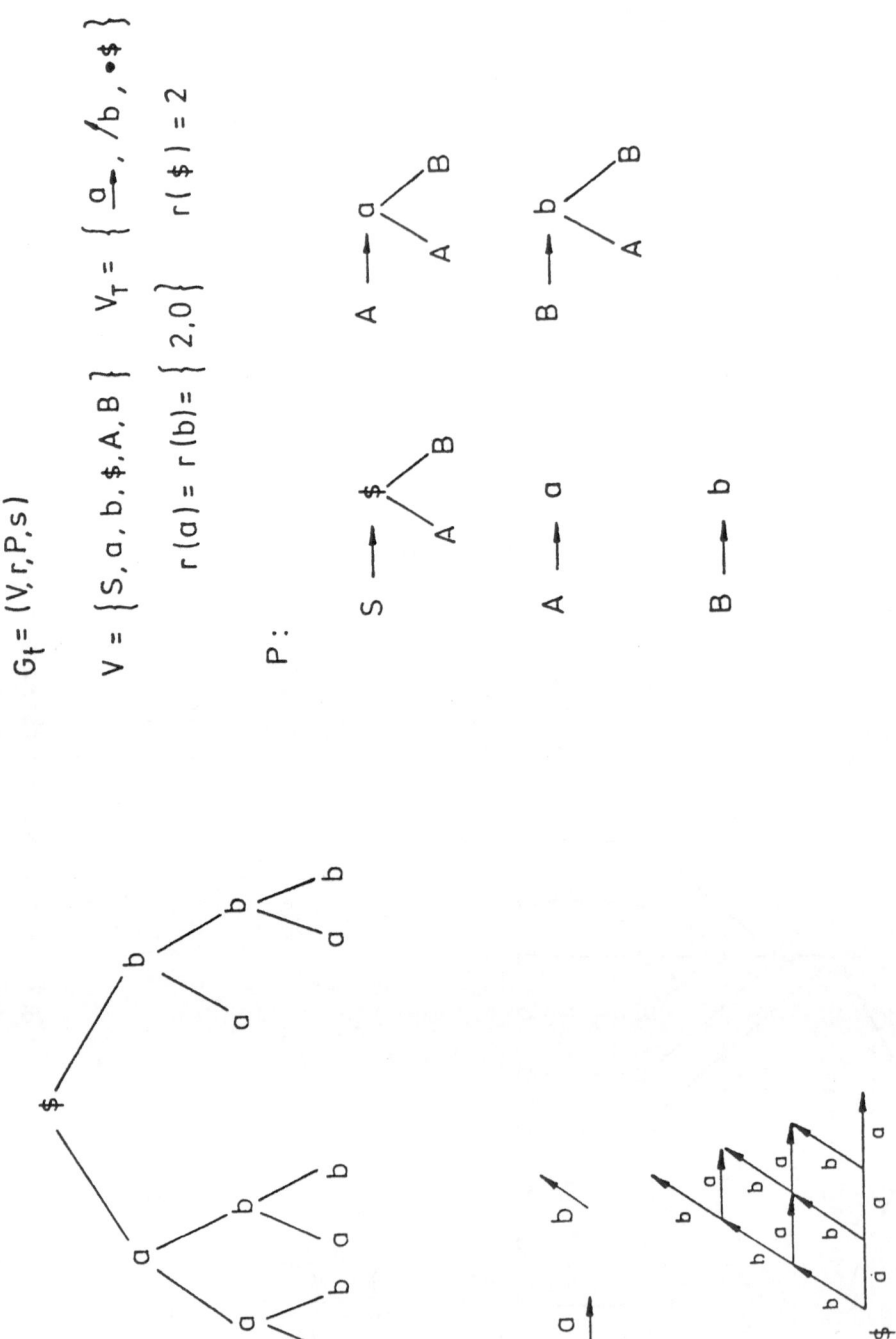

$G_t = (V, r, P, s)$

$V = \{S, a, b, \$, A, B\}$ $V_T = \{\overset{a}{\underset{\longrightarrow}{}}, \overset{b}{\underset{}{}}, \bullet\$\}$

$r(a) = r(b) = \{2, 0\}$ $r(\$) = 2$

P :

Fig.5 : A grammar for a tree structure

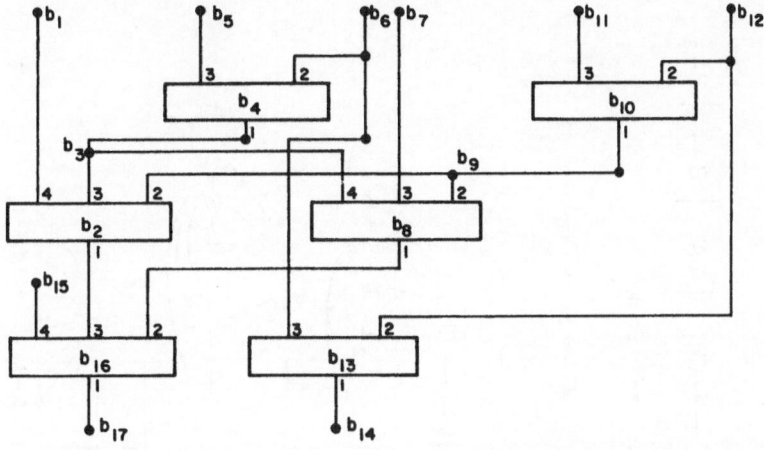

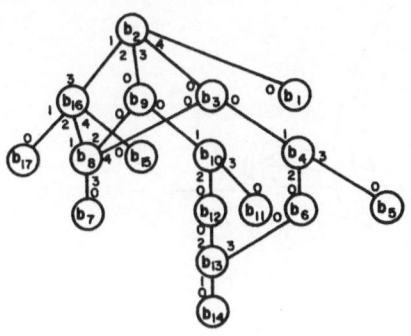

Grammar:

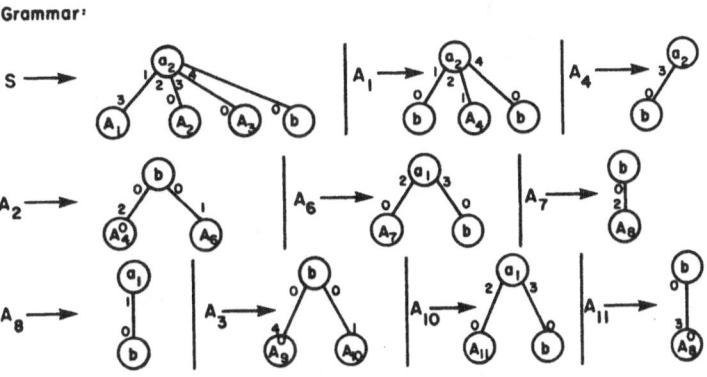

Fig.6 : A tree-graph grammar for a nonhierarchical graph structure

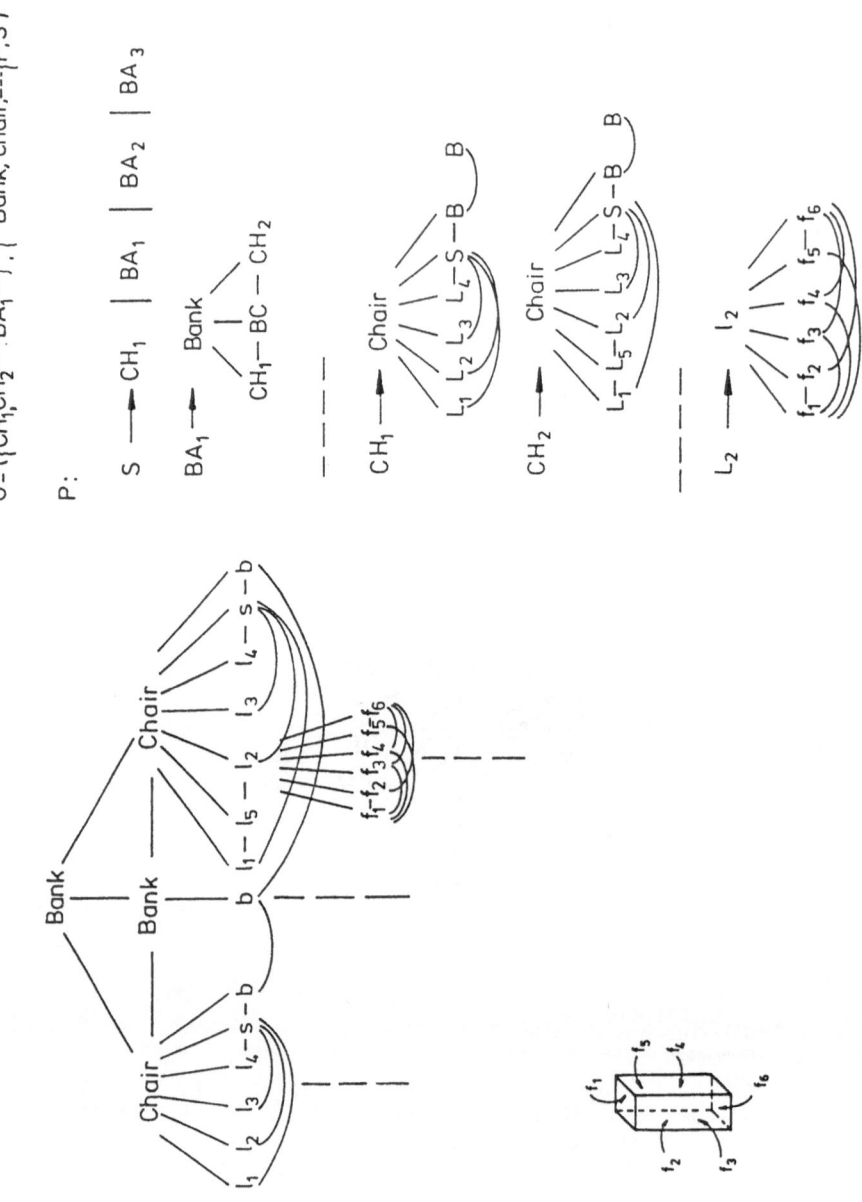

$$G = \left\{ \left\{ CH_1, CH_2 - - BA_1 - - \right\}, \left\{ Bank, chair, - - \right\}, P, S \right\}$$

P :

$$S \longrightarrow CH_1 \mid BA_1 \mid BA_2 \mid BA_3$$

$$BA_1 \longrightarrow \begin{array}{c} Bank \\ \mid \\ CH_1 - BC - CH_2 \end{array}$$

$$CH_1 \longrightarrow \begin{array}{c} Chair \\ L_1 \; L_2 \; L_3 \; L_4 - S - B - B \end{array}$$

$$CH_2 \longrightarrow \begin{array}{c} Chair \\ L_1 - L_5 - L_2 \; L_3 \; L_4 - S - B - B \end{array}$$

$$L_2 \longrightarrow \begin{array}{c} l_2 \\ f_1 - f_2 \; f_3 \; f_4 \; f_5 \; f_6 \end{array}$$

Fig.7 : A tree-graph grammar for a hierarchical graph structure

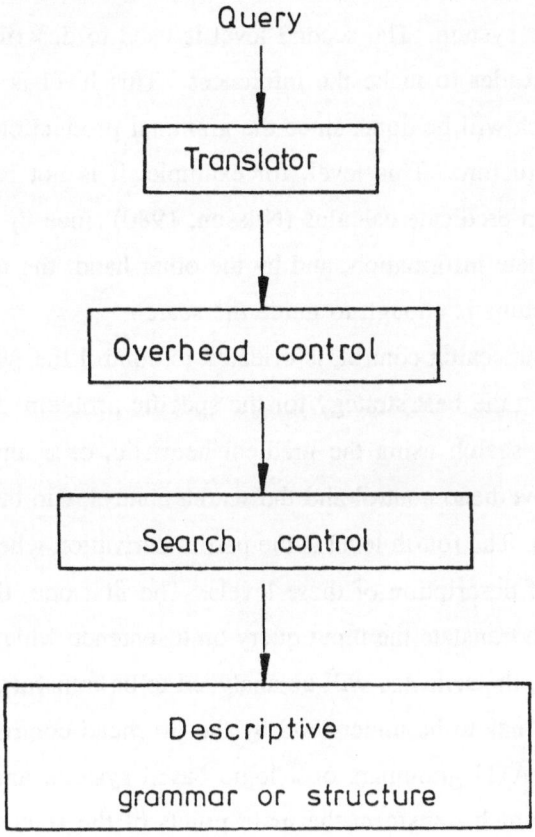

Fig.8 : The general methodology of the ITSDFG

the question for deduction and induction is posted, into a language which could be understood by the system. The second level is used to describe the overhead control of the system in order to make the inferences. This level is the core to specify the way that the search will be done, since the grammar production rules only have partial information of structure. This level, for example, it is not required in the semantic networks based on predicate calculus (Nilsson, 1980) since by one hand, the semantic net has the complete information, and by the other hand, the reasoning mechanism of the predicate calculus is enough to guide the search.

The third level, the search control, is oriented to control the way as the search will be followed, based on the best strategy for the specific problem. For example, the search could be a depth search using the gradient heuristic, or a uniform width search. In some cases, the overhead control and the search control, can be merged if the problem is simpler enough. The fourth level is the partial derivation where the search is applied. Let us see a brief description of these levels. The first one, the translator, consist of an analyzer which translate the input query on a sentence which will be understood by the system. Then, the sentence will be analyzed to be transformed into the appropriate commands which has to be understood by the overhead control. The first step can be done through an ATN grammars or a logic based system, and the second one, by a direct translator which transform the main points of the sentence on commands. The overhead control has the following goals :

- Start conditions and actions.

- End conditions and actions.

- The logic of the control and the computations.

- Scheduling.

The overhead control can be described by means of different types of procedures, for example by using a grammar for specifying the different steps. Fig.9 shows the overhead control described by a grammar of the example of Fig.7.

The search control is basically dedicate to guide the search in an efficient way. This control can be included at this level or at the level of the descriptive grammar. Depending where the control is included, the search could be in one of the followings places (Nilsson, 1980) :

- (a) In the selection process.

- (b) In the grammar production rules.

- (c) In a meta-control.

The search control is usually included in the selection process of a search, in the OPEN list of a basic search algorithm, where the next expandable nodes of the graph search are chosen. If the search control is included at the level of the descriptive grammar,case (b), then the search is guided by the proper grammar production rules. This case is possible if the questions are known a priori. The third case includes the control in the meta-control of the overall system, where for example, some general questions are described in the appropriate way.

4.2 Example

In order to clarify the complete system we show the following example :

Let us have a tree-graph-grammar (Sanfeliu and Fu, 1983) describing a bank following a hierarchical scheme (Fig.7)(Fig.12). Each one of the convex objects of the bank, (-a chair, a leg , $\cdots$-) is described by the nonterminals of the production rules, and the terminals are the different levels of description of the bank. Each convex object has its own reference base of coordinates, which means that in order to know the coordinates of a point with respect to the bank reference coordinates, the point has to be recomputed through the different relative reference coordinates.

We want to make some geometric deductions with respect to the relative position of the elements. For example we want to know the following :

" Is f_6 of L_1 of $Chair_1$ parallel to f_4 of L_2 of $Chair_2$? "

Since we have the geometric information of the points, the arcs, the surfaces, and the volumes of each part of the convex object, then we can answer this question by making inference (in this case deduction) through the grammar following the steps shown in the previous section. Let us see the complete process. The idea is to match the terminals

f_6 of L_1 of $Chair_1$, and f_4 of L_2 of $Chair_2$, with the ones existing in the grammar production rules, and then to calculate the normal vectors of f_6 and f_4 with respect to the bank reference coordinates. Once we have these vectors, we only need to compare them to find the parallelism.

The complete process is as follows. First, the query (Fig.9(a)) is translated into commands which can be understood by the system. Second, the commands are transformed into a sequence of actions which control the inference process. This step is shown in Fig.9(b) , where the overhead control has been expressed by means of a grammar which includes all the steps of the scheduling and control sequence. As is shown in Fig.9(b), in the derivation of the overhead control grammar, the sequence of actions start by matching the terminals and then continue through the production rules backwards until the bank production rule is found, and finally the two vectors are compared to look for the parallelism.

Third, a blind search is applied for finding the matching production rule and then the parallelism. The process, step by step, is shown in Fig.10, and the process stop when the production rule $S \rightarrow BA_1$ is found.

Another example is applied to the grammar shown in Fig.6. There, we have a tree-graph-grammar describing a diagram circuit by means of a nonhierarchical graph. The question is in this case :

" Is b_7 connected to b_{14} through b_4 ? "

Fig.11(a) shows some of the trails of the search over the graph structure, and in Fig.11(b) is shown, the step by step procedure of deduction through the tree-graph grammar to reach the final state (the goal is reached at the level of the right member of the A_8 production rule).

QUERY : Is L_1 of $Chair_1$ $\parallel$ F_6 of L_2 of $Chair_2$?

$\downarrow$ TRANSLATOR

$\parallel \{ L_1 - CH_1 , F_6 - L_2 - CH_2 \}$

(a)

OVERHEAD CONTROL :

$G = (V_N, V_T, \Delta, P, W)$
$V_N = \{< W >, < PC >, < PS >\}$
$V_T = \{\parallel, PS, PC, A, C, SC\}$
$\Delta = \{Bank, Chair_{i_1}, Back_{i_2}, B_{i_3}, S_{i_4}, L_{i_5}, Sup_{i_6}, Ar_{i_7}, V_{i_8}, \cdots\}$

$P :$

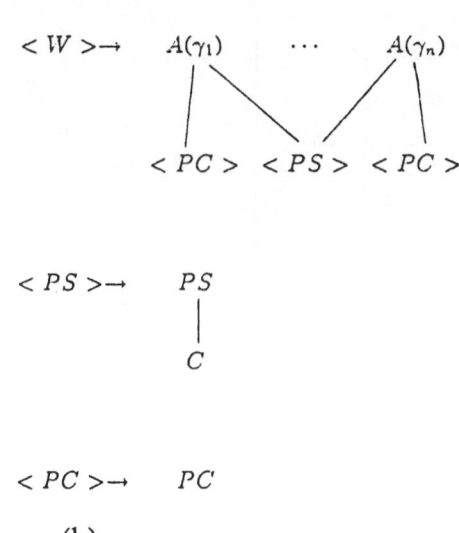

(b)

Fig.9 : (a) The query; (b) The overhead control;

where :

$$A \equiv \quad Activate \ nodes \ \{\gamma_1 \cdots \gamma_n\}$$

$$W \equiv \quad \| \ \{\gamma_1 \cdots \gamma_n\}$$

$$PC \equiv \quad To \ every \ new \ infered \ production \ rule \ do$$
$$[\vec{v'} = F'^{-1}\vec{u}]$$

$$PS \equiv \quad Stop \ in \ the \ top$$

$$C \equiv \quad Compare \ the \ results \ of \ \gamma_i \ to \ \|$$

Generated graph :

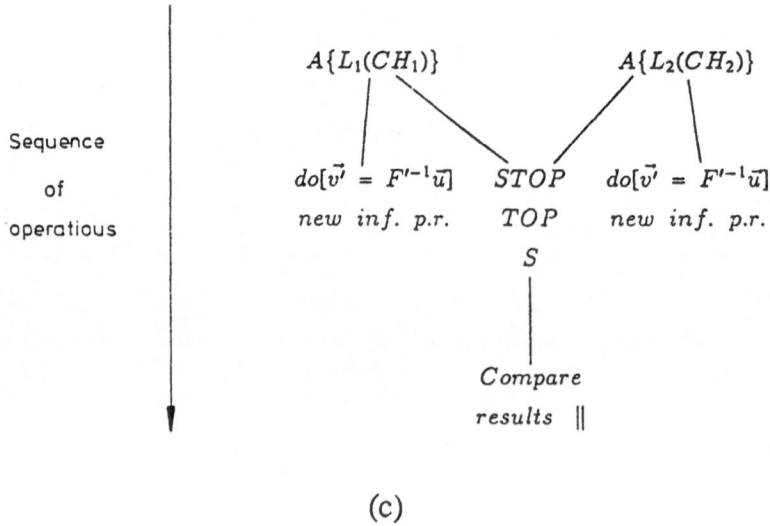

Sequence

of

operatious

(c)

Fig.9 : (c) The generated graph

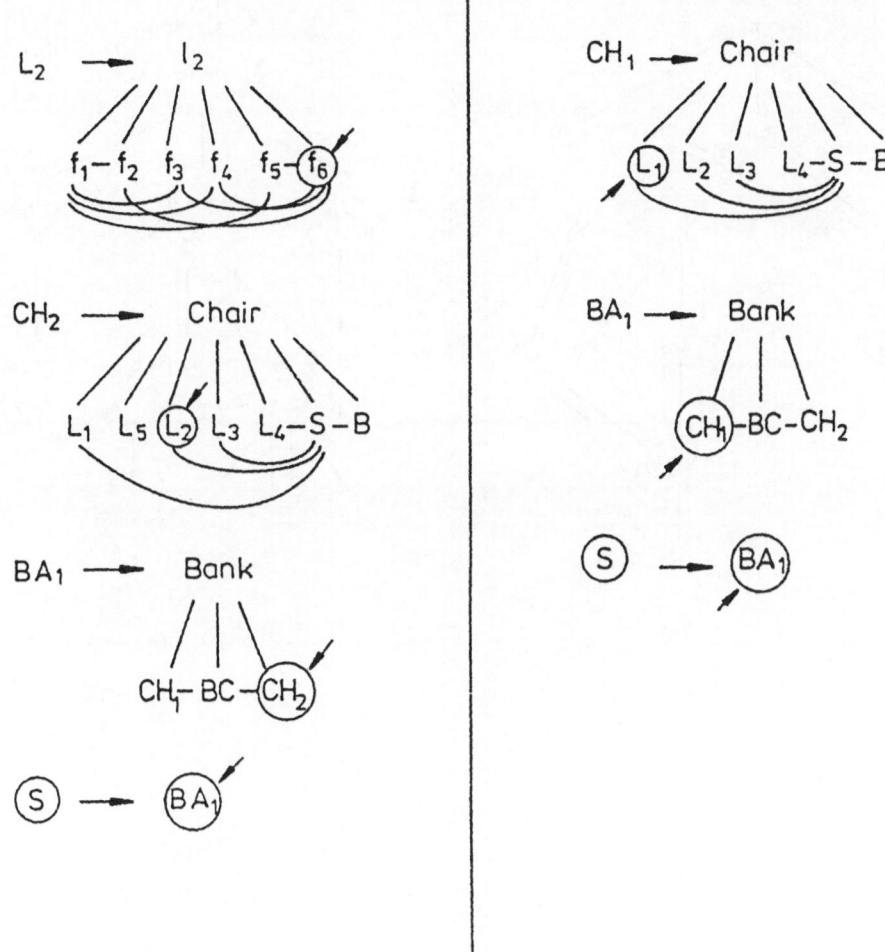

Fig.10 : The step by step process of the blind search

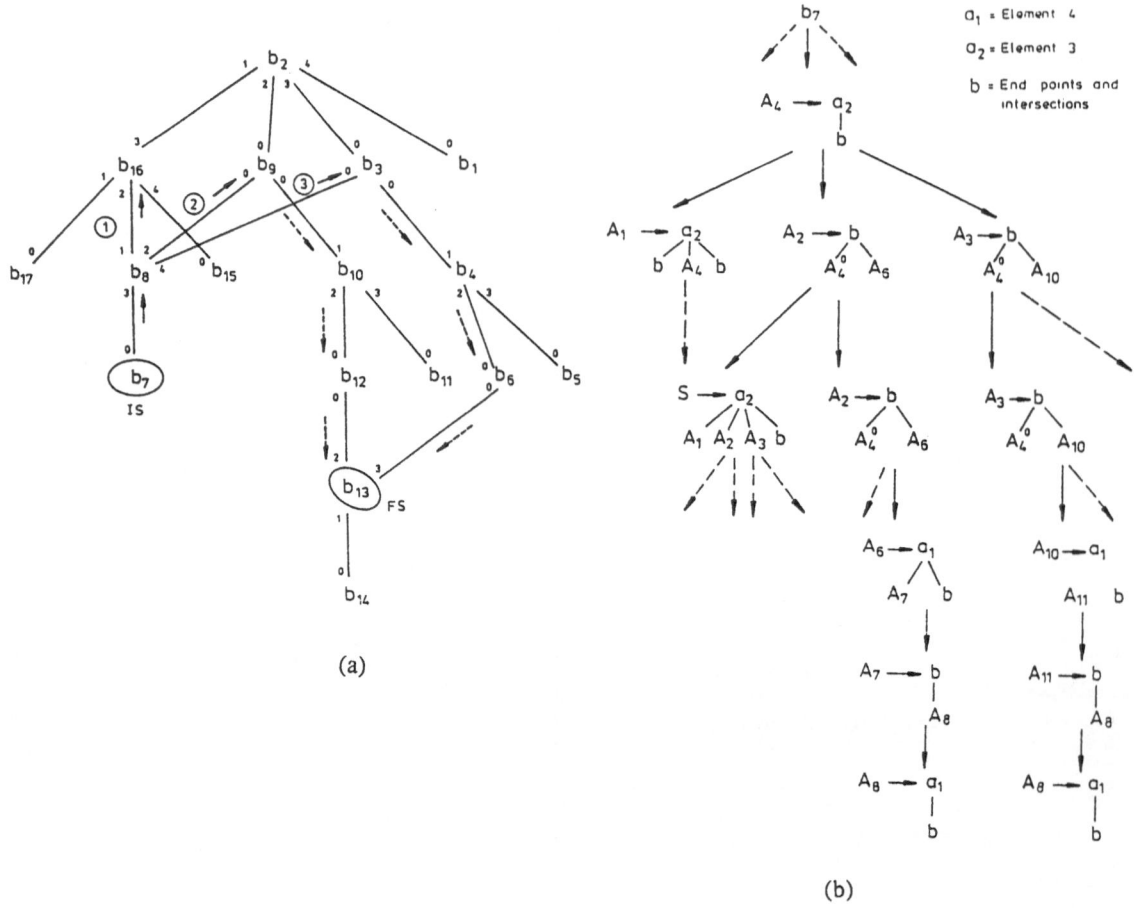

(a)

(b)

Fig.11 : (a) The graph structure for the example of Fig.6; (b) The deduction through the grammar for the mentioned graph

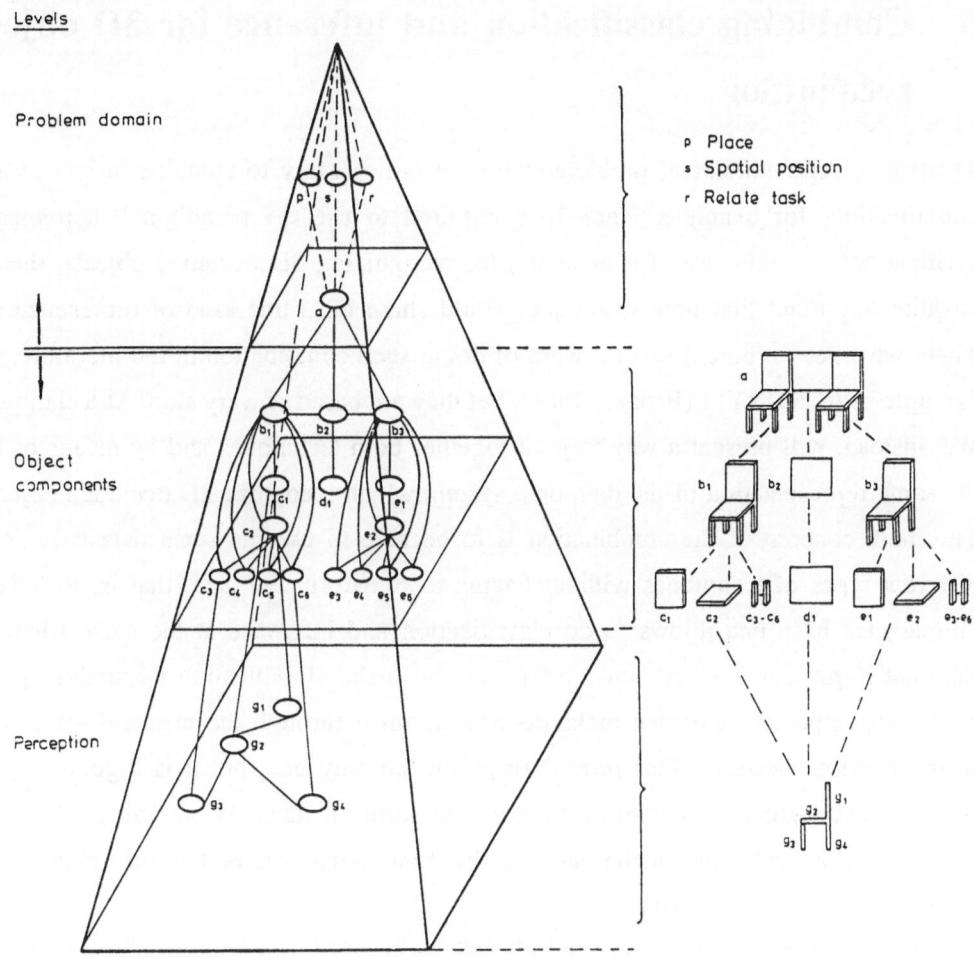

Fig.12 : The hierarchical structure of a 3D object

5 Combining classification and inference for 3D object recognition

There is a large number of problems where it is necessary to combine inference and classification, for example, when it is required to use the paradigm " hypothesis-verification". In this case, for example, for recognizing 3D occluded objects, then it is quite important that both techniques could share the same kind of representation. There have been reported several ways of doing such combination in the literature, for example in ACRONYM (Brooks, 1981), but they are based on very slow AI techniques. We, instead, will present a way to joint together both techniques and by means of the the same representation of the data on a system which recognize 3D occluded objects. The main concern of the combination is to be able to use the same data base from different types of techniques without having to transform the data, that is, to have a unique data base that allows to do classification and inference at the same time. If the data is presented as grammar rules, we can make classification by analyzing the data using a parser, and then make deductions on it through the mechanisms shown in the previous section. This procedure is not the only one, but it is a good way to integrate two techniques without having to transform the data. As we will see later on, the integration with other techniques can be done using part of this information and including the rest of the data.

As we have commented before, the classification and inferences procedures are the bases for the paradigm "hypothesis-verification". The classification is used to get the initial hypothesis and a coarse approximation of the hypothesis. The inference is applied in the verification process, that is, in the focus on attention process to improve the results obtained by the classifier. In order to apply such system in a 3D object recognition, the objects are represented by a hierarchy of levels (Fig.12)(Sanfeliu, 1984) where the "component object level", is structured in three levels : first level - the object -, second level - the convex objects - , and third level - the surfaces, the arcs and the vertices -.

The system works as follows : once the image is acquired, the primary process is applied (filtering and segmentation) and some special features and primitive structures are obtained. With this information the classifier get the first hypothesis. Then the orientation and position is obtained, and the classifier tries to find a coarse similarity measure between the reference object and the image object. In this process (Sanfeliu, 1984), the classifier not only obtains the value of similarity, but also what parts of the object have not been found and need a process of verification. Fig.14 shows some of the steps of the classifier and points out the parts of the object that have not been found. In Fig.13, the parts of the object that has not been found by the classifier and need a deeper attention (they are sent to the agenda) are shown.

The grade of similarity got from the classifier is ranked between 0 and 1 (0 is the best classification). If the reference object has a cost near 0, it means that has a high probability to be the correct one with respect to the recognition process. However, a cost higher than 0 implies that the reference object has not been completed recognized and that requires a verification process. This step is done by the inference process. With the information of the parts that has not been recognized , the system tries to find these parts through a reasoning process using the structure of the object. For example, if there is a missing arc, the inference can look on the reference object for features that can be used to detect the arc, i.e, to find the surfaces of each side of the arc. This process is done using the inference mechanisms. If the arc is found, then the similarity value will be diminished improving the grade of similarity. In these processes, the integration of the classifier and the inference is very important in order to get an efficient system.

In the next section we will show a system called PIRS which uses the mechanism shown in this section as the core of the process.

5.1   PIRS

The PIRS system (Perception Interpretation Robotic System) (Sanfeliu, 1986) is a system oriented to interpret 3D industrial scenes for applications on assembly and manipulation of objects. Moreover, the system is prepared to integrate the different sensors in order to reason with them towards the interpretation of a scene. The application domain is the recognition and identification of 3D objects which can be occluded or partially hidden in the scene.

The system shown in Fig.15, is based on a heterarchical structure and a blackboard architecture (Hayes-Roth,1985) which have two control flows, a bottom up and a

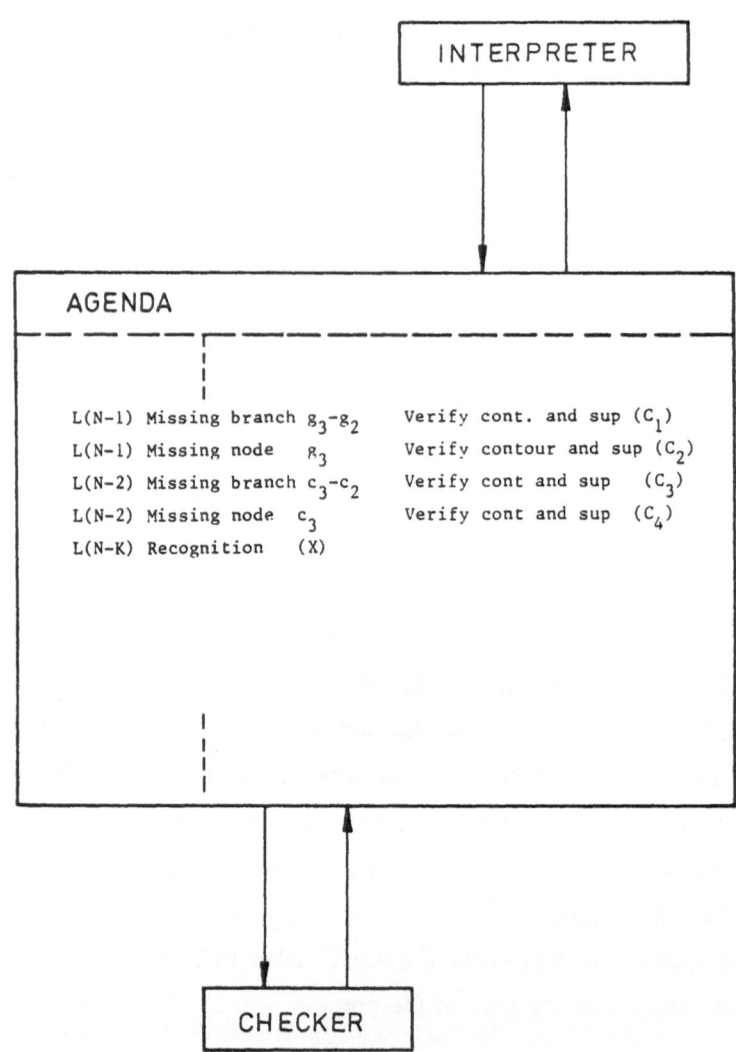

Fig.13 : Parts not found reported to the agenda

423

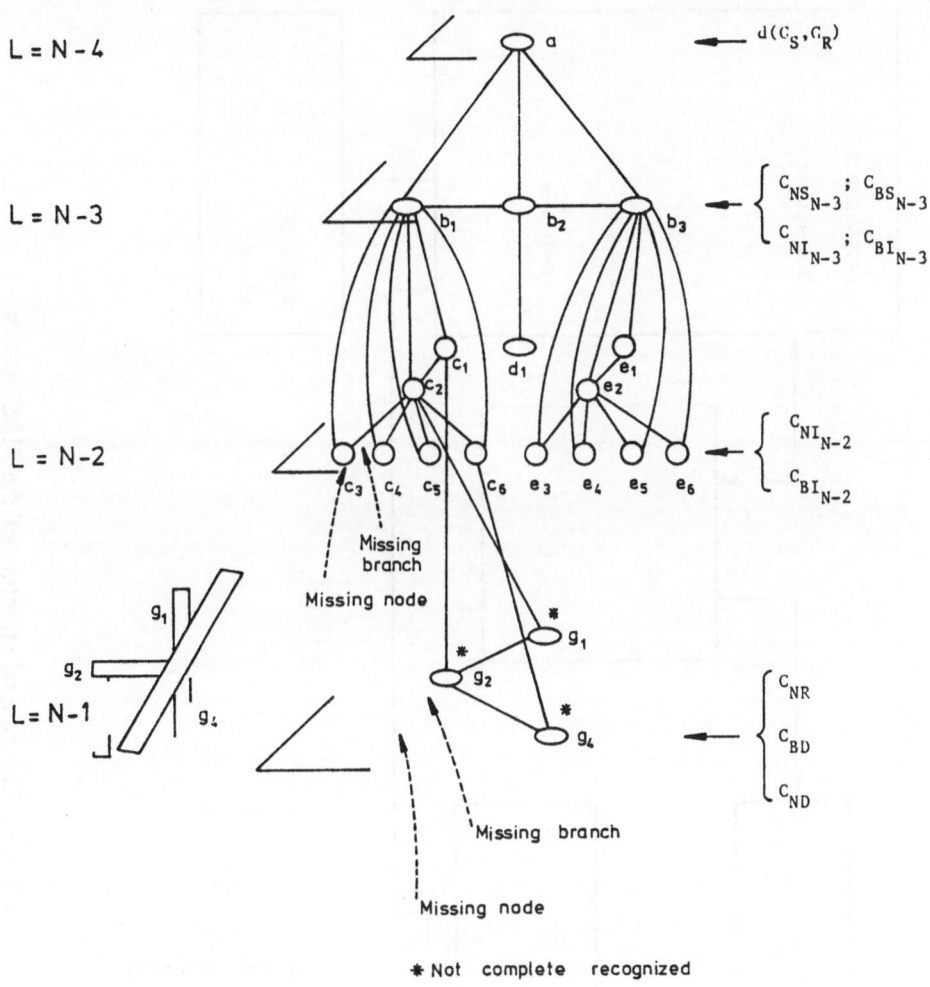

Fig.14 : The cost at the different levels of the hierarchy

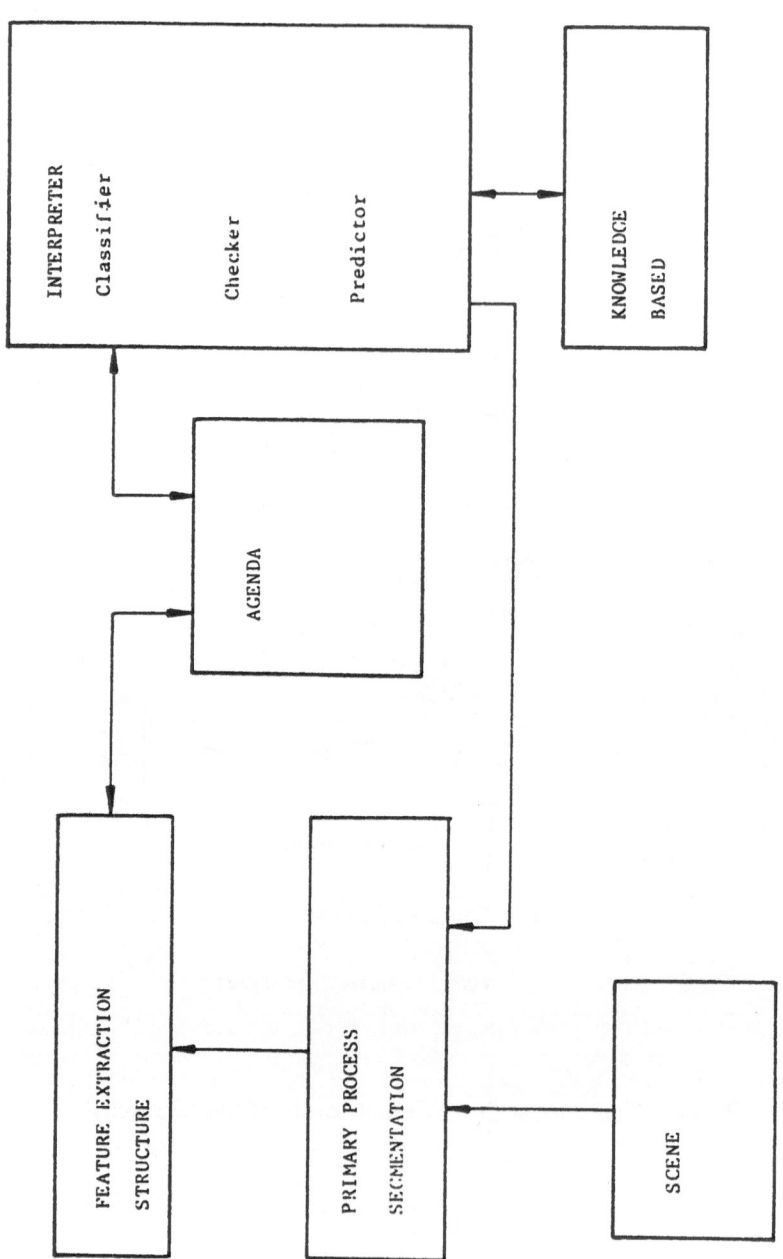

Fig.15 : The general structure of the PIRS system

top down, which are used in a joint or independently way. The system combines AI and pattern recognition techniques to perform the task in the way explained in previous section, and use a unique data base composed by three different data representations (semantic net, grammar structure and production rules)(Fig.16) (Brachman and Levesque, 1985; Goss and Hartmanis, 1980). The semantic net is used to describe the geometric structure of the objects and scenes, and their physical features. The grammar describes coarsely the object and is the base for doing the classification. Finally, the production rules have the information of the sensors, the domain, and the strategies.

In the first step, the system works in a similar way as the VISIONS (Hanson and Riseman, 1978) works, that is, if there is not information about the objects of the scene, PIRS does a primary segmentation and obtains some basic features. Once special features and primitive structures are obtained, the system proposes initial hypothesis, that later on, will be verified (hypothesis- verification paradigm). Then, a coarse classification is obtained through the application of the objects obtained in the initial hypothesis process, over the image. This classifier not only get a degree of similarity, but also tells what is missing in the object. With this information the checker has to try to improve the grade of similarity by "focus on attention" on the missing parts. This verification process is composed by several specialists, a geometric reasoning specialist, a symbolic reasoning specialist, and some planer specialists. These specialists have to reason with the information available of the object, the sensors and the domain.

If a priori of the scene objects is known, the "predictor" can be used to look for specific features in the scene, be using the knowledge of the domain and the sensors. This process allows to reduce in great deal the segmentation and the acquisition of features of the scene, that is, to reduce the time of interpretation. The predictor has a similar structure of the checker. The control of the three processes is done by the interpreter (Fig.17) by means of an agenda (the intermediate data base) where all the process interchange their data.

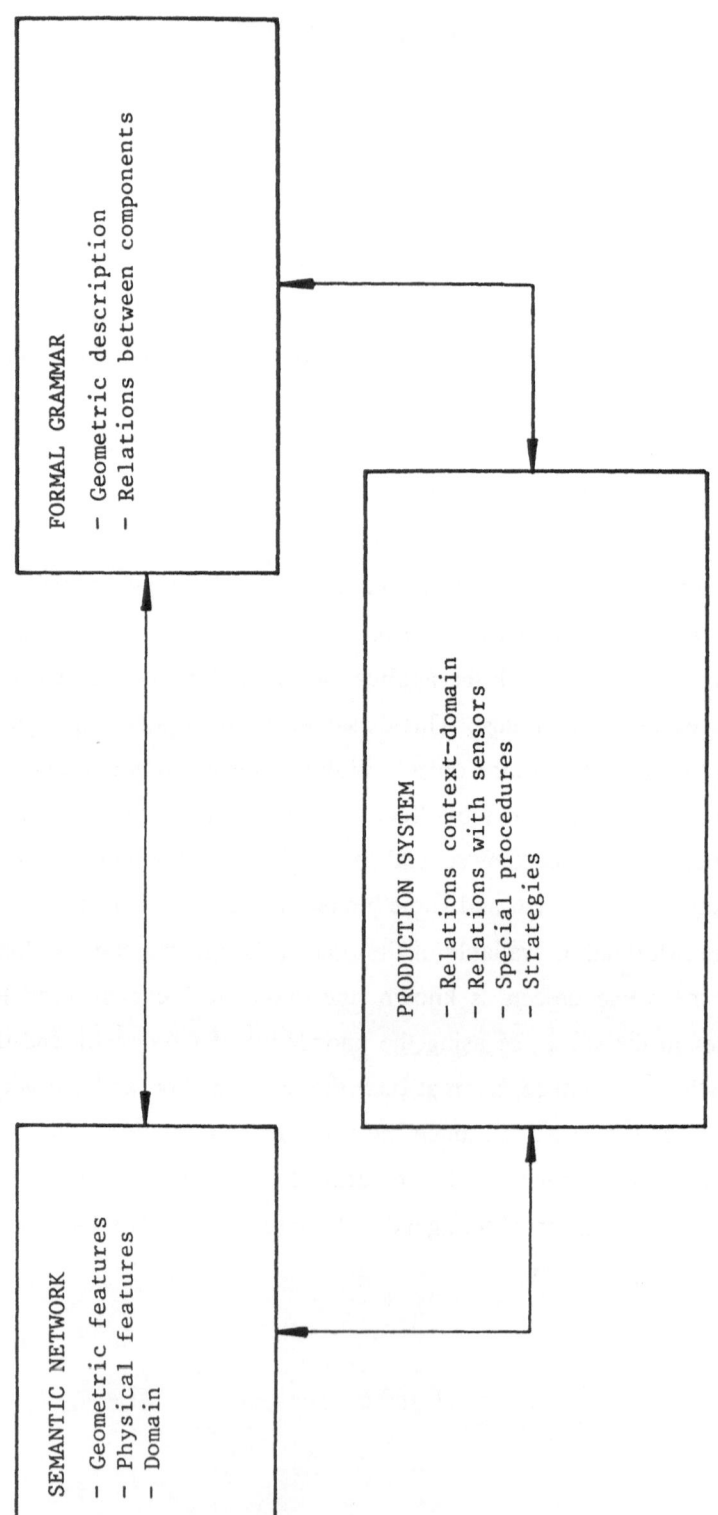

Fig.16 : The data knowledge base

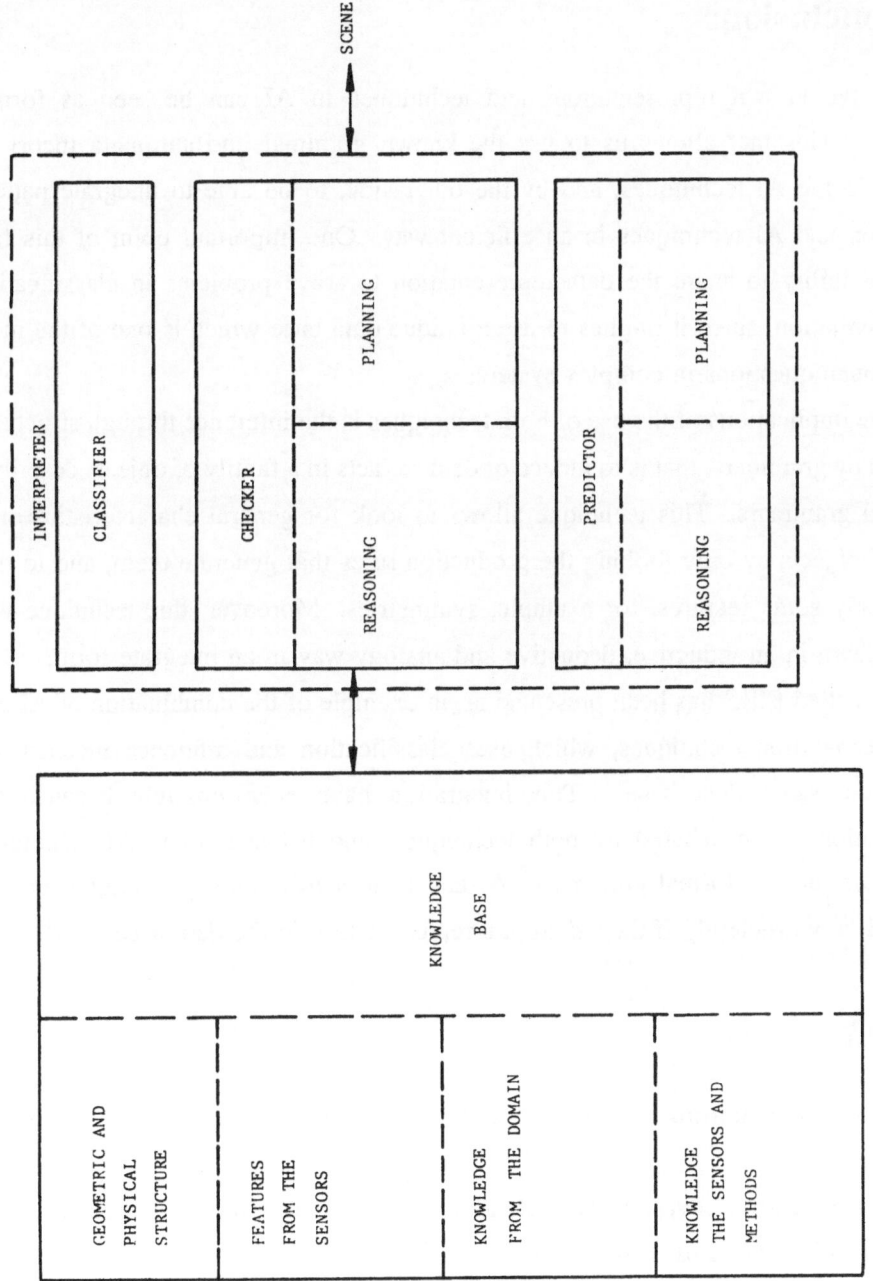

Fig.17 : The interpreter

6 Conclusions

Some of the known representations and techniques in AI can be seen as formal grammars. This fact allows us to use the known grammar and automata theory to formalize some AI techniques, and by the other side, to be able to integrate pattern recognition and AI techniques in an efficient way. One important point of this fact is the possibility to share the data representation to solve problems in classification and interpretation, since it implies to use a unique data base which is one of the most difficult open questions in complex systems.

One of the implications of the use of both techniques is the inference through structures described by grammars, that is to induce or deduce facts in a family of objects described by formal grammars. This technique allows to look for general characteristics on a family of objects by only looking the production rules that generate them, and to find very quickly some features, for example, symmetries. Moreover, this technique will allow to learn in an inductive, deductive and analogy way in an integrate form.

A system called PIRS has been presented as an example of the combination of AI and pattern recognition techniques, which uses classification and inference mechanisms through the same data base. This integration have been possible because the representation can be shared by both techniques, and because some AI techniques can be described as formal grammars. AI and pattern recognition techniques can be combined very efficiently if they share the representation in the data base.

7 References

Barr A. and Feigenbaum E.A. The handbook of artificial intelligence. *Pitman Books Limited, 1982.*

Brachman R.J. and Levesque H.J. Readings in knowledge representation. *Morgan Kaufmann Publ., Inc., Los Altos, 1985.*

Brooks R.A. Symbolic reasoning among 3D models and 2D images. *In Computer Vision (J.M.Brady ed.), North Holland, Amsterdam, New York, 1981.*

Fukunaga K. Introduction to statistical pattern recognition. *Academic Press, New York and London, 1972.*

Encarnacao J. Computer aided design-modelling, systems engineering, CAD-systems. *In Lecture Notes in Computer Science n. 89, Springer-Verlag, Berlin, Heidelberg, New York, 1980.*

Hanson A.R. and Riseman E.M Segmentation of natural scenes. *In Computer Vision Systems (A.R. Hanson and E.M. Riseman), Academic Press, New York, 1978.*

Hayes-Roth B. A blackboard architecture for control. *Artificial Intelligence 26, 1985, pp.251-321.*

Nilsson N.J. Principles of artificial intelligence. *Palo Alto, California, Tioga, 1980.*

Pavlidis T Structural pattern recognition. *Springer-Verlag, Berlin New York, 1977.*

Sanfeliu A. A distance measure based on tree-graph-grammars: A way of recognizing hidden and deformed 3D complex objects. *Technical report IC-DT-1984.01, Instituto de Cibernetica 1984. A summary in the 7th Int. Conference on Pattern Recognition, Montreal, Canada, July 30-Aug. 2, 1984.*

Sanfeliu A. PIRS: Sistema de percepción interpretación para robótica. *Technical Report IC-DT-1986.01 Instituto de Cibernética, 1986.*

Sanfeliu A. The inference problem in the syntactic and structural techniques. *To appear in Syntactic and Structural Pattern Recognition - Fundamentals, Advances and Applications, World Scientific Pub. Co. Pte. Ltd 1988.*

Sanfeliu A. and Fu K.S. Tree graph grammars for pattern recognition. *In Lecture Notes in Computer Science n. 153, Graph-Grammars and Their Application to Computer Science (Ehrig H, Nagl M., and Rozenberg G. eds.) Springer-Verlag, New York, 1983,pp. 349-368.*

Shortliffe E.H. Computer-based medical consultations: MYCIN. *Elsevier-North Holland, New York, 1976.*

Sickel S. Formal grammars as models of logic derivations. *5th Int. Joint Conference on Artificial Intelligence, IJCAI-77, Mass. Inst. of Technology, Cambridge, Massachusetts, USA, August 22-25, 1977.*

A SYNTACTIC APPROACH TO PLANNING

T.C. Henderson and E. Muehle

Dept. of Computer Science
3160 Merrill
Engineering Bldg.
The University of Utah
Salt Lake City, Utah 84112
U.S.A.

Abstract

Formal language theory concerns the study of two major issues:

1. **Generative Grammars**: a language is specified in terms of an alphabet, vocabulary symbols, rewrite rules and a start symbol; i.e., a method for generating all strings in the language, and

2. **Recognizers**: a language is specified by giving an alphabet, states, memory and state transitions; i.e., a method for deciding for any given string whether or not it is in the language.

Most of the syntactic pattern recognition work has exploited the recognition aspect of formal language theory. What we propose here is the use of generative grammars as a mechanism to help encode plans.

1. Introduction

Formal language theory permits both the analysis and synthesis of strings. Both of these aspects have been explored in the domain of shape analysis and pattern recognition [Bunke 82, Fu 73, Fu 74, Gonzalez 78, Lin 84, Rosenfeld 72]. Most of the work on synthesis has concerned the generation of regular patterns and textures. However, there has been little published on the use of grammars as a mechanism for solving the planning problem.

To solve a problem requires that the appropriate sequence of operations be performed in the correct order. Finding such a sequence is, in general, a difficult problem and many approaches have been proposed [Nilsson 71, Winston 84]. Most of

NATO ASI Series, Vol. F45
Syntactic and Structural Pattern Recognition
Edited by G. Ferraté et al.
© Springer-Verlag Berlin Heidelberg 1988

these methods lack the ability to focus well on a particular part of the problem. We propose that a generative grammar can provide such a mechanism.

2. A Syntactic Approach to Planning

We restrict our attention to planning in the context of a robot workcell. The task to be performed is light assembly. Thus, we are essentially concerned with plans for the assembly of small parts in a well-known environment.

First, it is necessary to have some way to model 3-D shapes and their structure. We have previously worked on the problem of 3-D shape representation and analysis [Davis 81, Henderson 81, Henderson 85], and we will use Stratified Shape Grammars as our representation scheme. Usually, a shape grammar is defined to solve the shape recognition problem. Here, however, we will define a shape and then take advantage of the shape grammar to help plan the sequence of operations.

Given a shape grammar, it can be used to help solve several aspects of the planning problem:

1. Rewrite rules impose an order on the operations. That is, given a parse tree of the 3-D structure to be built, it is straightforward to analyze the sequence in which the operations must be performed. In addition, it is also possible that opportunities for parallelism can be discovered.

2. Constraints on the positioning of parts can be recovered. Many such constraints are explicit in the rewrite rules (see Section 3 below). However, it is also possible to discover implicit constraints (e.g., by means of global analysis or constraint propagation).

3. Focus of attention is achieved. Since only the appropriate components appear together on the righthandside of a rewrite rule, it is possible to determine what parts of the shape are related. Moreover, one can use both ancestors, neighbors and decendants relations to focus attention.

In addition, the use of generative grammars permits a unified approach to the shape analysis problem. That is, the same underlying paradigm supports both the synthesis of the 3-D structures and the later analysis of those same structures. Thus, any change in the shape (i.e., a change in the grammar) is automatically reflected in the synthesis and analysis.

Finally, given the CAD-based context, it is possible that grammatical specifications of the 3-D structure can be synthesized from the CAD design information, thus getting around the difficult problem of grammar writing. At the very least, a graphical interface for grammar design would be reasonably easy to produce.

3. An Example

Consider a very simple example: the construction of Lincoln log houses. The 3-D shape primitives (terminal symbols) are the following:

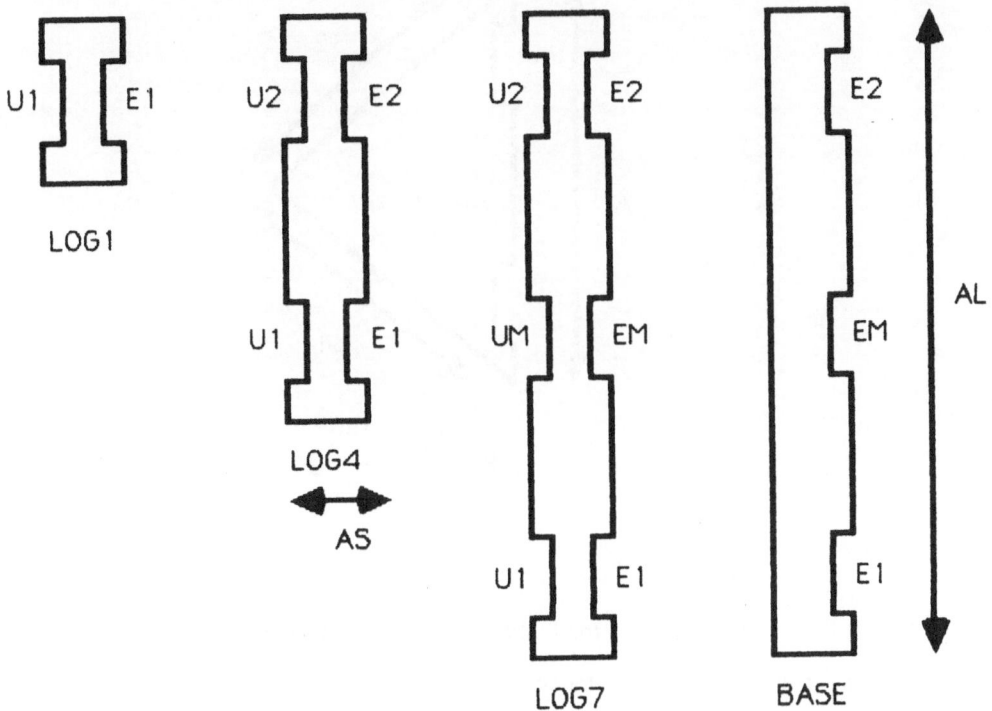

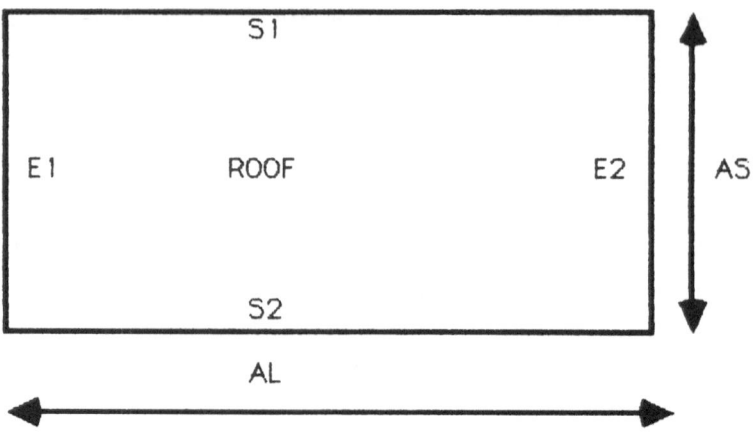

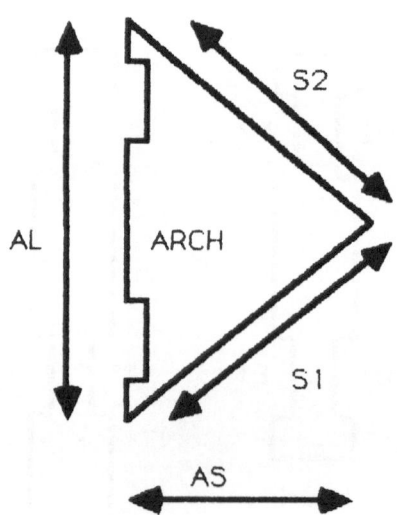

A stratified shape grammar for building a simple 4-walled house is:

Grammar for Lincoln Log House

```
SIGMA = {base, log1, log4, log7, arch, roof}
V = {house, top, bottom, 4wall, top-wall, bottom-wall}
```

```
base{E1,mE,E2}[Al,As]
log1{E,U}[Al,As]
log4{E1,E2,U1,U2}[Al,As]
log7{E1,Em,E2,U1,Um,U2}[Al,As]
roof{E1,S1,E2,S2}[Al,As]
arch{E1,E2,H1,H2}[Al,As,S1,S2]

house{}[h,w,l] :=  top{T1,T2,T3,T4}[Tl,Ts,Ht,Wt,Lt]
                    + bottom{B1,B2,B3,B4,B5,B6}[Bl,Bs,Hb,Wb,Lb]

C : Ti near Bi (i < 5)

S : Tl || Bl and Ts || Bs

Ga : 0

Gs : h = Ht + Hb; W = (Wt + Wb)/ 2; L = (Lt + Lb) / 2;

top{E1,E2,E3,E4}[Al,As,h,w,l] :=

                    arch{E1',E2',H1',H2'}[As',Al',S1',S2']
                  + arch{E1'',E2'',H1'',H2''}[As'',Al'',S1'',S2'']
                  + roof{E1''',S1''',E2''',S2'''}[Al''',As''']
                  + roof{E1'''',S1'''',E2'''',S2''''}[Al'''',As'''']

C : S1''' next-to S2'''' and
    E1''' touches E1'''' and
    E2''' touches E2'''' and
    S2''' touches H1'   and
    S2''' touches H1''  and
    S1'''' touches H2'  and
    S1'''' touches H2''

S : As''' || S1' and  As''' || S1'' and  As'''' || As'''' and
    As'''' || S2'' and Al' || Al''
    and  distance(H1',H1'') > length(log1)

Ga : E1 = E1'; E2 = E2'; E3 = E1''; E4 = E2''

Gs : Al = (Al''' + Al'''') / 2; As = (Al' + Al'') / 2; h = height(arch);
     l = length(Al);  w = length(As);

bottom{B1,B2,B3,B4,B5,B6}[Al,As,h,w,l] :=

     4wall{J1,J2,J3,J4,J5,J6,U1,U2,U3,U4,U5,U6}[Wl,Ws,h',w',l']
   + base{E1,M,E2}[Al',As']
```

```
       + base{E1',M',E2'}[Al'',As'']


C : U1 near E1  and  U2 near E2  and  U3 near E1'  and  U4 near E2'
    and  U5 near M  and  U6 near M'

S : Al'' || Al'  and  As'' || As'  and  distance(E1,E1') > length(log1)

Ga : Bi = Ji

Gs : Al = W1; As = Ws; h = h' + height(base); w = length(log7);
     l = length(log4);

4wall{E1,E2,E3,E4,M1,M2,U1,U2,U3,U4,U5,U6}[Al,As,h,w,l] :=

top-wall{E1',E2',E3',E4',M1',M2',U1',U2',U3',U4',U5',U6'}[Al',As',h',w',l']
+ bottom-wall{E1'',E2'',E3'',E4'',M1'',M2'',U1'',U2'',U3'',U4'',U5'',U6''}
           [Al'',As'',h'',w'',l'']

C : U1' near E1''  and  U2' near E2''  and  U3' near E3''
    and  U4' near E4'' and  U5' near M1''  and  U6' near M2''

S : Al' || Al''  and  As' || As''

Ga : Ei = Ei'; Mi = Mi';

Gs : Al = Al'; As = As'; h = h' + h'';

4wall{E1,E2,E3,E4,M1,M2,U1,U2,U3,U4,U5,U6}[Al,As,h,w,l] :=

 top-wall{E1',E2',E3',E4',M1',M2',U1',U2',U3',U4',U5',U6'}
        [Al',As',h',w',l']
+bottom-wall{E1'',E2'',E3'',E4'',M1'',M2'',U1'',U2'',U3'',U4'',U5'',U6''}
           [Al'',As'',h'',w'',l'']
+4wall{E1''',E2''',E3''',E4''',M1''',M2''',U1''',U2''',U3''',
      U4''',U5''',U6''}
        [Al''',As''',h''',w''',l''']


C : U1' near E1''  and  U2' near E2''  and  U3' near E3''
    and  U4' near E4'' and  U5' near M1''  and  U6' near M2''

S : Al' || Al'' || Al''' and  As' || As'' || As'''

Ga : Ei = Ei'''; Mi = Mi'''; Ui = Ui''';
```

```
Gs : h = h' + h'';


top-wall{E1,E2,E3,E4,M1,M2,U1,U2,U3,U4,U5,U6}[Al,As,h,w,l] :=

        log7[E1',Em',E2',U1',Um',U2'}[Al',As']
      + log7[E1'',Em'',E2'',U1'',Um'',U2''}[Al'',As'']

C : 0

S : Al' || Al''  and  As' || As''

Ga : E1 = E1'; E2 = E2'; E3 = E1''; E4 = E2''; M1 = Em'; M2 = Em'';
     U1 = U1'; U2 = U2'; U3 = U1''; U4 = U2''; U5 = Um'; U6 = Um'';

Gs : h = height(log7);


bottom-wall{E1,E2,E3,E4,M1,M2,U1,U2,U3,U4,U5,U6}[Al,As,h,w,l] :=

        log4{E1',E2',U1',U2'}[Al',As']
      + log4{E1'',E2'',U1'',U2''}[Al'',As'']
      + log1{E''',U'''}[Al''',As''']
      + log1{E'''',U''''}[Al'''',As'''']

C : 0

S : Ax' || Ax'' || Ax''' || Ax''''  x = (1,s)

Ga : E1 = E1'; E2 = E1''; E3 = E2'; E4 = E2''; M1 = E'''; M2 = E'''';
     U1 = U1'; U2 = U1''; U3 = U2'; U4 = U2''; U5 = U'''; U6 = U'''';

Gs : h = height(log4);
```

As can be seen, many of the constraints are explicit in the rewrite rules (e.g., Near, Parallel, etc.).

We are currently exploring the use of FROBS (frame objects [Muehle 86]) to express the shape grammar in an expert system format. For example, frobs can be defined for the vocabulary symbols:

```
*** Frobs ***

(def-class struct nil :slots (axis h w l))
```

```
(def-class house {struct} :slots (top bottom))

(define-class top {struct} :slots (joints))

(define-class bottom {struct} :slots (joints))

;;; assume that there are some instances of these 3 classes
```

and a rule can be expressed as:

```
(define-forward-rule identify-housea?ba)


                ^(parallel-p ?ta ?ba)
                (?top h ?th)
                (?top w ?tw)
                (?top l ?tl)
                (?bottom h ?bh)
                (?bottom w ?bw)
                (?bottom l ?bl)))
  (conclusion (and (?house top ?top)
                (?house bottom ?bottom)
                (?house axis ?ta)
                (?house joints ?j1)
                (?house h ^(+ th bh))
                (?house w ^(/ (+ tw bw) 2))
                (?house l ^(/ (+ tl bl) 2)))))
```

All structures have an axis frame, a height slot, a width slot, and a length slot. The axis frame contains the long and short axis of orientation. The height, width, and length slots contain those values for that particular structure. A house frame is a subclass of the structure frame, but also has slots for the bottom, and top frames of the house. The top and bottom frames are structures with a set of joints for reasoning about the connectivity between structures.

The identify-house rule says:

> If there exists a house with a top and bottom undefined,
> and there exists a top and bottom whose joints are near

each other, and the top and bottom axises are parallel to each other then there exists a house that has a top and bottom, with a set of joints, a set of axises, and a height width, and length.

Let us now see how this is useful in planning the construction of a Lincoln Log house. We describe a problem-reduction technique that successively reduces the state-space search by guiding the reduction through the use of rewrite rules.

Given a set of start states, S, a set of operators, F, that map states onto states, and a set of goal states, G, the rewrite rules of the grammar can be used to identify subgoals which must first be solved before the final goal can be achieved.

For example, given the goal of constructing a Lincoln Log house, the grammar gives us two immediate subgoals: build the top of the house and the bottom of the house. Given the geometric semantics of the relations on the attachment parts of the vocabulary symbols, it can be inferred that *top* is "OnTopOf" *bottom*, and therefore, that top should be built after bottom. We are currently exploring an implementation of these ideas to generate plans for a PUMA 560 to assemble Lincoln Log houses.

4. Conclusion

We propose that the syntactic approach may be used as the basis for planning. A grammar permits the natural recovery of sequence information, recovery of constraints and the focus of attention. We are presently exploring light assembly tasks in a robotics workcell.

5. Acknowledgments

This work was supported in part by NSF Grants MCS-8221750, DCR-8506393, and DMC-8502115.

REFERENCES

Bunke, H. (1982) Attributed Programmed Graph Grammars and Their Application to Schematic Diagram Interpretation. IEEE Transactions on Pattern Analysis and Machine Intelligence PAMI-4(6):574-582, November.

Davis, L. and T.C. Henderson. (1981) Hierarchical Constraint Processes for Shape Analysis. IEEE Transactions on Pattern Analysis and Machine Intelligence, PAMI-3(3):265-277, May.

Fu, K.S. and B.K. Bhargava. (1973) Tree Systems for Syntactic Pattern Recognition. IEEE Transactions oon Computers C-22:1087-1099, December.

Fu, K.S. (1974) Mathematics in Science and Engineering. Volume 112: Syntactic Methods in Pattern Recognition. Academic Press.

Gonzalez, Rafael C. and Michael G. Thomason. (1978) Applied Mathematics and Computation. Volume 14: Syntactic Pattern Recognition. Addison-Wesley Pub. Co.

Henderson, T. and L. Davis. (1981) Hierarchical Models and Analysis of Shape. Pattern Recognition 14(1-6):197-206, May.

Henderson, T.C. and Ashok Samal. (1985) Shape Grammar Compilers. Pattern Recognition 19(4):279-288.

Lin, W.C. and K.S. Fu. (1984) A Syntactic Approach to 3-D Object Representation. IEEE Transactions on Pattern Analysis and Machine Intelligence PAMI 6(3):351-364, May.

Muehle, Eric. (1986) FROBS Manual. Technical Report PASS-note-86-11, U. of Utah.

Nilsson, N. (1971) Problem Solving Methods in AI. McGraw-Hill.

Rosenfeld, A. and D. Milgram. (1972) Web Automata and Web Grammars. In B. Melzer and D. Michie (editors), Machine Intelligence, pages 3307-324. Edinburgh U. Press, Edinburgh.

Winston, P. (1984) Artificial Intelligence. Addison-Wesley, Reading, Mass.

IX. WORKING SESSIONS

Working Group A: 2D and 3D Image Understanding

Chairman: *Roger Mohr*

Participants: *Hans-Peter Biland (Reporter)*

Christopher I. Connolly

William I. Grosky

Klaus Kayser

Rajiv Mehrotra

J. Ross Stenstrom

Preface

In this Working Group we tried to be constructive, to clearly state unsolved problems, and to synthesize ideas on the subject. Since the discussion time was limited to less than three hours, we had to restrict the field of interest in order not to bring out ideas that were too general and evasive.

Features and Models

Consider the problem of recognizing complex objects in situations where only *incomplete and / or imperfect* information is available due to light conditions (e.g., shades, shadows), viewing condition (e.g. , perspective, hidden parts, mutual occlusions), noise, optical distortions, etc.

- For our recognition process, we have to take advantage of local and global information that is as robust and reliable as possible: we base the recognition on *features*. We do not want to restrict the kinds of features. From the wide variety of features the following examples are included:

  - the elevation of a vertex / corner

  - the area and the position of a hole

  - an edge line or a surface boundary line

  - the position of the illumination source

- Because of the imperfection of the information, we need a *redundant* model. The redundancy helps to improve the signal-to-noise ratio know from information

NATO ASI Series, Vol. F45
Syntactic and Structural Pattern Recognition
Edited by G. Ferraté et al.
© Springer-Verlag Berlin Heidelberg 1988

theory. The model of an object indicates which features are present, and describes their interrelations. Examples for such relations between features are:

- the dependence of a shadow line on the position of the illumination source.

- the dependence of the angle between edges in the projection on the eye-point.

• The relations in a model can be used in a verification mechanism: given two features a and b test whether $q(a,b)$ holds. This is a consistency checking procedure. A more active usage is to have a procedure that reduces the dimensionality of the parameter space of the unknown or partially known features. For example, the procedure.

- *predict _ line _ y(line _ x)* might predict the position of a line y given the position of a line x, and

- *predict _ angle _ of _ light _ source (body _ line, shadow _ line)* might predict the angle of the light source position given an edge and its corresponding shadow.

The Recognition Problem

The assumed inputs to the recognition task are:

1. The *models* of all the possible objects, i.e. all the objects that have to be expected,

2. A *set of features* possibly having superfluous, missing, and more or less distorted elements due to noise.

In the recognition process we have to decide to which model a subset of features most probably belongs. This is usually done by matching the features sequentially to every model. This usually done by matching the features sequentially to every model, and judging the quality of each match. It has to be pointed out that a *group of features* may be restrictive enough to select a small subset of models, optimally only a single model (Mehrotra).

This leads to an open problem: Given a set of models and given information about the robustness of the feature extraction methods for each kind of feature, determine the set of features that should be first extracted in order to have a concise and robust selection of model subset (indexing).

About the Matching

Matching can be done either in a global manner using, for instance, the Hough transform, or by finding directly some kind of near-homomorphism between the features and the model. Such a direct method is sometimes impractical. Instead, matching with a "good" subset of feature data is first performed, which can lead to the indexing mentioned above. To proceed, a strategy is needed. We should like to distinguish between two cases:

1. If the models allow only checking for consistency, we can only select input features (we might also consider using *missing* features) and check the models for sufficient consistency. Various pruning techniques may be helpful in this process. For instance, we can use the "branch and bound" technique when looking for the best match. Moreover, the strategy for choosing the feature to match can also be included in the model.

2. If the models allow prediction of features, the recognition process can reduce the search space of the features it tries to match. This allows tuning the search by selecting a specialized feature extractor. This algorithm can be described as follows:

```
WHILE   model is not confirmed nor rejected DO
        predict a feature
        check for its presence
        update the model-object match
```

Ayache and Faugeras (1986) give a good illustration of this process. However, their work could be extended towards finding a better first match throughh good indexing.

The preceding description of the matching process was very abstract. For a more concrete formulation, we should have to answer many more questions, for example:

- *What rules can guide us to find the optimal first set of features to match*
  Initially matching a larger number of features will have several effects:

- The models which will survive to be candidates for verification will be fewer.

- The prediction of location will be more precise.

- However, performing the first match will then be harder; where is the best balance? Can it be found automatically?

- *Can we build models using a general frame for predicting feature location from partial and imprecise information*

This is one of the key points concerning cooperation from multiple knowledge sources. Geometric techniques and Kalman filtering are inappropriate tools here, since they are valid only when the geometric uncertainty is small. More general tools, such as the one introduced by Brooks in ACRONYM do not seem to be suitable for real applications.

- *How can we extend "static" model (model of completely defined and fixed shape) towards "generic" one?*

REFERENCES

R. Mehrota and W.I. Grosky: "SMITH: An Efficient Model-Based Two-Dimensional Shape Matching Technique" (in these Proceedings).

N. Ayache and O. Faugeras: "HYPER, A New Approach for the Recognition and Position of Two-Dimensional Objects", IEEE PAMI, Vol. 8, pp 44-54, (1986).

Working Group B: Waveform and Speech Recognition

Chairman: R. de Mori

Participants: E. Vidal (Reporter)

                F. Casacuberta

                T. Chang

                H.M. Rulot

                E. Tanaka

                L.D. Wu

Introduction

Waveform Recognition is usually known as the task of giving certain type of *interpretations* to a (set of) unidimensional signal(s) which, without loss of generality, will be assumed to be defined in the time domain.

Depending on the specific area of interest, waveforms can represent quite different (natural) physical events. For instance, they can represent speech (utterances), electrocardiogram or electroencephalogram (ECG, EEC) recordings, or many other biological signals. Also, waveforms can represent earth vibrations (earthquakes) and, furthermore, they can represent the response of system to an artificial excitation. Examples of this later class are radar signals, and artificially originated seismic recordings.

The problems involved in the recognition of the different kinds of waveforms are very diverse; however, all of them share a common difficulty of requiring some type of *model* of the underlying physical system to allow for convenient analysis and/or interpretation. In some cases, like (certain) seismic signals, the model can be fairly simple, for the signals can be properly considered as the result of the convolution of a (one shot) mechanical excitation with the time response of the mechanical system composed of the different ground layers. At the other extreme we may face problems, such as speech understanding, in which the signals are the result of a hierarchycal process in which psychological, pragmatic, semantic, linguistic, mechanical, and acoustic contributions are merged together to produce the observed outcome. In such cases, no single model can account adequately for the corresponding waveforms; rather, a hierarchy of models, associated with the above mentioned abstraction levels, are required to achieve interpretations to one of these levels.

Despite the great differences existing among the different areas related with waveform recognition, some essential common aspects can be identified. This has led us to the agreement on a common canonical block diagram for the process of waveform recognition (Fig.1). Let us analize with some detail what each block of this diagram is assumed to do, and what are considered to be its inputs and outputs.

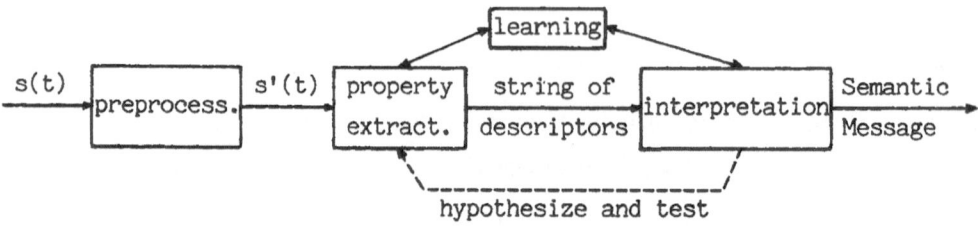

Fig 1.: Canonical black diagram for waveform recognition

Preprocessing

The input waveform s(t) is submitted to a number of processes (devices) including analog filtering, sampling and digitizing, and possibly further (digital) filtering. One important characteristic of these processes is that the time domain where the original signal was defined is not (essentially) changed. But perhaps more important is the fact that very little, if any, explicit task-specific knowledge (model) is used. The output of this block is therefore assumed to be a series of numbers s'(t) t=1,2,... representing the amplitude of the (filtered) original waveform at the appropriate instants of time.

Property Extraction

The signal s'(t) is known to convey information about a number of "features" or (elementary) properties of the underlying system (or system's state). Extracting this information is thus the main function of this block. The types of properties, and the ways property-items of a given waveform can be related to each other, are diverse, as are thus the possible outputs of the block. However, due to the sequential nature of the waveform "segments" where properties are to be extracted, most types of outputs are structured as *strings of Descriptors*. Depending on how the segments are defined, we may have either *fixed duration or variable duration* descriptors and, in any case, the whole string representation can be either *fixed length* (fixed number of segments, regardless of the actual duration of the whole waveform), or *variable length* (number of segments on the duration of the waveform). Fixed duration segments are the easiest to obtain since they are defined independently of the actual waveform instance being analyzed. Variable length segments, on the other hand, are more attractive because they can potentially lead to a more concise (compact) representation case can be complex and troublesome.

The decriptors of the waveform representation just discussed can be of three basic types: Numeric, Symbolic, or Hybrid. A *Numeric descriptor* is a set of numbers, or more specifically, a *vector* which is the result of applying certain numerical transformations to the waveform segment considered ("signal processing"). A *Symbolic descriptor* is the result of assigning, to a waveform segment (or to some numeric descriptor obtained from it), one (or more) label(s) which provide a "closed form" description of the segment. There are many ways is which this labeling can be carried out, but perhaps one of the most widely adopted is that know as "Vector Quantization". In this case, the set of labels is often referred to as "codebook", and it can be obtained automatically by means of clustering techniques. The labelling process itself is then usually based on non-parametric techniques of Pattern Recognition. Instead of using pure symbolic or numeric descriptors, a *hybrid representation* can be conveniently adopted, in which some numeric attributes are attached to the symbolic descriptors in order to augment their descriptive capabilities.

Interpretation

The input of this block is a string of descriptors which is assumed to convey information that is both concise and relevant with respect to the model of the underlying system implicitly or explicitly assumed. The essential function can be stated in terms of finding convenient (consistent with the model) *relations* among the different segments of the original waveform or, more specificaly, among their descriptors. The complexity or this block greatly depends on the type of interpretations which are desired as outputs. In any case, the interpretation results can be understood as "semantic messages" (SM) out of a (possibly infinite) set of messages which constitute the "semantic universe"

(SU) of the problem considered. Therefore, it is the complexity (size) of this universe which determines the complexity of the interpretation task. If the SU is "small" (i.e., a small number of SM's are expected), the interpretation can be considered as a simple process of *classification,* and the output can consist of just a *label* specifying the classification result. Examples of this type are Isolated Word Recognition, (Optical) Character Recognition, etc. On the other hand, if the SU is "large" a simple label can not completly specify the SM; rather, this message should consist of some type of *structured information.* For instance, in (speech) Natural Language systems, the SM must be adequate to drive complex actions (e.g.: a phrase of a Query Language to drive a Data Base access). In these cases, the Interpretation process rather than being a simple process of classification, should be considered as a process of *"understanding".*

When SU's of increasing complexity are involved, Syntactic Pattern Recognition and/or Artificial Intelligence methodologies are usually adopted. In particular, Continuous Speech Recognition with limited vocabulary and simple protocol (syntax), is currently being approached with certain success by means of Stochastic Finite-State Automata or the closely related Hidden Markov Modelling. On the other hand, for Speech Recognition Systems that aim at dealing with (pseudo-) Natural Language, it would be advantageous (and even necessary) to make use of Artificial Intelligence methodologies, and especially those concerning with Natural Language Processing. It should be noted, however, that, in this case, the noise (errors) and imprecision always present in the input signals seems to render inappropriate most usual Natural Language parsing techniques.

Learning

Given the great dependency of the whole Interpretation process on some a priori model of the underlying system, the (implicit or explicit) establishment of this model is therefore one of the most important aspects of any proper approach to Waveform Recognition. This aspect is usually referred to as "Learning" and, also, as "Model Parameter Estimation" or "Knowledge Acquisition". Learning can be performed in two basic (non-excluding) ways: compiling human possessed ("deductive") knowledge from specific examples of the underlying system behaviour. In any case, Automatic Learning is presently one of the most challenging problems of Waveform Recognition. "Deductive" learning is usually approached through expert-system-like knowledge acquisition methodologies, while "inductive" learning is approached by methods of

Concept Learning (Artificial Intelligence), Grammatical Inference (Syntactic Pattern Recognition), and/or Parameter Estimation (Statistical Pattern Recognition).

Hypothesize and Test

There are two basic ways in which the acquired knowledge (model) can be used to obtain appropriate interpretations of the given waveform: top-down or model-driven, and bottom-up or data-driven. However, when complex interpretation tasks are afforded, an dequate combination of theses is usually required, leading to what have been called *active systems* in which a "hypothesize and test" paradigm is adopted. In this way, Property Extraction can be (partially) guided by the (set of) hypotheses which are generated as partial interpretation result(s). This allows us to perform an interesting "successive refinement" loop which can lead to a proper final result while keeping the resource requirements within reasonable limits.

Conclusion

Some current views on Waveform and Speech Recognition have been briefly outlined. However, there seems to be a need of further investigation for conceiving a common (formal) framework in which different techniques and methodologies could be adequately accomodated and conveniently integrated within a most general framework of Pattern Recognition and Artificial Intelligence.

Working Group C: Hybrid Techniques

Chairman: H. Bunke

Participants: T.C. Henderson (Reporter)

                H. Baird

                G. Cristobal

                R.M. Haralick

                J. Kittler

                S. Ressler

                A. Sanfeliu

                R. Siromoney

                K.G. Subramanian

Hybrid methods range from system based on the well-integrated use of several otherwise distinct methods to systems simply comprised of juxtaposed modules of different techniques. The most common motivation for choosing a hybrid method is the failure of narrower classical methods. When this is so, documenting that failure can be helpful to other researcher. A frequent penalty of choosing a hybrid approach is estrangement from standard analytical methods. This suggest that hybrids should be cautious, minimal extensions to well-known prior work. In order to categorize hybrid systems, it is useful to use three properties:

1. *Integral Purity* - this a mesure of the succinctness and inherent nature of the technique applied; i.e., whether it is derived from the problem or is some more general technique borrowed from another problem domain.

2. *Number of Techniques* - a hybrid system must be a blend of more than one approach; and

3. *Number of Representations* - a hybrid system is based on several representation which describe the problem from different points of view; these representations have the property that they can be transformed depending on the application.

In terms of these three axes, classical techniques lie near the origin. It may also be useful to designate as *heterogeneous* all systems which involve multiple techniques and/or representation. Subclassification within heterogeneous systems includes then:

NATO ASI Series, Vol. F45
Syntactic and Structural Pattern Recognition
Edited by G. Ferraté et al.
© Springer-Verlag Berlin Heidelberg 1988

1. *Multi-method Systems*: these involve a loose coupling of techniques and/or representations which are focused to resolve a problem in an independent way, and

2. *Hybrid Techniques*: these are well-integrated approaches which coherently combine several methods and/or representations.

Figure 1 shows these ideas pictorially.

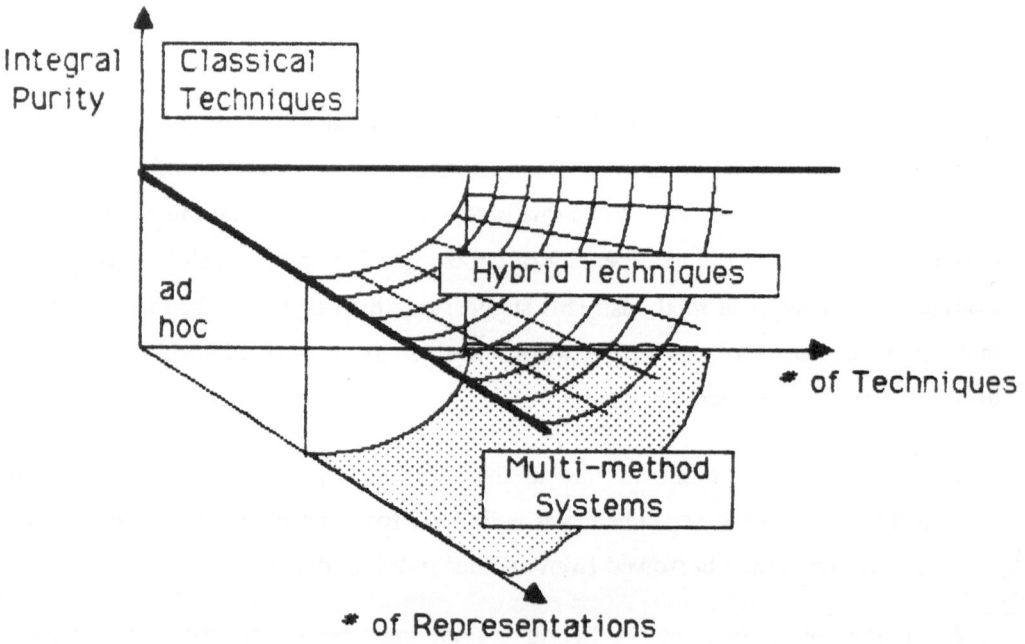

Figure 1. An Organizational Scheme for Heterogeneous Methods

Major issues involved in the use of hybrid techniques include:

* *Control* - overall control of the process must be consistent with the technique,

* *Interfacing* - the data structures and techniques must be appropriately interconnected with respect to the application,

* *Transformation* - the data structures should be modifiable to fit the problem domain, and

* *Performance Measures* - there must be some way to determine the cost of increasing the complexity with respect to the results obtained

It is essential that the overall control of the method be well-integrated with the underlyne techniques.

It is perhaps useful to consider computer vision as one particular example. Figure 2 shows one approach to computer vision. The goal is

Computer Vision Paradigm

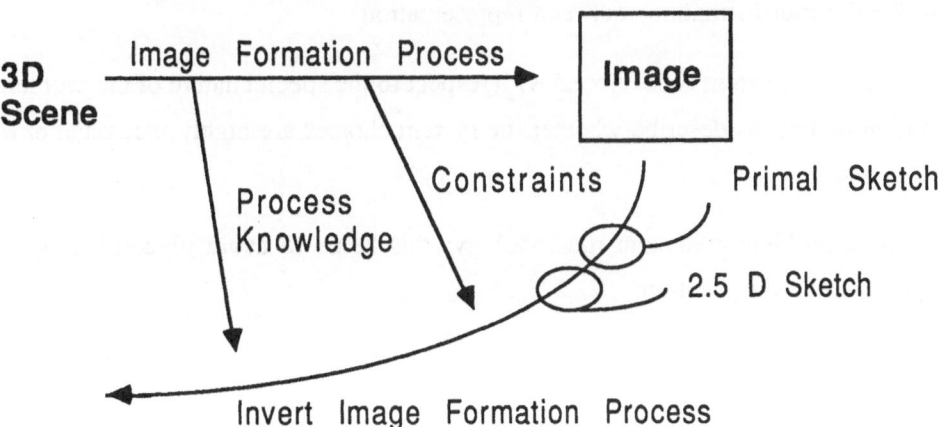

Figure 2. Computer Vision Paradigm

to recover the 3-D scene information from a 2-D image. The steps involved in achieving that goal include building several intermediate data structures which encode information recovered fron the image based on constraints and knowledge derived from the image formation process. Pattern recognition is a wider problem, honewer. We are now dealing with encoded data which can be produced from a variety of approaches. To date, most attempts to recover the pattern have been based on one of four approaches: statistical, structural, syntactic, or logical. In general, anything goes in the selection of the technique and the pattern recovery process. This has tended to make the study of pattern recognition much more *ad hoc* than might be desirable. Hybrid techniques attempt to exploit the parallel analysis of the data through the use of more than one technique in performing pattern analysis.

Some examples of hybrid techniques include stochastic grammars and knowledge-based systems which incorporate several different methods. It is essential that workers in this area:

* strive for clarity

* develop performance measures to estimate the cost and the efficiency of the system versus the increase in complexity.

* describe the transformations between representation

* discuss how the system is developed with respect to the special nature of the problem, and in particular, describe whether the system choices are highly motivated or *ad hoc*, and

* required a problem statement methodology; this helps to avoid pigeon-holing the problem and the solution.

Working Group D: Models and Inference

Chairman   *T. Pavlidis*

Participants *M. Kaul (Reporter)*

               K. Aoki

               N. Perez de Blanca

               R. Molina

               J. Quinqueton

               L. Shapiro

               C. Sielaff

Summary

First, the title of the working group was discussed. Models are automata, grammars, etc. There is a trade-off between the level of primitives and the complexity of recognition. For the statistical approach the problem of extrancting the best discriminating feature is solved, but not for the syntactic approach. There are some NP-completeness results about grammatical inference problems. So a major questions is, whether one has to start with a human made grammar, guess it and test it. Another suggestion is, restrict to a finite set of rules and try to find the best grammar within this finite search space. On the other hand one has not to take NP-completeness results into account if the relevant strings do not tend to grow very large. Otherwise one has chosen the wrong primitives or wrong relations between them. But how can one describe the essential features of the letter E e.g., taking into account, that.

is recognized by humand. Perhaps the theory of catastrophes can explain, why syntactic pattern recognition procedures are so sensible againts small phenomena. There are several existing successful pattern recognition systems, in which high-level primitives are used, which involve a lot of knowledge.

T. Pavlidis closed the working group with a summary: Models depend on the goal.

NATO ASI Series, Vol. F45
Syntactic and Structural Pattern Recognition
Edited by G. Ferraté et al.
© Springer-Verlag Berlin Heidelberg 1988

X. PANEL

PANEL : ARTIFICIAL INTELLIGENCE VERSUS SYNTACTIC TECHNIQUES: THEORETICAL AND PRACTICAL ISSUES

Panelists :

G. Ferraté (Chairman)

R. De Mori

T.C. Henderson

R. Mohr

T. Pavlidis

A. Sanfeliu (Reporter)

L. Shapiro

One of the first issues raised by the panel was the proper definition of the field of artificial intelligence. Some of the panelists shared the opinion that there exist two levels in this field :

1. AI engineering, also called weak AI, which is focused to find the solution of specified problems usually associated with human intelligence, for example, scene recognition or speech recognition. Many of these problems share several disciplines, e.g., signal processing, information theory, pattern recognition or logic.

2. Art or science AI, also called strong AI, which is oriented towards philosophic and metaphysic processes of natural intelligence, i.e., to understand the meaning of the concepts or to emulate the human intelligence process.

AI engineering could be considered as what has been done in pattern recognition research work with respect to classification problems, for different reasons. On the one hand, the solution of an application depends on the required knowledge and context, and as far as pattern recognition is concerned there are no limits on the techniques and the types of representations that can be applied. On the other, many techniques and types of knowledge representation used by people of both fields, for example

NATO ASI Series, Vol. F45
Syntactic and Structural Pattern Recognition
Edited by G. Ferraté et al.
© Springer-Verlag Berlin Heidelberg 1988

matching or searching, have been named in different ways, but they are basically the same (which are basic techniques of structural and syntactic pattern recognition). This point was raised by several panelists, and one of the examples shown was the matching between graphs, which is being used in different ways by pattern recognition and artificial intelligence people. It was reported that, at the present time, there are several techniques and representation methods which are being used by both group of reseachers, but with different names, so that they are creating some confusion. However, there is a great deal of effort in the pattern recognition field, in the theoretical bases and the implementation problems of matching non exact strings, trees and graphs, and in the problem of obtaining a distance measure between those structures and the language of a grammar.

Other panelist discussed the problem of the integration of different knowledge bases, called hybrid systems, and the relationship with pattern recognition and artificial intelligence (basically logic) techniques. It was pointed out that the representation of knowledge can be structured and unified in such a way (for example trough formal grammars), that only the appropriate knowledge could be considered depending on the application. That is, to say for example, that an explicit knowledge base of an object may be used, and depending on the application, the data can be transformed in the best way for doing, for example, classification or reasoning. One way of this integration was shown in an early presentation, where formal grammars where used to describe the objects in a scene and the same grammar was used for classifying an object, for obtaining the subgraphs trough reasoning and for getting some properties of the objects. This case was explained for scene representation.

Other hybrid systems were discussed, and their problems described. In the case of speech recognition, it was reported that the integration of several kinds of knowledge and techniques are being used, and that the techniques of the aforementioned fields were mixed together. Some of the problems come from the consistency of the knowledge base and the distortions of the real world with respect to references. The first problem needs true maintenance procedures for solving the consistency problem, and the second one requires the application of information theory for the solution. Moreover it was reported, that in order to include some knowledge which was not procedural, an expert system was used.

The utility of expert system shells to be applied in pattern recognition problems was also discussed. The point was to get to know which are the requirements to be considered for applying expert systems shells in pattern recognition. Several answers were reported. Some panelists point out that many problems in classification can be expressed as algorithms, so there is no need for an expert system. They also commented that once a classification problem is expressed in such a way, it is very difficult to verify the outcome of the system, and that the control in many of the existing tools, cannot be modified. However other panelists said that at the present time, for some kind of knowledge, the only way to include this knowledge is through expert systems, if we want to have flexible systems, and this could be a way of integrating different kinds of knowledge bases.

XI. LIST OF PARTICIPANTS

AMAT J.
FACULTAD DE INFORMATICA
PAU GARGALLO 5
08028 BARCELONA
SPAIN

BAIRD H.
AT AND T BELL LABORATORIES
COMPUTING SCIENCE RESEARCH CE
600 MOUNTAIN AVE. 2C-557
MURRAY HILL, NJ 07974 U.S.A.

BASAÑEZ L.
INSTITUTO DE CIBERNETICA
DIAGONAL 647
08028 BARCELONA   SPAIN

BUNKE H.
INSTITUT FÜR INFORMATIK UND
LANGEWANDTE MATHEMATIK
LÄNGGASSSTRASSE 51
CH-3012 BERN - SWITZERLAND

CONNOLLY C.
GE CORPORATE RESEARCH AND DEVELOPMENT
BOX 8 ROOM 4C1, BLDS. K-1
SCHENECTADY, NY 12345   - USA

DE MORI, R.
MCGILL UNIVERSITY
SCHOOL OF COMPUTER SCIENCE
805 SHERBROOKE ST.W.
MONTREAL, QUEBEC H3A 2K6 - CANADA

GROSKY, W.I.
COMPUTER SCIENCE DEPT.
WAYNE STATE UNIVERSITY
DETROIT, MICHIGAN 48202 - U.S.A.

HATTICH, W.
FRAUNHOFER-INST. FÜR INFORMATIONS-
UND DATENVERARBEITUNG-IITB
SEBASTIAN-KNEIPP STRASSE, 12-14
7500 KARLSRUHE 1 - GERMANY

AOKI, K.
UTSUNOMIYA UNIVERSITY
DEPT. OF INFORMATION SCIENCE
FACULTY OF ENGINEERING
UTSUNOMIYA 321 - JAPAN

BARRENA, K.
CALIXTO DIEZ N.11 4B
48012 BILBAO
SPAIN

BILAND, H.P.
COMMUNICATIONS AND COMPUTER SCIENCE DEPT.
IBM ZURICH RESEARCH LABORATORY
8803 RUSCHLIKON - SWITZERLAND

CASACUBERTA, F.
FACULTAD DE INFORMATICA
DEPT. SISTEMAS INFORMATICOS Y COMPUTACION
46071 VALENCIA
SPAIN

CRISTOBAL, G.
FACULTAD DE INFORMATICA
LAB. DE INTELIGENCIA ARTIFICIAL
CTRA. DE VALENCIA, Km. 7 - 28031 MADRID - SPAIN

FERRATE, G.
INSTITUTO DE CIBERNETICA
DIAGONAL, 647
08028 BARCELONA
SPAIN

HARALICK, R.M.
DEPT. OF ELECTRICAL ENGINEERING
UNIV. OF WASHINGTON
SEATTLE, WA 98195 - U.S.A.

HENDERSON, T.C.
DEPT. OF COMPUTER SCIENCE
3160 MERRILL
ENGINEERING BLDG. SALT LAKE CITY
UTAH 84112 - U.S.A.

JUAN, J.
INSTITUTO DE CIBERNETICA
DIAGONAL, 647
08028 BARCELONA
SPAIN

KAUL, M.
UNIVERSITAT PASSAU FUR INFORMATIK
POSTFACH 2540
8390 PASSAU
GERMANY

KAYSER, K.
DEPT. OF PATHOLOGY
HOSPITAL FOR THORACIC DISEASES
AMALIENSTR, 5
6900 HEIDELBERG - GERMANY

KITTLER, J.
DEPT. OF ELECTRONIC AND ELECTRICAL ENGINEERING
UNIVERSITY OF SURREY
GUILDFORD, GU2 5XH
UNITED KINGDOM

LAGUNA, P.
DEPT. DE ESTADISTICA
FACULTAD DE CIENCIAS
UNIVERSIDAD DE GRANADA
GRANADA - SPAIN

LAPRESTRE, J.T.
ELECTRONICS LAB, UA 830 OF CNRS
UNIVERSITY OF CLERMONT II
B.P. 45
63170 AUBIERE - FRANCE

LI-DE WU
DEPT. OF COMPUTER SCIENCE
FUDAN UNIVERSITY
SHANGHAI - P.R. CHINA

MEHROTRA, R.
COMPUTER SCIENCE DEPARTMENT
UNIVERSITY OF SOUTH FLORIDA
TAMPA, FLORIDA 33620 - U.S.A.

MORH, R.
CRIN CAMPUS SCIENTIFIQUE
BOITE POSTALE 239
54506 VANDOEUVRE-LES-NANCY CEDEX
FRANCE

NOLTEMEIER, H.
LEHRSTUHL F. INFORMATIK I
UNIVERSITAT WURZBURS
AM HUBLAND
D-8700 WURZBURS - GERMANY

PAVLIDIS, T.
DEPT. OF ELECTRICAL ENGINEERING
SUNY
STONY BROOK
NY 11794 - U.S.A.

PEREZ DE LA BLANCA, N.
DEPT. ESTADISTICA
FACULTAD DE CIENCIAS
UNIVERSIDAD DE GRANADA
18071 GRANADA - SPAIN

QUINQUETON, J.
CRIM
860 MONTE DE ST. PRIEST
34100 MONTPELLIER - FRANCE

RESSLER, S.
NATIONAL BUREAU OF STANDARDS
GAITHERSBURG
MD 20899, - U.S.A.

RULOT, H.
CENTRO DE INFORMATICA
DR. MOLINER, S/N
46100 BURJASOT (Valencia) - SPAIN

SANFELIU, A.
INSTITUTO DE CIBERNETICA
DIAGONAL, 647
08028 BARCELONA - SPAIN

SERRA, T.
DEPT. ELECTRICIDAD Y ELECTRONICA
FACULTAD DE CIENCIAS
CTRA. DE VALLDEMOSA, Km. 7,5
PALMA DE MALLORCA
SPAIN

SHAPIRO, L.
DEPT. OF ELECTRICAL ENGINEERING
UNIVERSITY OF WASHINGTON
SEATTLE, WA 98195
U.S.A.

SIELAFF, C.
UNIVERSITAT HAMBURG
FACHBEREICH INFORMATIK
BODENSTEDTSTR, 16
2000 HAMBURG 50
GERMANY

SIROMONEY, R.
DEPT. OF MATHEMATICS
MADRAS CHRISTIAN COLLEGE
TAMBARAN
MADRAS 600 059, INDIA

TANAKA, E.
UTSUNOMIYA UNIVERSITY
DEPT. OF INFORMATION SCIENCE
FACULTY OF ENGINEERING
UTSUNOMIYA 321-31 - JAPAN

TORRAS, C.
INSTITUTO DE CIBERNETICA
DIAGONAL, 647
08028 BARCELONA
SPAIN

SIROMONEY, G.
DEPT. OF MATHEMATICS
MADRAS CHRISTIAN COLLEGE
TAMBARAM
MADRAS 600 059
INDIA

STENSTROM, J.R.
GENERAL ELECTRIC CO. RESEARCH
AND DEVELOPMENT CENTER
KW C625 ONE RIVER ROAD
SCHENECTADY, NY 12345 - U.S.A.

TONG CHANG, G.
ACADEMY OF SCIENCES OF CHINA
PROF. OF DEPT. of AUTOMATION
TSINGUA UNIVERSITY
PEKING -P.R. CHINA

VIDAL, E.
FACULTAD DE INFORMATICA
DEPT. SISTEMAS INFORMATICOS Y COMPUTACION
46071 VALENCIA - SPAIN

NATO ASI Series F

NATO ASI Series F

Vol. 23: Designing Computer-Based Learning Materials. Edited by H. Weinstock and A. Bork. IX, 285 pages. 1986.

Vol. 24: Database Machines. Modern Trends and Applications. Edited by A.K. Sood and A.H. Qureshi. VIII, 570 pages. 1986.

Vol. 25: Pyramidal Systems for Computer Vision. Edited by V. Cantoni and S. Levialdi. VIII, 392 pages. 1986.

Vol. 26: Modelling and Analysis in Arms Control. Edited by R. Avenhaus, R.K. Huber and J.D. Kettelle. VIII, 488 pages. 1986.

Vol. 27: Computer Aided Optimal Design: Structural and Mechanical Systems. Edited by C.A. Mota Soares. XIII, 1029 pages. 1987.

Vol. 28: Distributed Operating Systems. Theory und Practice. Edited by Y. Paker, J.-P. Banatre and M. Bozyiğit. X, 379 pages. 1987.

Vol. 29: Languages for Sensor-Based Control in Robotics. Edited by U. Rembold and K. Hörmann. IX, 625 pages. 1987.

Vol. 30: Pattern Recognition Theory and Applications. Edited by P.A. Devijver and J. Kittler. XI, 543 pages. 1987.

Vol. 31: Decision Support Systems: Theory and Application. Edited by C.W. Holsapple and A.B. Whinston. X, 500 pages. 1987.

Vol. 32: Information Systems: Failure Analysis. Edited by J.A. Wise and A. Debons. XV, 338 pages. 1987.

Vol. 33: Machine Intelligence and Knowledge Engineering for Robotic Applications. Edited by A.K.C. Wong and A. Pugh. XIV, 486 pages. 1987.

Vol. 34: Modelling, Robustness and Sensitivity Reduction in Control Systems. Edited by R.F. Curtain. IX, 492 pages. 1987.

Vol. 35: Expert Judgment and Expert Systems. Edited by J.L. Mumpower, L.D. Phillips, O. Renn and V.R.R. Uppuluri. VIII, 361 pages. 1987.

Vol. 36: Logic of Programming and Calculi of Discrete Design. Edited by M. Broy. VII, 415 pages. 1987.

Vol. 37: Dynamics of Infinite Dimensional Systems. Edited by S.-N. Chow and J.K. Hale. IX, 514 pages. 1987.

Vol. 38: Flow Control of Congested Networks. Edited by A.R. Odoni, L. Bianco and G. Szegö. XII, 355 pages. 1987.

Vol. 39: Mathematics and Computer Science in Medical Imaging. Edited by M.A. Viergever and A. Todd-Pokropek. VIII, 546 pages. 1988.

Vol. 40: Theoretical Foundations of Computer Graphics and CAD. Edited by R.A. Earnshaw. XX, 1246 pages. 1988.

Vol. 41: Neural Computers. Edited by R. Eckmiller and Ch. v. d. Malsburg. XIII, 566 pages. 1988.

Vol. 42: Real-Time Object Measurement and Classification. Edited by A.K. Jain. VIII, 407 pages. 1988.

NATO ASI Series F